# THE LUSITANIA CONTROVERSIES
## BY GARY GENTILE

On May 15, 1915, the British liner *Lusitania* fell victim to a German torpedo off the south coast of Ireland. In scarcely twenty minutes the crack Cunard steamship sank in 300 feet of water, leaving on the surface of the troubled green sea a blight of human flotsam: a handful of lifeboats, the cries of drowning passengers and crew, and the bodies of the dead. Among the 1,198 fatalities were scores of children and babes in arms, and 123 American citizens.

Death struck some with merciful swiftness. Most, however, languished in the frigid water and suffered the terrible numbing cold until their strength and endurance gave way. The silence that followed was more telling of the tragedy than the direful screams of the wretched.

The sinking of the *Lusitania* was the latest in a list of German atrocities that shocked the civilized world.

Shock quickly yielded to heated controversy as investigations and inquiries sought suitable answers to a host of provocative questions about the legality of unrestricted submarine warfare and the torpedoing of merchant vessels carrying none but innocent civilians. Was the *Lusitania* armed? Was she carrying a cargo of high explosives? Had her captain disregarded recent Admiralty warnings about U-boats operating in the vicinity? Did a minor infraction of international law make the *Lusitania* a legitimate target? And, philosophically, can an inhuman act be condoned or justified within the context of all-out war?

*The Lusitania Controversies* takes a bold and enlightened look at these and other issues that are still being argued today.

But there is more. The story then becomes an exciting tale of adventure: an impious trek through the history of deep wreck-diving from its meager beginnings in the 1950's, through miscarried commercial salvage operations which sought to wrest from the *Lusitania* the secrets that some people believe to exist, to the highly acclaimed mixed-gas diving expedition in 1994, in which the author took an active part.

The author brings original insight to this historical retelling because he has been deeply involved in deep wreck-diving and shipwreck research and writing for more than twenty-five years. He is considered a pioneer in the field of technical diving, and was instrumental in introducing mixed-gas diving technologies to the realm of shipwreck exploration.

A saga of such incredible breadth and dimension could not be fitted between the covers of a single volume without excluding particulars that are compelling in the extreme. In this chronological account, Book One begins with the *Lusitania's* construction in 1905; covers her career, sinking, aftermath, and global repercussions; examines the liner's lasting impressions and publication history; discusses aborted and ineffective attempts to dive the wreck; then details the commencement and evolution of deep wreck-diving as far as 1979.

The untold story of early wreck-diving is rich in passion, portrayal, betrayal, and adventure. The author relates the exploits of his predecessors through unproven times when equipment and wreck exploration techniques were primitive. He follows the evolution of the activity through the slow maturation process until a greater understanding of the special methods and hazards was wrought after years of trial and effort.

Book Two picks up where Book One leaves off, carrying the story of deep wreck-diving forward from 1980. The second volume recounts a host of exciting explorations into the dark foreboding interior of the Italian liner *Andrea Doria*, which came to be a proving ground for deep wreck-divers the world over. Other exciting episodes abound as the author pursues ever-deeper challenges in the search for sunken shipwrecks.

Then came the development of helium breathing mixtures, which permits divers to explore shipwrecks at depths that were previously undreamt of. The author takes the reader on the first-ever descent to the German battleship *Ostfriesland*, at 380 feet, which he made in 1990. He takes the reader on discovery dives to tankers, freighters, submarines, and World War One U-boats that lie off America's east coast. He evokes the thrill of what deep wreck-diving is all about - with insightful observations that can come only from one who has done it all.

Finally, Book Two covers in untold depth the successful 1994 mixed-gas *Lusitania* diving expedition - as well as the legal wrangles beyond. This behind-the-scenes look includes sinister elements that sound more like the plot devices of a grade B movie than abject reality: spies, sabotage, threats of lawsuit, even a bomb which destroyed a car in the author's driveway on the day before he left to testify in a federal hearing against the *Lusitania's* rival salvor. All the dirty details of the courtroom drama are included.

*The Lusitania Controversies* is a fresh, unexpurgated view of one of the most notorious shipwrecks in history. The *Lusitania* is a wreck whose infamy will not be forgotten.

# The Lusitania Controversies
## by Gary Gentile

# BOOK TWO
## Dangerous Descents
### into
## Shipwrecks and Law

5 - Part 6 - 1980's: Down to Deeper Depths

106 - Part 7 - Early 1990's: Mixed-Gas Revolution

      177-208 - Photographic Insert

209 - Part 8 - Flashback: Historic Interlude

224 - Part 9 - Grand Adventure on the Lusitania

302 - Part 10 - After Shocks

376 - Whither

384 - Index            392 - Books by the Author

Gary Gentile Productions
P.O. Box 57137
Philadelphia, PA 19111
1999

Copyright 1999 by Gary Gentile

All rights reserved. Except for the use of brief quotations embodied in critical articles and reviews, this book may not be reproduced in part or in whole, in any manner (including mechanical, electronic, photographic, and photocopy means), transmitted in any form, or recorded by any data storage and/or retrieval device, without express written permission from the author. Address all queries to:

Gary Gentile Productions
P.O. Box 57137
Philadelphia, PA 19111

Additional copies of this book may be purchased from the same address by sending a check or money order in the amount of $25 U.S. for each copy (postage paid).

Website: http://www.pilot.infi.net/~boring/gentile.html

Picture Credits

All uncredited photographs were taken by the author. The front cover top image is a contemporary postcard from the author's collection. The front cover bottom image is an underwater photo of the first letter "A" in the *Lusitania's* name on the high side of the bow at 270 feet. Every attempt has been made to contact the photographers or artists whose work appears in this book, if known, and to ascertain their names if unknown; in some cases, copies of pictures have been in public circulation for so long that the name of the photographer or artist has been lost, or the present whereabouts are impossible to trace. Any information in this regard forwarded to the author will be appreciated. Apologies are made to those whose work must under such circumstances go unrecognized.

The author wishes to acknowledge Jon Hulburt, Drew Maser, and Gene and Joanie Peterson, for reviewing the manuscript.

International Standard Book Number (ISBN) 1-883056-07-1

First Edition

Printed in Hong Kong

# Part 6
# 1980's: Down to Deeper Depths

## Return to the *Andrea Doria*

In the stirring saga of wreck-diving, the 1980's could be called the decade of the *Doria*.

I spent six years clamoring to charter a boat for a return trip to the Grand Dame of the Sea. I beseeched and cajoled a number of captains, all to no avail. It was not unknown, therefore, that I wanted to go back to the wreck and was looking for a boat to take me. When the opportunity finally presented itself, however, I didn't incur the responsibility of filling the boat. It was a captain's charter, and I went simply as a customer.

Trips to the *Andrea Doria* might never have come into prominence had it not been for the *Sea Hunter*, a boat introduced to me through the Eastern Divers Association, and which I chartered after EDA folded. The original owners were Ron Burdewick, an EDA member who made the 1974 *Andrea Doria* trip, John Lachenmeyer, an electrical engineer, and truck driver Sal Arena. By 1980, Burdewick had sold out his share of the boat and quit diving, leaving Lachenmeyer and Arena as partners.

I helped to spread the word about the trip, and in very short order there were enough people interested in the venture to fill *two* trips. I signed up for both.

This was strictly a bare-bones charter. There were six divers on the 39-foot boat, one captain, and no crew. The divers were expected to share the duties of steering and standing watch. Today, after some fifty trips to the *Doria*, I can't remember everyone who was on every trip - they all blend together in my mind. In 1980, my dive buddy was Bill Nagle. Also aboard were John Barnett, vice president of City Bank, and fellow electrician Jim Murtha. The captain was John Lachenmeyer.

If ever a charter could be described as a trip to hell, this was the one.

The *Sea Hunter* was ill-equipped for long-range overnight trips. The cramped quarters below deck housed only four bunks, which shared space with the engine and head, making the compartment interminably hot and unbelievably noisy and suffusing it with the scent of diesel fuel and other malodorous fumes. The head was more cramped than a phone booth: so small that even Superman would have had trouble changing costumes in it. The door was misaligned

## 6 – The *Lusitania* Controversies

with the partition so the latch wouldn't catch. In anything other than flat calm seas, the door swung open and exposed the squatter to his bunk mates.

And those were the good points.

There was no galley, no cold storage, no air-conditioning or any way to fan out heat from the engine, and no electricity other than that provided by storage batteries, which was sufficient only to start the engine and to run the bilge pumps, cabin lights, and electronics - for a limited time. The major concession to safety was an inflatable chase boat which, because there was no room for it on the after deck, had to be stowed topside (uninflated) on the bow, which was reached by means of a narrow walk-around that was accessed only by climbing over scuba gear packed against the cabin bulkhead.

To make up for these deficiencies we brought food in coolers that occupied valuable deck space. Artie Kirchner, who went on the second trip, supplied a stove and an outboard motor for both trips. The stove was a home-made three-burner propane affair that Kirchner knocked together from loose lumber and spare parts. The outboard motor for the chase boat had no pull-cord and wouldn't start electrically on its own, so Kirchner included a motorcycle battery to jump-start it.

The after deck was jam-packed with double tanks, camera boxes, gear bags, decompression reels, pony bottles, weight belts, and miscellaneous cases of assorted paraphernalia, to say nothing of boat accessories such as anchor ropes, a spare grapnel, rescue and trail lines, a ball float, a crate full of tackle, and so on. There was no room to walk and hardly a place to stand. The only sure way to cross the deck was to wait for a lull in the wave action and step quickly over the blockade into an open spot, and repeat the process till reaching the destination. I attempted circumnavigation by balancing on the gunwale while holding onto the low railing, but nearly got pitched overboard for my efforts; afterward I went by way of the obstacle course.

Seven people making four dives each required twenty-eight sets of doubles. Half the tanks were stowed below on the deck between the bunks. Valve tops stuck up like sharpened punji stakes, a trap for the restless sleeper who flung his arm out of the bunk or who got up during the night to pee or take his turn on watch. Halfway through the trip, all the full tanks below were lugged up through the narrow doorway into the cabin, the empty tanks were hauled below, then the tanks in the cabin were distributed outside on the after deck. It was like playing Chinese checkers with gangling, hundred-pound marbles.

Lachenmeyer slept in the wheelhouse on the flying bridge which, because there was no ladder, was reached by jumping up from the railing onto the narrow ledge behind the door. Whoever lost the bunk lottery snoozed on the portside bench. Because the bench was only three-quarters as long as the human body, the snoozer's feet and lower legs hung over nothingness; the bench wasn't wide enough to curl up on.

With accommodations so limited, I voluntarily slept on the flush deck in the cabin. Every morning I had to roll up my sleeping bag and fill the floor with luggage that had been stashed under the table or piled in the corners. Every once in a while a bad roll brought someone's suitcase down on my head or across my body. Sometimes it roused me, but more often then not I awoke in the morning buried under drysuits and bags of food.

The cabin doors couldn't be closed because heavy gear boxes were stacked against them. Not only did the ocean spray waft into the cabin continuously, but there was no sill to keep out water that rolled across the deck. Compared to the luxury and dimensions of the *Atlantic Twin*, diving off the *Sea Hunter* was like traveling in the bilge beneath steerage. But who cared about minor discomforts? We were going to the *Doria*!

Lachenmeyer and Arena drove the boat from Freeport to Montauk, which was about 110 miles from the wreck site. This set the trend for most trips in the years to follow. We loaded our gear in great excitement and with some trepidation, for the wind was howling and the seas were breaking hard upon the shore. The plan called for us to depart late in the afternoon, travel all night, reach the wreck site in the morning, dive that day and the next, then return overnight for a morning arrival to the dock. As they say, the best laid plans . . .

We ran into mountainous waves right outside the inlet, with never a thought of turning back. The forecast called for the winds and seas to moderate. They *had* to moderate, because according to the laws of physics and gravity they couldn't get any worse. Long heaving rollers conspired with driving rain to beat us back to shore, yet we persisted. The boat fell off the tops of whitecaps like an overloaded raft in a whitewater river.

On the after deck, coolers and gear boxes slid back and forth like clothes agitating in a washing machine, and tanks slipped out from under bungee cords to add to the confusion. There was no hope of securing the gear without risk of losing life and limb. I braced myself between bulkheads and watched the equipment reach its lowest level of potential energy - flat on the deck, and compacted so tight that nothing could move.

Nothing stayed dry. Whatever wasn't soaked by spray blowing in through the open doors was slowly saturated by the humidity. Everything was dampened but our spirits and enthusiasm.

Sitting alone on the enclosed wheelhouse, Lachenmeyer maintained a steady course through seas that we estimated rose as high as fifteen feet and never as low as ten. It was no night to be out in a small boat. Still, the *Doria* lured us on.

Despite our "gung ho" attitude toward the adverse weather, we

## 8 – The *Lusitania* Controversies

were forced to turn back because of mechanical problems: the harsh pounding loosened a bearing block that supported the propeller shaft. The uneven torque that was imparted to the propeller caused a shimmy that threatened to damage the transmission or engine or both. During a short lull between wave heights, Lachenmeyer turned the boat onto a reciprocal course and headed back to port.

It was long after midnight before we reached calmer water near shore. We drove through the inlet into the protected harbor within. Not until we approached the dock, however, did we learn that excessive vibration had jammed the throttle linkage. Lachenmeyer was unable to reduce the speed of the engine. He could shift the transmission into neutral and into reverse and back into forward, but only at full rpm's. This made docking a difficult maneuver at best.

We raced toward the dock at breakneck speed in the absolute blackness of night. The only way to reduce headway was to throw the engine into neutral and let friction with the water slow the boat, then brake by shifting into reverse. While Lachenmeyer manipulated the controls, the rest of us arranged ourselves on the bow in order to fend off the wooden piles and planks as we converged on the dock with incredible celerity. Because the boat drove forward and backward in spurts, we couldn't stand upright on the slippery, rain-swept deck. We had to lean over and hold onto the rail.

It was madness to back into a slip between boats. The only alternative was to tie up broadside to the wharf. In the Cyclopean eye of a hand-held dive light Lachenmeyer misjudged the speed of the boat with the proximity of the dock. Too late he thrust the engine into reverse. I suddenly saw the silhouette of the platform at eye level and within arm's reach. The *Sea Hunter's* stem slammed into a pier so hard that it knocked me off my feet, and the boat bounced off like a rubber ball before anyone could throw a line around a pile or cleat. The sudden reverse surge flattened us in a heap among a tangle of loose lines. The boat charged backward into the harbor at flank speed.

After the boat recoiled from the collision, Lachenmeyer shifted to neutral then to forward. The boat leaped ahead like a bronco on the attack, once again knocking us off our feet on the spray-slickened deck, like a handful of dropped pickup sticks. At ramming speed we crashed into the wharf the second time. Someone lassoed a pile as I jumped for the platform. I ran to another pier and someone tossed a stern line which I caught and snubbed around a cleat. The Keystone Kops had landed.

After dawn, Lachenmeyer inspected the engine compartment and discovered the loose bearing block that was the cause of all the trouble. He called Arena and told him to bring the necessary parts to Montauk. Arena had to work (he delivered home heating oil), so he didn't arrive until late afternoon. Together they effected repairs, and that night we were ready once again to brave the foibles of the broad Atlantic. According to the weather report, wave heights had dropped from ten-to-fifteen feet to eight-to-ten. We decided to go for it, hoping for further moderation.

No one got much rest that night. We each took a stint at the wheel, steering blind by compass through the dark starless overcast. Climbing to the wheelhouse required grip, agility, and foul weather gear. We were afraid of losing someone overboard during watch changes, so we initiated a strict protocol which required each person coming on watch to declare his intention by rapping on the cabin ceiling, which was the wheelhouse floor, and when going off watch to announce his safe arrival on deck. Down below, we had to shout to be heard over the roar of the engine and the crashing of the waves. To sleep, I had to wedge myself tight with clothing and gear bags. Otherwise I rolled continuously from side to side like a marble on a seesaw.

The newly installed radar was the only way to observe the passage of ships and fishing vessels that lay strewn across our route. The *Sea Hunter* pitched and yawed like a roller coaster ride gone wild. In the cabin, clothes and sleeping bags were damp and already beginning to mildew. The smell of diesel fumes permeated the boat. We were cold, wet, sick, and tired - and the trip had hardly begun.

I have few memories of the dives to separate them in my mind from the many dives I made on subsequent trips. Mostly I recall the awful time on the boat. Morning brought little relief from the weather and pounding seas. Nagle and I tied the hook onto a promenade deck stanchion. During that dive and the next we explored the high side of the boat deck and promenade deck, recovered several windows, decompressed like a pair of sneakers in a clothes dryer, and somehow managed not to puke our guts out. I also photographed the wreck.

Between dives and afterward my energy reserves were nil. I ate cold canned food not because I couldn't light the stove but because I couldn't catch it. The stove slid across the deck like a sled on ice, crashing first into one gunwale and then into the other. I watched the bouncing stove the way a spectator watches a tennis match. Tanks and gear formed a solid pack, requiring strength and determination to negotiate and to set up for the next dive.

That night we all took turns standing radar watch. Even though the *Andrea Doria* lay between the incoming and outgoing shipping lanes, vessels sometimes strayed from the recommended routes and fishing trawlers plowed the seas in crisscrossing patterns in an around-the-clock quest to stock their holds with catch. An anchor light was no sure proof against being run down.

After my watch I fell into a deep sleep. The seas were still running five to six feet, but the boat didn't seem to roll quite as much as it had been. At six o'clock I was brusquely brought back from dreamland by Lachenmeyer's alarming shout: "We're sinking! We're sinking!" If he had yelled "Incoming!" I might have snoozed on, waiting for enemy shells to land nearby, but going down so far from shore was a serious proposition for one who didn't know how to swim. My eyes popped open and I was fully alert in an instant.

Lachenmeyer ripped open the hatch cover at my feet. Sea water sloshed over the engine mounts. "Start bailing!"

I didn't need a second invitation. I never got out of a bed roll so fast in my life, even under fire. I snatched up the trash bucket and unceremoniously dumped the carefully collected refuse over the side. Murtha rolled off the bench to lend a hand. Lying on my belly I lowered the bucket into the bilge, filled it with oily water, and thrust it up to Murtha, who emptied the bucket overboard and swung it back to me. Desperation struck fear in our hearts and lent strength to our arms.

The situation was serious. Nagle had dozed off during watch and had failed to rouse his relief. Water leaking through glands and seams tripped the automatic switches. The bilge pumps came on and ejected the water efficiently. But since the bilge pumps worked on battery power, the engine had to be run periodically in order to recharge the batteries. When the boat was under way, a centrifugal pump operated by the engine kept the water level down, and the batteries were constantly being charged. But with no one keeping an eye on conditions during the latter part of the night, the bilge pumps functioned until they drained the batteries dry. Then the water rose steadily until it overflowed the floor boards.

A boat captain sleeps like a cat: with both eyes closed but with subconscious senses alert. Lachenmeyer detected the boat's sluggish motion, emerged from the wheelhouse to find the anchor light unlit, and went below to determine the cause. He descended into the darkness expecting to find that a circuit breaker had tripped. Instead he splashed more than knee-deep into cold, briny water, and saw wavelets lapping the edge of the bunk in which Nagle was asleep. That's when he started screaming.

The reason that the boat no longer rolled like it had been despite tempestuous seas was that it was half full of water: weighted down like a punching doll whose base was packed with sand. And that wasn't the worst of it. Dead batteries meant more than not having power for the pumps. It meant there was no electricity to spin the starter motor. There wasn't even enough energy to operate the radio and send a call for help. Lachenmeyer tried both: the starter groaned a couple of times before he shut it down to conserve what little power remained, and the radio wouldn't come on at all. We were dead in the water in more ways than one.

Lachenmeyer ransacked his brain for a way out of our predicament, Murtha and I kept bailing, Nagle looked on helplessly, Barnett and the rest pulled up another hatch cover and began a bucket brigade. In a few minutes of hard work we were able to lower the water level. The boat wouldn't sink as long as we kept bailing. For what it was worth, we stemmed the flood before it reached the electrical contacts on the batteries and the starter motor.

Then Lachenmeyer had a brainstorm that bordered on the absurd, with no chance at all of succeeding - but what did we have to lose? Due to the rough weather we hadn't been able to inflate the raft much less launch it and mount the motor. Still unused, therefore, and fully charged, was the six-volt motorcycle battery Kirchner had supplied to start the tiny outboard. Lachenmeyer connected jumper cables from the baby battery to the boat's giant wet cells. He climbed to the wheelhouse, took a deep breath, and pushed the starter button. There was only enough juice in the motorcycle battery to give one jolt to the starter motor. The engine turned over a single revolution, sputtered, caught, and thrummed loudly to life.

A thunderous cheer arose from the deck as we praised our new salvation.

The engine purred for several minutes before reaching its ideal operating temperature. Then Lachenmeyer cut in the centrifugal pump. Great jets of water spurted from the hull pipes and before long the water level dropped considerably. Soon the cabin was habitable again, except for the oil-slickened floor boards which made walking a hazardous duty. The engine also recharged the batteries, giving us electricity for the lights, radar, and - of greater import - the radio. Lachenmeyer intercepted a message that was transmitted from the *Nantucket* lightship. In fact, the *Sea Hunter's* radio didn't have the signal strength to reach land; our contingency plan for Coast Guard assistance required that messages be relayed through the lightship that guarded the Nantucket shoals, or through other ships in the vicinity.

We each breathed a sigh of relief. But now that the danger was over, I saw no reason not to resume our diving as long as the engine was kept running. Few captains would have considered such a proposal after the close call we'd had with catastrophe. But Lachenmeyer was an easy-going person. Besides that, after having his head in the bilge and breathing noxious diesel fumes, he was too busy throwing up over the rail to register too strong a protest. I preyed on him in his sickened state, and he consented to let us have one more dive before heading the boat for home.

That dive was the best one yet, with visibility better than seventy-five feet despite the early hour: it was only seven a.m. I went in by myself, shot a roll of film, and recovered another window. For seven-

## 12 - The *Lusitania* Controversies

teen minutes bottom time at 180 feet I decompressed for thirty-five minutes. By eight o'clock all the divers were aboard and the boat was under way.

With the excitement of the moment over, my strength and enthusiasm waned. By then I was suffering from malnutrition as well as from dehydration. The ice in the coolers had long since melted and most of the food had spoiled. Other victuals that had to be cooked could not be prepared on the still itinerant stove. We couldn't even boil water for coffee or heat a can of soup. I was reduced to adding cold water to a cup of instant rice. After half an hour of partial reconstitution I chewed the hardened grains; they crunched between my teeth like moldy mouthfuls of sand.

Like everyone else I took my turn at the wheel. Cloudy skies and rough seas chased us all the way home. The boat didn't reach Montauk until late that night, after the restaurants had closed, so we slept on empty stomachs. We were up early and eager for a hearty breakfast. When Lachenmeyer finally stepped from the boat to the dock, he was so weak and disoriented that he stumbled around in circles and would have walked off the edge of the platform had I not grasped his arm in time. Copious amounts of steaming coffee soon revived us all, and helped put in perspective the uncommon thrill of shipwreck exploration with the occasional vicissitudes of oceanic travel.

As I look back on events, I behold that trip as a turning point in the growth of deep wreck-diving - not because of anything extraordinary that occurred during the dives, which were ordinary by comparison with many later dives, but because of what the trip inaugurated: annual excursions whose continuity has gone unbroken to this very day. What might have been little more than an isolated occurrence became the first step on a treadmill with no end in sight.

Accessibility to the *Andrea Doria* gave divers a goal to attain, a firm objective to reach. Those who were not otherwise daunted by depth or dissuaded by contrary conditions, could now set their sights on achieving new purpose in the underwater world of deep wreck-diving.

The second trip that year was less work and more fun than the first but, alas, not as memorable for lack of adversity. Alas?

### Tragedy on the *Andrea Doria*

The following year the *Sea Hunter* ran two more trips to the *Doria*. Again I was full of excitement, but the euphoria of returning to the Grand Dame of the Sea was quickly diminished by tragedy.

We arrived over the wreck in the late afternoon. Nagle and I accepted the onerous task of tying in the hook. A strong current was running. We fought hard to pull down the anchor line through dark, murky water. At 200 feet, through the gloom, we detected the shadowy curve of the lower hull some ten feet away. The grapnel had slid off the smooth metal plates and lay somewhere below.

We let go of the line, kicked furiously toward the wreck against the current, and landed on the remains of a twisted trawler net. I spent only a few seconds clawing my way forward and upward before I decided that it was crazy to continue. The anchor line was already out of sight, the vast hull presented no points of reference, and the possibility of entanglement was considerable. I signaled to Nagle that I was calling it quits. He agreed. We aborted the dive after seven minutes.

We spent the night anchored in the sand nearby. The next morning the tie-in went more smoothly. Visibility had increased to thirty feet, the current had slacked off, and the grapnel caught on the high side of the promenade deck. Nagle and I did the job quickly and, after releasing a Styrofoam cup to let those on the boat know that the grapnel was secure, had time left over for exploration.

Later that morning came the bad news. Stan Smith had gotten separated from John Barnett on the wreck, and never saw him on the anchor line. Was there a liftbag floating that Barnett could be decompressing under? There was not. As we scanned the horizon with binoculars for signs of a stricken diver, I was overcome with a sense of deja vu - I remembered waiting and praying for John Pletnik to surface from the *Pinta*. We gradually gave up on the idea that Barnett was doing a drift decompression.

That afternoon I went down with Smith to look for Barnett's body. We started at the forward end of the promenade deck and worked our way aft. Smith stood safety for me as I dropped into each doorway and cast my light around inside. As I descended into the large explosion hole through which the *Topcat* divers had removed the statue of Admiral Andrea Doria in 1963, Smith drifted a few feet aft and peered down the next doorway about fifteen feet away. He spotted a pair of fins protruding from a room about ten feet below. He raced back and signaled to me with his light.

I came up out of the hole, followed Smith to the doorway, and looked down to where his light illuminated the fins. I descended into the darkness. Barnett lay face down on a steel plate that was the side partition of a rest room which, like the wreck, lay on its side. The height of the opening was the width of a door. I worked my way into the room by Barnett's side to examine the situation.

The regulator had fallen out of his mouth. According to his pressure gauge his tanks contained plenty of air. His pony bottle regulator was in place around his neck. There was no blood in his mask or signs of struggle. He was not entangled in any way, nor were there any

## 14 - The *Lusitania* Controversies

lines, wires, or projections that could have prevented his backing out. Oddly, a promenade deck window was lodged in the back of the room, within inches of his face. It seemed as if he was looking at the window when, for some unknown reason, he lost consciousness - but this is only speculation.

Smith and I had no difficulty in pulling out the body. We carried it up to the promenade deck. I put Barnett's second stage into my mouth and took a couple of breaths. The regulator delivered air effortlessly. We sent the body to the surface on a liftbag after taking the precaution of tying it to a decompression line. The trip was then terminated. It was a long ride home with the body of a friend lying on the after deck under a tarp. We could hide his face from sight, but we couldn't hide it from our imaginations.

The autopsy revealed massive embolism - exactly what you would expect after shooting a body to the surface from a depth of 190 feet. The medical examiner was not a hyperbaric physician and understood nothing about the expansion of gases in a non-breathing diver during an uncontrolled ascent. Why Barnett passed out or why he stopped breathing was never ascertained. Circumstances suggested that he overbreathed his regulator and passed out from a build-up of carbon dioxide. (We didn't know about oxygen toxicity at the time, so the possibility of a seizure was never considered.) Barnett was survived by a wife and four children.

Arena was concerned that the Coast Guard would find his charter arrangement out of accord with regulations. Technically, any chartered boat that spent more than twelve hours at sea was required to have two captains in addition to a mate. He pleaded with me to let him put me down on the investigation report as crew. I had nothing to gain by this other than being co-named in a legal action should Barnett's wife decide to sue, but in the end I let him use my name in order to get him out of trouble with the Coast Guard, and so he didn't lose his ocean operator's license.

## China from the *Andrea Doria*

Barnett's death did not put a damper on diving the *Andrea Doria*, any more than a highway fatality disinclines motorists from traveling interstate highways after driving past the site of an accident before the wreckage is cleared off the road. Death is a way of life. After the mourning, the survivors pick up the shattered pieces of their lives and carry on. A week after Barnett's death I was back on the *Doria*, exploring new territory and making more finds.

I returned again the following year, but not until 1983 did the *Andrea Doria* receive the promotional boost that fostered the allure the wreck so richly deserved. Prior to that time, divers clambered over the *Doria's* huge hull like fleas on an elephant's back - and visualized about as much of the wreck as fleas suppose about an elephant's behavior. All that was about to change.

Peter Gimbel's documentary on the salvage of a safe from the foyer deck also depicted a jumbled pile of china in the first class dining room, from which his team recovered quite a number of pieces. Although the deck plans clearly showed the location of the room, I wanted to know if any china was left, and, if so, what special conditions might apply in reaching the spot. Forewarned, as they say.

At a party after a symposium where Gimbel featured the film, I asked him about the china. Both he and team member Nick Caloyianis described the place for me. They cautioned me that getting there on scuba could be risky - Gimbel explored the wreck at the end of an umbilical hose that fed a never-ending supply of gas, and that also doubled as a guideline to the point of entry. The respect I hold for Gimbel's many accomplishments is not diminished by the concession that a wreck-diver he was not. By that I mean that his temperament and goals differed from those of wreck-divers. Rather, Gimbel was an adventurer and filmmaker who spent some time on wrecks. Exploring dark interiors was a specialty in which wreck-divers excelled.

I had already made incursions into the *Doria's* deep recesses along the boat deck and promenade deck. I had entered the very compartment where Gimbel found the china - in 1974, when I dropped through the small square cut-out. Since then, Gimbel's commercial divers had enlarged that opening to the size of a garage door. Entrance and egress were now easy and unrestricted.

Which is not to say that I contemplated the proposition without trepidation. It would be smug and presumptuous to sound audacious in retrospect. At the time, I anticipated the deed with great regard.

My buddy on this trip was Steve Gatto, a non-union electrician soon to gain affiliation and begin his apprenticeship. We met through the Dive Shop of New Jersey when our relationship was that between dive master and novice. Now he had reached his majority and, although this was his first trip to the *Doria*, he had paid his dues in the Mud Hole. Nine years earlier, when John Starace was my buddy on a similar *Doria* venture, we were dubbed "the Ford and the Ferrari." Starace was the Ford. Now, sad to say, I had become the Ford. Gatto was the Ferrari.

Because Gatto and I had a specific objective in mind, we wanted to establish the mooring line close to the foyer deck opening. I spread the deck plans across the table so Arena could see where we wanted to be. He dropped buoys on the bow and stern of the wreck, then positioned the *Sea Hunter* over the estimated location. When Gatto and I dropped down the anchor line we found the grapnel within sight of the hole. We shackled the down-line eight feet from the entry point.

## 16 - The *Lusitania* Controversies

I won't elaborate on the incidents of the dive since I've already given a detailed account in my book, *Andrea Doria: Dive to an Era.* Suffice it to say that we entered the black abyss, found the mother lode, suffered a few scary moments, and recovered a couple of bags full of china. The significance of the event lay in the abrupt change in perception experienced by the deep diving community toward the holy wreck of holies, the *Andrea Doria.*

Whereas previously divers strove for years for the opportunity to simply touch her hallowed hull, aspiring to bring back some trinket they chanced to find on her rusting exterior, hereafter the way was opened for them to explore the wreck's vast innards. No longer were the pitch-black compartments looked upon as unpenetrable catacombs from which one couldn't hope to return. The innermost recesses were similar to those within many other wrecks - a little deeper of course, more convoluted perhaps - but accessible all the same to those with an adequate base of experience and a respectable dash of daring.

Add the lure of fine Italia china, imprinted with the ornamental crown logo, and the *Andrea Doria* became a more desirable dive than ever: a Siren calling to wreck-divers who had a penchant for collecting lost relics. Literally overnight a new era was ushered in as interest in the wreck reached a fever pitch that, still today, shows not the slightest sign of diminishing.

### NEW KID ON THE *DORIA* BLOCK

Waiting impatiently to jump on the bandwagon was Steve Bielenda, owner and operator of the dive boat *Wahoo*, which ran out of Cap Tree, Long Island. Bielenda was running a weekend shuttle service to the *Oregon* and *San Diego* almost as if it were a commuter route. No sooner did the grapevine buzz with word of the great china redemption than he asked me to lead a charter to the appropriate location.

After finding a trove whose value could be called incalculable, one might wonder why I accepted his proposition. By keeping the spot and the route a secret I could have filled my bags with plates, cups, and saucers galore. To what end?

I suppose if I'd found a box of hundred dollar bills or negotiable bonds I'd have kept them all for myself - or at least have shared them with Gatto. But the value of an artifact is purely intrinsic: its worth is in the finding, in the retrieving, in the restoring, not in the possessing or in the amassing of unlimited quantities. One can display only so many plates of different designs; the extras go into storage.

Just as I found satisfaction as a dive master in offering opportunities for upcoming divers, I felt deep gratification in providing the means for others to achieve their wreck-diving ambitions. If they wanted artifacts that I had already found, I was willing to point the way. So I accepted Bielenda's offer to go on his trip as a non-paying crew member.

The *Wahoo* was a luxury yacht compared to the *Sea Hunter*. The 55-foot-long fiberglass hull sported a broad beam and plenty of bunks and deck space. The boat was amply operated by at least two captains and three or four mates. A cook prepared hot meals for the passengers and crew in a galley that boasted conventional and microwave ovens, freezer, refrigerator, and coffee maker, and hot and cold running water. The boat had a shower, television, and VCR - in fact, all the comforts of home. A dozen people could gather in the main salon for meals.

I showed Bielenda how to grapnel the wreck between two well-placed markers, instead of letting the hook fall where it may - the normal practice. This made my job easier under water. Helping to tie in was John Lachenmeyer. He had parted ways with the *Sea Hunter*, leaving Arena the sole owner. Lachenmeyer now owned the *Sea Hawk* in partnership with Frank Persico. Lachenmeyer's captain's license was available for trips to the *Andrea Doria*.

I could recognize subtle features of the wreck from the plans or from previous dives, and this enabled me to determine where on the hull I landed and to orient myself accordingly. Having made such a statement, I am forced to admit that on my tie-in dive with Lachenmeyer I found myself disoriented when the wreck first came into view. So providentially did the grapnel fall where I wanted that it dropped straight through the opening into the foyer deck, where the china was located. I had never seen the hole from above, so it took a few seconds to get oriented. We quickly shackled the chain about eight feet away and descended into the gloom, to emerge a few minutes later with a stack of patterned china.

Diving the "china hole" in the early days was like being on a game show in which the contestants won a specified amount of time in which to fill a shopping cart with whatever items they wanted as they raced along the aisles of a store whose shelves were stocked with rare merchandise.

Back on the boat I began a routine that I've continued to this very day. I drew a sketch of the wreck and the anchor line and its relationship to the china hole. Then I showed the customers how to find their way to the hot spot. I also warned them about any pitfalls they might encounter. No one came back empty-handed. Satisfied customers meant repeat business for the *Wahoo* and newfound friends for me.

By leading people to the china trough I had the rest of the wreck to myself, so I went exploring unmolested while the others bagged up. This led me to discover the riches of two gift shops and the wonderful items they contained: jewelry, figurines, and souvenir spoons with a painting of the *Andrea Doria* in the bowl.

Rarest and most unique of all my finds were unique pieces of art which once graced the inner wall of the first class cocktail lounge. These were glazed ceramic panels created by the Italian artist and pottery maker Romano Rui. They were worth more to me than a hundred bags full of china.

Bielenda's offer to act as guide was extended through subsequent years. This was a fortuitous opportunity for me, as Sal Arena relieved me of my duties on the *Sea Hunter* the following year. His behavior was always unacceptable, but finally he went too far. The story is worth telling.

## RELIEVED OF *DORIA* DUTY

The prospect of picking up a piece of china attracted a growing number of divers. My job was to set the hook within reach of the foyer deck opening. However, Arena managed to drop the grapnel at the wrong end of the wreck, three hundred feet from the china hole. I recognized the stern wing and enclosed promenade deck, and knew what I had to do. Dave Stroh helped me drag the line along the hull. We crawled and hauled the length of a football field, every yard of progress more difficult than the last. If Arena was doing his job properly, he should have been feeding us slack. But despite a current that was negligible, we felt as if we were being pulled backward.

There were moments when I was breathing so hard that I had to stop and rest. I'd partially catch my breath, exchange glances with Stroh, and struggle on - using one hand to hold the chain and the other to pull myself forward. We were only thirty feet from our goal, and nearly exhausted, when there was no more line to be had. We secured the chain to a stanchion on the promenade deck from which the china hole was visible. Then the two of us lay down on the hull and gasped, completely spent. Our bottom time was nearly over, but after a couple minutes of heavy breathing we felt strong enough to proceed, so we ducked into the dining room and scooped up several pieces of china that were exposed above the mud. All the other divers brought back souvenirs, too. From that perspective the trip was a rousing success.

In another way those days at sea were a nightmare. Arena got nastier as time went on. He cursed constantly and abused the customers verbally, he humiliated them at every opportunity, he interrupted their conversation with ridicule, and he ordered them about as if they were indentured servants. He behaved so dictatorially that he's been known ever since as Little Hitler (and worse).

Although none of Arena's vituperation was aimed directly at me - not because I was his crew but because I always replied in kind - I

## 1980's: Down to Deeper Depths - 19

finally got fed up with his foul-mouthed gruff and refused to hear any more. My equanimity had reached its end because I couldn't stand to see the customers so ill-treated. In the customers' defense, I lit into him so fiercely that he retreated to the sanctity of the flying bridge. He must have thought he was safe there because he continued to scream down at me.

I motioned him down to the deck with my fist. "When I was in the army I took all the chicken shit I'm ever going to take. So you come down here and we'll settle this once and for all."

The customers - my friends - were shocked. No one had ever heard me raise my voice much less call someone out for a fight. I was so well known for my tact that I was often asked to intercede in disputes in which I wasn't involved - cool heads were now and then at a premium. My easy-going temperament had earned me the name of Nonchalant Lamont. And besides that, I hadn't been in a fist fight since Kenneth Stoker beat me up in grade school. I admit it: I was overcome with anger. It happens to me every decade or so.

Arena must have thought I was bent on murder. He sequestered himself on the flying bridge for the remainder of the day. When he finally worked up the nerve to mingle with the troops, he told me that I was fired from my position. From a personal standpoint my newly appointed limbo status was beneficial. I no longer had to stand watches, I didn't have to help with boat chores, I could go diving whenever I wanted, and I wasn't required to pull the hook - that job was delegated to the customers, who had to pay good money for the privilege. Nor did I have to help steer the boat. More important, my opposition to his subjugation brought about the desired effect: Arena modified his behavior for the remainder of the trip. He continued to act tyrannically, but he no longer screamed invectives. (The change was not permanent: he resumed his obnoxious behavior on the very next charter.)

My long association with Arena and the *Sea Hunter* was over. Our parting was no sweet sorrow. But the story didn't end there. Arena asked Jon Hulburt to drive him home from Montauk. During the drive, Arena bragged to Hulburt how he had intentionally dropped the grapnel in the stern in order to prevent anyone from recovering china. I suppose he might have reasoned that an empty-handed diver would come back the next year to obtain the souvenir that he hadn't gotten the first time. Later, he freely acknowledged to me that he delighted in thwarting people from achieving their aspirations. It must have given him a sense of power and domination. Thus when he finally figured out from my bubbles that I was hauling the chain to the bow, he snubbed the line around a cleat in order to stop me.

Making admissions to Hulburt was a dimwitted conceit. Hulburt was one of those very customers who had paid so much to be raped. Hulburt also appreciated the potential harm in Arena's viciousness, and communicated this to me. Now I was not just angry, but incensed. By making me huff and puff along the bottom, Arena put Stroh and me at considerable risk: we could have passed out from the strain, and drowned. I can forgive a deed that is done through ignorance, but never through intent.

## The Lure of the *Andrea Doria*

Diving the *Andrea Doria* became firmly established as an annual event that many people looked forward to, first-timers as well as those who couldn't get enough of what the wreck had to offer. Because of this attraction, and because the wreck had become so readily accessible, divers with intermediate experience were inspired to gain the deep-water skills and proficiency which they needed to possess in order to make the dive with relative safety, and to be comfortable enough at depth to make the dive a rewarding experience. Thus the *Doria* became the carrot on the stick that encouraged many underwater adventurers to achieve their full potential.

## Repetitive Dive Extrapolation

The china in the *Doria* lay at a depth that exceeded 200 feet. The Navy Tables did not permit repetitive dives beyond 190. This was not a function of physiology but of testing and mathematics. At those depths, the U.S. Navy found it more expedient to put divers in the water for a single long exposure instead of for multiple exposures of shorter duration. Consequently, for exposures deeper than 190 feet they didn't bother to produce repetitive group designations needed to calculate residual nitrogen times. This didn't mean that repetitive dives to depths of 200 feet and beyond couldn't be made, only that no one had ever done it.

I overcame this shortcoming in the Navy Tables by a method of extrapolation. For example, the total ascent time (decompression time) required by a 20-minute dive to 210 feet was 40 minutes. The next larger total ascent time at 190 feet was 44 minutes, from which I drew the repetitive group letter N. After five hours N became C, which yielded six minutes bad time in the repetitive dive timetable for 190 feet. I added six minutes to my planned repetitive dive time, rounded up to the next five minute increment as an additional safety or fudge factor, and decompressed according to the timetable recommended for a 210-foot dive of the resultant duration. This meant that on a repetitive dive I could do 14 minutes at 210 feet, followed by a decompression of 56 minutes, which was the time required for a non-repetitive dive for 25 minutes at the same depth. I did it for years, and it worked.

After originating this repetitive dive methodology and field testing it myself, I didn't think it would hurt to have scientific verification after the fact. I called the University of Pennsylvania, and wound up fomenting fierce disaccord with Chris Lambertson, head of the hyperbaric department. He had a fit when I told him that I was diving in excess of 200 feet on scuba. He said it couldn't be done. I thought he must have misunderstood, and heard me to say that I was *contemplating* such a dive, so I repeated myself casually, emphasizing that I had made many such dives over the years, and that repetitive deep dives were routine. I wanted to know about repetitive dives to depths deeper than 200 feet.

He refused to hear it. He told me gruffly how much experience he had working with animals in the laboratory. He rambled on about decompression models and hyperbaric experiments and theoretical assumptions. When I finally had a chance to interject, I reminded him that I commonly made repetitive deep dives without any ill effects. Ergo, it couldn't be impossible and there was nothing theoretical about it. I've done it, therefore it's done.

"Then you're no sport diver," he shouted angrily. "You're a commercial diver."

"Commercial divers get paid for diving," I rebutted. "For me it's just a hobby."

"There are no other categories," he insisted. "You're either a sport diver or a commercial diver."

I couldn't understand his compulsive need to categorize my activity, to make it fit into a niche that he had long ago preconceived. Why couldn't he accept what I did because I did it? After all, I wasn't bandying idle speculation, I was presenting empirical data. He couldn't accept my data because they hadn't been observed in his precious laboratory - my diving activities occurred in the real world where he exercised no control. Eventually I came to realize that Lambertson possessed an unalterably closed mind to all matters of fact that contradicted his unscientific presumptions. Lambertson was the kind of scientist impersonator who, five hundred years earlier, would have crucified Galileo for his heretical observations, or insisted that the world was flat despite Magellan's circumnavigation.

I didn't let Lambertson's blinders prevent me from making repetitive deep dives. Nor, I suspect, did he let my observations of truth shake his prejudices. Down with empiricism!

## Passing of the Old Guard

Digressing for the nonce from the Grand Dame of the Sea, whose influence on the evolution of deep wreck-diving cannot be overemphasized, other wrecks and occurrences played influential roles whose value should not be overlooked. Most prominent among events was a death dealing dive on the *Sommerstad*, at 170 feet. The date was July 12, 1982.

I was nineteen minutes into a twenty minute dive and about to turn for the anchor line when I saw a strange tableau in the distance. Something was wrong with the picture that I couldn't at first comprehend. John Dudas was crouched on his knees by a steel beam on which was tied a sisal line that angled up toward the surface. He appeared to be examining the knot. Facing him from across the beam was Bill Nagle, who seemed to be waiting in abeyance. Approaching Dudas from the opposite side was Kathy Warehouse. She drew to a halt a couple of arm's-lengths away and made no closer approach. She just watched.

My attention was distracted momentarily by movement overhead. Adjacent to the nearly vertical sisal line was a falling decompression reel, unwinding in extreme slow motion as if someone on the surface or at a shallower depth had dropped the reel but grabbed the end of the line. Visibility was fifty feet laterally, and better when looking up. The reel at that moment was forty feet off the bottom, and I could see parallel lines of sisal another fifty feet above.

When I looked back at the trio in the wreckage it came to me what was amiss: no exhaust bubbles were erupting periodically over Dudas's head. His regulator hung limp on his chest.

I charged forward in sudden anger, shouting mentally, "Don't you people see what has happened? Why don't you give him some air?"

I shoved a regulator between Dudas's lips and desperately pushed the purge. The air that went into his mouth came right back out, as if his throat were sealed off from his lungs. I placed a hand on the back of his head so I could hold the regulator tight between his lips. But no matter how long I held down the purge button, the air simply filled his oral cavity and burped out around his mouthpiece. The needle on his pressure gauge was pegged on zero. When I squeezed his arm I felt that awful muscle limpness that I'd encountered too often in the past.

John Dudas was dead.

Once I accepted the dire reality I shifted mental gears. Rescue devolved to body recovery and time was running out. I tried to get Nagle to help but he was useless. He was traumatized and completely nonresponsive to my entreaties. His eyes maintained that dull, vacant stare of the dispossessed. Not only was Dudas his best friend, but he held Dudas's skills in such high regard that he was overwhelmed by an inability to acknowledge the unthinkable truth that Dudas had committed some deadly mistake. Warehouse, once I got her attention, was willing to take direction.

I yanked a nylon choker from my mesh bag and handed it to Warehouse while I unrolled a liftbag. She slipped the choker under Dudas's tank strap, but I removed it and looped it around his chest. I didn't want to send the tanks to the surface only to have the body slip out of the harness. I put enough air in the liftbag to pull the choker taut. Then I took my decompression reel off the back of my tanks and tied the end of the line to the loop in the choker.

With the rigging secure, I pulled off a few feet of sisal and handed the reel to Nagle. He was still in shock and didn't even know I was there. I shook him by the shoulder. He transfixed me with his sightless gaze. I punched him hard and shoved the reel against his chest. He finally took hold of it. I indicated what I wanted him to do: the simple procedure of grasping the dowel by the ends so the line unreeled as the liftbag rose to the surface. He performed the task like an automaton.

I ascended to the liftbag floating above Dudas's head, held my regulator under the opening, and purged air into the opening until the liftbag started to rise. I dropped back to the bottom as the body soared away. Nagle did his job mechanically. The sisal spun off faster and faster then came to a sudden halt. I cut the line and tied it to the same beam to which the other line was tied.

Now I was in desperate straits. My tanks were down to 500 psi. My planned twenty minute dive had stretched to twenty-five. And my legs were horribly wrapped in thick strands of monofilament with which the wreck was enshrouded. Warehouse came to my rescue. She dropped down around my feet and slashed away with her knife until I could kick my legs free. Then I grabbed Nagle and yanked him toward the anchor line, which was out of sight beyond the limit of visibility. I prayed that my sense of direction would not fail me now.

Kicking hard, we reached the grapnel about a minute later. I pulled myself fast up the line for the first seventy feet, then slowed to the proper rate of ascent (which at that time the Navy considered to be sixty feet per minute). I was breathing like a banshee from the exertion of the recovery and the fast swim. No current was running, so I concentrated on breathing deep and slow. I soon got my breathing rate down to three breaths per minute - my usual rate at rest during decompression. I can cut it down to two breaths per minute in an emergency, but I never feel satisfied with the quantity of air which that rate provides. But no matter how conservative I was with my air, I still wasn't going to make it through the extended decompression.

I switched to my pony bottle while I still had some reserve. Then, rather than trying to tough it out on my own, I indicated to Gene Peterson that I was low on air and that I needed to use his back-up. Several other divers were on the line, but there was no way to communicate to them the dreadful events that had transpired. After forty-five minutes of decompression I surfaced to find the body floating impiously at the end of a rope, and a boat full of numbed divers bursting with tears, sorrow, and mournful curiosity. They needed reassur-

## 24 - The *Lusitania* Controversies

ances that no one could provide.

The burning question that rippled through the boat that day, and through the diving community later, was: how could an accident possibly occur to the diver who was the most skilled, the most experienced, and the most respected of them all? And: if Dudas was not infallible, what were my chances of long-term survival?

John Dudas was a friend with whom I shared mutual respect. I had been to his house for meals and he had been to mine. We had been on many trips together - although we had never dived as buddies because we both preferred diving alone. We had spent many hours discussing shipwreck research. We had sat together at social gatherings. Yet I would be remiss if I tried to make excuses for his shortcomings. No disparagement to his memory is intended by analyzing the mistakes he made that resulted in his demise.

After the sordid ordeal was over I reconstructed from available evidence the most likely scenario to account for his final moments. Pre-dive conversation played an ironic part in establishing Dudas's antiquated methods and mind set.

I quipped with Dudas as we were getting dressed for the dive. When he said with mock annoyance that his tanks weren't filled to capacity, I suggested jokingly that he complain to the person who had shorted him on air, knowing full well that *he* had filled the tanks. He was an instructor who owned a backyard dive shop. He responded on cue, "I can't. I filled the tanks myself."

During a discussion about back-ups he showed me the octopus rig he used for teaching students. I asked him if he wanted to borrow my spare regulator to put on his pony bottle. He said he didn't believe in pony bottles; he carried enough air on his back. He also mentioned that he hadn't done any deep diving in years, having to spend all his available time doing check-outs in the quarry.

Nagle complained about a problem he was having with a regulator he had bought from Dudas. The second stage venturi kept icing up and wouldn't deliver air; he had to breathe off his alternate regulator until the ball of ice melted. Dudas scoffed at Nagle's complaint because he was using the same brand, and he had never experienced icing. Peterson confirmed Nagle's diagnosis, adding that because that brand of regulator was so unreliable both he and Norman Lichtman refused to sell it at The Dive Shop of New Jersey.

After donning his wetsuit Dudas took a buoyancy compensator from his gear box. He looked at me, scowled, and said, "I only wear this when I have to look good for the students." He rolled up the BC and put it away.

Dudas entered the water five minutes before I did. His bottom timer stopped automatically when his body hit the surface; it showed thirty minutes. That was precisely when I looked at my own timer.

Thus I was able to calculate that when I came upon him on the bottom he had been dead for only five minutes. But dead is dead, whether it's for a minute or a month.

When we first reached the wreck site a fishing boat was already there. Arena spent more than an hour grappling the wreck because the anglers were working along a bridle. A bridle consists of two anchors connected by a long loop of rope. One anchor is placed at either end of the wreck, and the connecting rope is reeved through a ring secured to the boat. Pulling the boat back and forth along the rope enables the anglers to fish the entire length of the wreck. One diver mentioned that he passed the bridle rope at a depth of ninety feet. This would explain why Dudas's reel was unwinding slowly when I spotted it: he must have passed the ninety foot mark when he let go of the reel in his last desperate bid to swim to the surface, and the reel fell down on the other side of the bridle, which hung up the sisal.

What person who is nearly out of air takes the time to tie off a decompression line, knowing that he doesn't have enough air to reach the surface much less to decompress? Dudas must have been so complacent on the bottom that he didn't monitor his air consumption throughout the dive, and didn't bother to check his pressure gauge prior to ascent. He either forgot that air is consumed in greater quantities at depth, or he lost his respect for the hostility of the environment.

Couple this information with what Dudas had told me years before: that because of neoprene compression he had to kick his way off the bottom, and that his wetsuit didn't gain positive buoyancy until he reached fifty feet. I often used to watch him decompress upside down, clinging to the anchor line so he wouldn't float to the surface.

From all these facts I inferred that somewhere above ninety feet but deeper than fifty, Dudas ran out of air. Unable to power his way to the point where neoprene expansion would float him, he lost consciousness, then fell back down to the bottom, coincidentally alighting at his tie-off point. Moments later, Nagle happened along, realized that his best friend was dead, and went catatonic at the sight.

What caused Dudas's death? In the back of my mind lies the horrible thought that Nagle's offhanded comment might have confused him for a precious couple of seconds. If Dudas thought that his regulator had iced up, he might have wasted valuable time groping for his other second stage - then wondered how *that* one could have iced up too if he hadn't even used it. But I don't truly believe that it was a contributing factor.

Dudas's death was more the result of subconscious machismo, of reliance upon strength, of a desire to preserve the old ways of diving, of his disapproving attitude toward progress. He carried no pony bottle, wore no buoyancy compensator, and repudiated the reliance upon back-up plans and procedures. He was a dinosaur among mammals. The legacy that he left behind - for those who paid attention - was an eye-opening reinforcement that the sea is a cruel courtesan who pays no tribute to those who ignore the hard-earned lessons of the past.

These sentiments are not intended as an indictment. Dudas was a product of his times, with a mental disposition that resisted change. The earliest divers were pioneers who necessarily felt comfortable in accepting more risk than the divers of later-day refinement were willing to accept. Because pioneers are used to doing things the hard way, they feel little drive to alter their behavior or to adapt to new and improved methods of doing things. They often repudiate progress.

In the final analysis, many divers - myself included - found that bad habits and unsafe practices were deceptively easy to acquire, especially after too long a period of security. Life is a risk, but diving is riskier. And for that reason a diver can't afford to let down his guard, can't let vanity overrule common sense, and can't let peer pressure prompt daring deeds that one is not presently equipped to handle.

Dudas left more of a mark on the world than three children and a pregnant wife. If his death can be viewed in any light that is positive, it is this: his passing raised the level of awareness among deep divers everywhere that strict attention must be paid to every facet of a dive, from planning to equipping to execution, with particular heed paid to contingency back-ups for unanticipated occurrences. Disdain and arrogance have no place under water.

## THROWN OFF THE *SEA LION*

At Dudas's funeral, George Hoffman was a pall bearer on the handle opposite mine. Across the coffin I said solemnly, "Hello, George." He looked me up and down, from necktie to black shined shoes, and replied grimly, "Nice suit." That didn't leave me much of an opening. Those were the last words we exchanged for the next ten years.

The cause of Hoffman's enmity went back to the previous year, when we had a falling out. From late autumn to early spring, when the water was exceptionally cold, the *Sea Lion* sat idle at the dock. Each club had a few hard-core divers, but not enough to fill the boat. Since I had contacts with all the major clubs, and knew the divers scattered throughout the realm who wanted to dive deep wrecks, I was able to organize charters that no one else could fill. Some of these trips I ran myself, others I scheduled through The Dive Shop of New Jersey.

I made special arrangements with Hoffman to charter the *Sea Lion* out-of-season on one condition: we dived deep or we didn't dive at all. This meant that if the weather was so bad that we couldn't reach the Mud Hole, the trip would be canceled. There would be no Money Wreck shenanigans. I couldn't entice hard-core divers to drive long distances at a time when the weather was the least predictable, if they thought they might get stuck leaving the dock just so Hoffman could justify charging a charter fee for the day. These people didn't just want to get wet, they wanted to make a dive that was challenging. Hoffman agreed because the alternative was to have no charter at all, instead of the possibility of picking up a few extra trips on those weekends when conditions were favorable. As a corollary, I checked with Hoffman the day before a dive so we could decide then and there whether or not to proceed.

The system worked well until Hoffman purchased a new boat to replace the clunker he had been herding people onto for years. A storm was on the horizon and it wasn't all due to the weather: a new and larger mortgage contributed the most to the upcoming fury. The wind blew hard one day before a late autumn dive, and small craft advisories were posted by the Coast Guard. Under the conditions of our agreement the trip should have been blown out. But when I called Hoffman on the phone to confirm, he demurred. He suddenly needed lots of cash and he needed it right away. He wanted to take us to the *Delaware*. When I reminded him of our special arrangement not only did he renege on the deal, but he threatened retaliations if I did not show up with a check. Reason works only with reasonable people.

Since this dive was on the Dive Shop schedule, I called Norman Lichtman and discussed the matter with him before taking appropriate action. I also foretold what measures Hoffman would adopt if he didn't get his way. Despite the consequences, neither Lichtman nor I were willing to be intimidated into backing down from a position of rightfulness. After all, we had obligations to divers who depended on us to make proper decisions in their behalf. I couldn't see people driving from as far away as Pittsburgh and Washington, DC to make a leisurely dive on the *Delaware*. Pittsburgh was seven hours away by car, and Washington, DC was four.

So who should call Hoffman and give him the bad news?

I wanted Lichtman to do it as the owner of the shop. He thought it was my responsibility as the dive master. I acquiesced and made the fatal call. Hoffman reacted predictably in keeping with his character: in resounding dialogue replete with anti-euphemisms, he canceled all twelve Dive Shop trips for the upcoming season, he refused to take other Dive Shop charters for the rest of his natural life, and he banned me forever from setting foot again on the *Sea Lion*.

Now I know how Demeter must have felt in the moment before he was killed by Democrates for bringing him the news of Sparta's defeat. (This incident inspired the phrase that has come down through history as "Don't shoot the messenger.") Hoffman's reaction confirmed my long-held aphorism: George Hoffman had no friends, he had only paying customers.

It might seem to some that my unwillingness to accept disgraceful conduct was counterproductive in that it diminished the transportation opportunities available to me, and reduced accessibility to certain wrecks. Be that as it may, I've always believed that one whose standards are flexible really has no standards at all. Many things in my life are more important than diving. Principle is outstanding among them.

### ANOTHER FATALITY ON THE *SOMMERSTAD*

Although I don't like to dwell on morbidity, the *Sommerstad* was the site of another misfortune which occurred coincidentally on the first anniversary of Dudas's decease. I was diving with Jon Hulburt when he called my attention to a flashlight beam in the distance. I figured that someone had dropped a light. As we got closer, however, we saw that the still-burning light was attached to the wrist of a diver who wasn't blowing bubbles.

Any prospect that he might still be alive was dashed when we saw that the mask was half-flooded with blood-stained water and that the vacant eyes were open and immersed. Perfunctorily I tried to give him air, but Jerry Rosenberg had breathed his last. His tanks were bone dry. Hulburt and I had no need to communicate intent: we both knew what had to be done, and we did it. In a trice we had the body rigged with a liftbag and a decompression line, and we sent it to the surface. We tied the sisal to a piece of wreckage at the very edge of the wreck. On the way back to the anchor line, unflappable Hulburt retrieved a porthole.

Unknown to us as we returned to the anchor line, the air in Rosenberg's drysuit expanded so much during ascent that the body overtook the liftbag, which flipped upside down and dumped out all its air. Then the neck seal blew out when the body reached the surface. Arena jumped overboard fully clothed when he saw the largely distended drysuit purging itself of air. The body sank before he reached it. Soaking wet, Arena radioed a report to the Coast Guard. He had no way yet of knowing the circumstances of the calamity and he couldn't assume the worst. It was distinctly possible that the diver who had surfaced and then resubmerged was only unconscious, and that he might resurface downcurrent of the boat. The Coast Guard dispatched not only a search-and-rescue helicopter but a 150-foot buoy tender that happened to be nearby.

The full story did not emerge until all surviving divers returned to the boat. Jerry Rosenberg teamed up with Mike McGarvey and Brad Sheard. When Rosenberg indicated that he was low on air, both

McGarvey and Sheard tied off decompression lines about ten feet apart, Sheard figuring that three divers on a single line might create complications. No sooner had they left the bottom - ascending together in parallel - than Rosenberg ran out of air. McGarvey handed his pony bottle regulator to Rosenberg, who then clung to McGarvey's tanks and began to hyperventilate. Air that should have lasted twenty minutes or more Rosenberg consumed in two.

McGarvey was now task-loaded with controlling his ascent while unreeling the line and fighting the extra weight on his back. Rosenberg yanked the regulator from McGarvey's mouth and put it into his own. McGarvey swung around to try to dislodge Rosenberg and reclaim his only remaining source of air. They fought over the regulator. McGarvey wrested the mouthpiece away from Rosenberg, took two breaths, then passed the regulator back, but in those few seconds Rosenberg lost consciousness and slowly sank out of sight. Sheard witnessed the horror-filled drama from a position of relative safety.

Hulburt and I provided the finishing touch by recounting our part in the affair. Arena then called the Coast Guard and advised them to call off the search. He gave two sound reasons: the diver was dead (not adrift downcurrent) and we knew where the body was - tied to the wreck at the end of 170 feet of sisal. The Coast Guard willingly provides lifesaving service, but once it is ascertained that a person is dead they cease all search and rescue operations. They will not conduct a body search. This time, however, they countered with a proposal that seemed completely out of character: they dispatched another rescue vessel whose underwater recovery team intended to bring up the body for possible resuscitation.

If this sounds like the harebrained scheme of a modern Dr. Frankenstein, it isn't far from the mark. A physician who was advised of the situation had his own recommendation to make: he thought that Rosenberg - although clinically dead - could be revived as a victim of cold water drowning. Granted that the bottom temperature was twelve degrees above freezing, there were other factors to consider that common sense conceded. Hours must pass before the body could be brought to the surface and transported to a hospital, during which time the lungs could not be respirated. Even though no more nitrogen was being absorbed by the tissues, a significant decompression penalty would be violated by the sudden reduction of pressure. And the body had skyrocketed once to the surface, which must have embolized the lungs beyond recuperation. The Coast Guard (or the doctor advising them) refused to accept the knowledgeable advice of people in the field.

Death is forever.

What put a kink in the immediate plan was Arena's offhanded mention over the radio that the body lay 170 feet deep. This was a

## 30 - The *Lusitania* Controversies

depth that the Coast Guard divers were not capable of dealing with because it exceeded by 40 feet the arbitrary depth limitation established by the certifying training agencies. The rescue vessel turned back. Timidity earned our scorn.

The buoy tender continued to prowl around the *Sea Hunter*, and the helicopter ran a zigzag pattern behind our stern - just in case. The Coast Guard stubbornly refused to accept our analysis of the situation. In fairness, the Coast Guard didn't know our expertise. Nor did they give up readily if there was a chance of effecting a rescue.

The Coast Guard then offered suggestions with the force of demands. Arena played the uncomfortable middle man as a confrontation developed between us and the Coast Guard. "Us" in this case meant Hulburt and me. The Coast Guard was likely driven by consultation with the doctor. Arena listened to the Coast Guard radio operator, shouted down from the wheel house what they wanted us to do, then we responded, and Arena told the operator what we had said. I didn't envy Arena his position, for he had to take headstrong gruff in both ears, and pass it along to someone who didn't want to hear it.

The Coast Guard argued fiercely for us to dive again at once and to refloat the body so it could be rushed by air to the hospital. Hulburt and I were willing to do the job, but not at the expense of incurring unnecessary risk. With only a few minutes surface interval and desaturation time, it would be near suicide to make a repetitive dive to 170 feet for the length of time that was necessary to complete the task. The doctor, sitting comfortably away from danger in his temperature-controlled office, and completely ignorant of hyperbaric medicine, either didn't understand or didn't care about the potential consequences of the actions he was prescribing.

I was reminded of the army officer whose troops suffered great losses while storming a hill held by the enemy. He was forced by overwhelming resistance to retreat and leave his casualties behind. So he ordered another attack just to recover the dead. "I don't care how many more men we lose, I don't leave bodies behind."

Sacrificing lives for so meaningless a cause is an asinine attitude that I can't comprehend. Especially when the life to be sacrificed was mine. Hulburt agreed with me wholeheartedly. Yet the doctor insisted more volubly as the minutes continued to pass.

From the deck I shouted up to Arena in the wheelhouse, "Tell the doctor if he's so damned determined to have the body, to come out here and bring it up himself." Hulburt was more explicit.

Meanwhile, we poured over the U.S. Navy Air Decompression Tables to determine how soon it was reasonably safe to go back down. The bottom time of our first dive was twenty-one minutes, after which, with a fudge factor of conservatism plugged into the formula (Navy Tables were perilously short) we decompressed for thirty-nine.

We wanted to have at least fifteen minutes to carry out the recovery - and not have to hang until our skins shriveled off. The shortest surface interval that we were willing to accept was two hours and twenty minutes.

This did not sit well with the doctor. We argued back and forth with Arena acting as a two-directional mouthpiece. Hulburt and I refused to budge from our position - we wouldn't go in a minute sooner. We lounged on the boat while the silent bubbles of nitrogen percolated through our bloodstreams to be offgassed through our lungs. The process is invisible but very real. The doctor lacked our intellectual awareness.

Finally the moment came. The helicopter flew circles around the boat, ready to charge to the rescue the instant the body hit the surface. We rolled overboard with extra liftbags in hand and a firm plan in mind. Rosenberg's drysuit should have been drained of air so overexpansion shouldn't recur. We intended to remove his weight belt in order to add to the overall buoyancy that was provided by his drysuit.

The water was clear and cold below the thermocline: fifty feet of visibility and 44° F. We oriented ourselves at the grapnel, then struck out straight across the wreck toward the tie-off point. (On the first dive we had skirted the perimeter - a longer and more circuitous route.) Bright ambient light enabled us to recognize features along the way and to correct our course accordingly. We found the tie-off within minutes, but we never would have relocated the body if we hadn't had a trail of sisal to follow across the sandy desert. It had come to rest some one hundred fifty feet off the wreck.

We examined the body closely. By some eerie coincidence the weight belt had come unbuckled and lay on the bottom beneath Rosenberg's waist - as if someone else had gotten there before we did and removed it. Hulburt and I exchanged uneasy looks. We disconnected the original liftbag - which was held up at the end of the choker by a trapped pocket of air - as a precautionary measure of mistrust. Then we secured another liftbag to the loop. It required only a couple of puffs of air to lift the body off the sand and to send it soaring to the surface.

Now occurred an unanticipated consequence. As the body soared upward the sisal line was picked up off the bottom like the shaft of an inverted pendulum. We kicked as hard as we could but within seconds our guideline to the wreck was out of sight. Hulburt took a compass bearing and I aligned myself with the direction in which the line disappeared. We had a few anxious moments as we swam across the featureless white sand, but managed to find the wreck at the precise location of the tie-off point.

The sisal stood vertical and taut, but I tugged on it anyway to make sure there was tension on the line. Then we oriented our way across the wreck. Ever alert, Hulburt retrieved another porthole on the return to the anchor line. We began our ascent at thirteen minutes; we decompressed for forty-two.

Unknown to us, there now transpired a debacle of epic proportions. Sheard donned his drysuit and fins, swam to the body, and, after ascertaining that the weight belt had been removed and that the body was held afloat by the drysuit and not the liftbag, he cut the sisal line that secured the body to the bottom.

The buoy tender launched a small boat to aid in picking up the body (I refuse to write patient). Sheard swam back to the *Sea Hunter* after the boat arrived. The Coast Guard rescuers gathered in the liftbag and held on to the body while the helicopter positioned itself for the lift. The men in the boat couldn't figure out the quick release mechanism on the tank harness, and they didn't know how to disconnect the drysuit inflator. Diving systems were alien to them. Holding on to Rosenberg by the arms and the head, they secured the downline from the helicopter's winch to the torso with the tanks still attached.

The helicopter had trouble maintaining position. It rocked back and forth as it rose and fell. Finally given the go-ahead to haul away, the pilot didn't compensate properly for the additional load. The liftbag was still secured to the body, and was now weighted down with two hundred pounds of water. The body was dragged through the wave tops for a couple of dozen feet, lifted up to spin crazily on the wire, then dropped back into the sea where it submerged - again, and again, and again. After being pummeled in this fashion for several hundred feet, the winch operator succeeded in pulling in enough wire to lift the body clear of the surface. Soon the body was winched up until it was even with the cargo deck. The helicopter hovered as the crew tried to get the body into the cabin, but the tank straps and the liftbag hung up on the protruding edge of the platform. Then the situation got really out of control.

The performance would have been comic if it hadn't been obscene. The body hung half in and half out of the cargo bay, trussed up like side of beef in a butcher shop. The flyboys didn't know any more about scuba gear configurations than their waterborne counterparts. Unable to extricate the body from the tank harness and drysuit inflator, they slashed the straps and hose with a knife and dragged Rosenberg and his equipment aboard piecemeal. The liftbag remained flapping outside the helicopter. Hulburt had fitted it with Kevlar straps that couldn't be cut. After the flight, the liftbag had more than fifty holes in it.

Worse was yet to come. The helicopter stayed so long on site, and maneuvered so long while the body was being loaded, that it was running low on fuel. The pilot transmitted a mayday to Coast Guard headquarters that he might have to ditch the million-dollar aircraft at sea. As it raced toward land, another helicopter was dispatched on an intercepting course in order to pluck the survivors out of the water. It was touch and go for a while, but with the second helicopter flying escort, the first one barely managed to make it back to base without crashing.

Jerry Rosenberg was pronounced dead at the hospital.

Poor Rosenberg was a likable fellow who deserved better than he got. He also shouldn't have been diving in water that was so far over his head.

His wife was a rare woman. Although she was crushed by the loss of her husband at so tender an age, and so early in their lifelong commitment, she was consoled by the fact that he died doing something that he loved so dearly. Her eulogy was a fitting finish to a tragic human drama.

### . . . Back at the Ranch

Meanwhile, I was thankfully alive and abiding in uncomplaining poverty.

None of my friends or relatives understood my reason for renouncing a secure and well-paying job to endure a life of financial hardship. I admit that it doesn't seem like a sensible choice to those impelled by the sole desire to accumulate material wealth. Only an author or artist can truly appreciate the inner urge and motivation that make self-expression a prerogative.

Nor did my friends understand my dedication to writing. They believed that I sat around all day watching television. Little did they know that I didn't even own a set. I lost the majority of my worldly possessions in divorce, and most I never replaced. I subsisted quite well without a radio or television, without knowing the latest songs and shows or the hottest jingles and commercials. This gap in my knowledge of current events I managed to survive.

I used to quip that I was so broke that I couldn't pay attention, and sometimes that wasn't far from the truth. Yet I always knew where my next dollar was coming from - and exactly how much and when it was coming, since dividends were regular and paid on time. I lived on a minuscule budget that couldn't tolerate much in the way of unexpected outlays such as home or auto repairs and medical bills. The only charity I donated money to was myself.

By the time I paid my fixed expenses - taxes, mortgage, utility bills, health and automobile insurance, and child support - I was left with $2,400 per year to spend on other essentials such as food, clothing, gasoline and auto maintenance and repairs, books, diving, and so on. Impecuniosity forced some major revisions in my lifestyle. I became conservative in my spending (some said cheap) and scrimped however possible. Childhood companion Tom Gmitter knew that I kept the house heat low; he stopped by for a visit one wintry afternoon with his overcoat draped over his arm - so he could put it on after coming inside.

Long gone were the costly TV dinners of my rich electrician days. I had to learn to cook the cheapest foods the market had to offer. I never wasted so much as a single grain of rice. Once, when a box of crackers got soaked in my cooler on the boat, I spread the crackers on a towel in my yard in order to dry them in the sun. I couldn't afford to throw them out because I didn't have the cash at the time to buy another box.

Clothing I didn't care about as long as I was warm. They may say that clothes make the man, but I say that clothes merely cover him up; it's what's underneath that counts. I had a penchant for wearing garments till there was nothing left to wear. In this regard I was a trend setter: I dressed in jeans with jaggedly torn knees long before it was popular to do so. When my trouser legs got threadbare I cut them off above the knees and converted them to shorts: the first "cut-offs." And "threads" to me wasn't a beatnik phrase but an accurate description of what I sometimes dressed in.

I kept clothes long past their usefulness, almost until they fell off my body. On the way home from a Long Island dive that was blown out by a summer storm, a group of us stopped at Coney Island. I didn't go on any rides, but I was astonished to find William Beebe's original bathysphere on display. While walking through the park, Steve Gatto and Tom Packer took note of the well-worn seat of my shorts, which was not just thin but was showing through the bottoms. They ran up behind me and yanked down on the tattered legs, exposing me completely to a posterior exposure. There was not enough material left to patch.

Under my drysuit I wore longjohns leftover from my electrician days. They were pretty thin, so I donned one pair over another for added warmth. Once, dressed in this attire between dives, my private parts dropped into public view through the remnants of the crotch - to the embarrassment of the women on board. To solve the problem I wore one pair backward. There was no truth to the rumor that I got my clothes from CARE packages. But I accepted hand-me-downs and throw-aways from anyone. I had to. I didn't give to charity because no charity would accept what I had to offer.

I made household rags out of old tee shirts that I thought were too worn and torn to wear. One day after washing the laundry, while sorting and folding my clothes, I discovered a distressing fact: I couldn't tell the difference between the tee shirts and the rags; they all looked the same. I didn't know which to wear and which to use for scrubbing the floor.

For typing paper I used the backs of letter-sized sheets that were printed only on one side: discarded correspondence, duplicate photocopies, advertisements received in the mail, and any old papers that were slated for the trash.

If poverty builds character then I had my share of both. But I disagree with the philosophy of starving artists that hunger accents creativity. I've never been able to write when my stomach was feeling pangs.

I paid income tax to the State, and school tax, personal property tax, and real estate tax to the city and county, but with allowable deductions I didn't have enough earned income to meet the minimum taxation threshold according to the rules of the Internal Revenue Service. Not only did I not have to pay federal income tax, one year I got a rebate! As usual, the amount due on my tax form was zero, so I didn't send any money. But because of a new (and temporary) rule, I fell into the category of a disadvantaged household. The IRS paid *me* instead of me paying them.

I paid my bills like a juggler with at least one ball in the air, using a system that I called "defrayed prestidigitation." Say a bill was due on the twenty-eighth of the month. I didn't have enough money in the bank to pay it. But a direct deposit was due to be made into my account on the first. I mailed the check so it would arrive on the last possible due date, calculating that by the time it was handled, deposited, and cleared by the bank, my account would have sufficient funds to cover it. I didn't necessarily live from hand to mouth; I lived from credit to debit. I used credit cards as a convenience but paid them off at the end of the month so as not to incur wasteful interest charges.

While this self-imposed indigence had its drawbacks, the reader should by no means feel pity for my plight. There were great boons and benefits to my new way of life. I didn't have to get out of bed to the clamor of an alarm clock, I didn't have to fight commuter traffic or wait in the snow for a bus, I didn't have to deal with the quirks of narrow-minded co-workers, I didn't have to take orders from a domineering boss, and I was always home in time for dinner.

There was a marvelous sense of freedom in having every day to myself in which to do anything I wanted. And while I was wont to quip that an author is always up at the crack on noon, in reality I put in more hours per year at my chosen occupation than I ever did as an electrician. Outsiders found this difficult to believe because of the number of activities in which I was engaged.

Unlike those who work from nine to five for fifty weeks a year, and have their evenings and weekends off, my hours were infinitely flexible. At home I labored seven days a week, often till late at night, and accomplished nearly a fortnight's work in the course of a single week.

Then I might take a two-week trip without losing any income. By the end of the year I worked many more hours than the average person who had a full-time job.

I skewed my schedule to fit the diving season. In winter I became a recluse and slaved away at my desk. I had few distractions and not much of a social life, although there was the occasional girlfriend and, rarely, a relationship. One day merged with the next through consecutive writing sessions till I lost all track of time. "What day is it?" I might ask someone who happened to call on the phone.

Once I started a writing binge I disapproved of interruptions. One time I worked for eleven days straight without ever leaving the house or even opening the door. I didn't suffer from cabin fever; I enjoyed it. It's a good thing, for without a master to crack the whip, a person lacking will and self-discipline will lose his sense of purpose. An author who doesn't have a strong work ethic will not be very productive.

My adjustable work schedule enabled me to travel for weeks or months at a time, then make up for it by working round the clock.

During the most active part of the dive season I was constantly on the go. One summer I had so many trips scheduled back-to-back that I was away from home for three straight months. I emptied the refrigerator and turned off the electricity and gas. Then I went diving, backpacking, diving, canoeing, and diving. Another time I wrote for seven months straight, without a single day off. From seclusion at home to camping under the stars, my world was one of extremes.

## A Blessing in Disguise

The army doctors had warned me that the nerve damage in my shoulder and leg was permanent, and that the pain would get worse with age.

The pain I felt after a hard day's construction work was excruciating. The dull ache in my shoulder became a sharp twinge that forced me to hang my arm limp at my side. My leg felt as if it were being twisted like a towel that was being wrung dry, then drawn back and forth through a ringer. Having to stand on a crowded bus only made it worse. When I got home I lay on the sofa and raised my leg straight up against the wall, then lowered it to an elevated position on the cushion. This seemed to let the congested blood flow out. Taking analgesics would have been a simpler treatment but I preferred the psychosomatic approach; I didn't want to become dependent on artificial means. I accepted pain as a way of life.

An unanticipated blessing that resulted from my change in occupation was a marked reduction in chronic pain. Once relieved of the constant standing that was required by construction work, I found the pain diminishing. The foot that was almost completely numb gradu-

## 1980's: Down to Deeper Depths - 37

ally regained some feeling, and the hypersensitivity of my arm and leg gradually receded.

Within months I felt better than I had at any time since my injury. A desk job was great medicine. I had - and continue to have - trouble sitting, but that is easily overcome: I squirm constantly - not due to nervousness, but because changing position relieves the throbbing that's so distracting. Mere discomfort is so much better to have than constant pain.

Thus a whole new world was opened to me: a pain-free world in which I could follow my own pursuits.

### Expanded Shipwreck Research Program

A welcome relief from the mental strain of creative writing was my interest in shipwreck research. Research involves no creativity, just dogged determination. In the 1970's it was difficult to visit the major repositories of information for more than a day at a time - and that at the cost of eight hours pay. I got out of bed at 5 a.m., drove for three hours to Washington, DC, then perused documents all day at the National Archives, the Library of Congress, or the Naval Historical Center. I didn't get home until eight or nine at night. Sometimes I made the trip alone, sometimes I carpooled with Eric Garay, a fellow diver and helicopter design engineer.

Now, with more time to spare and no wages to lose, I expanded my research objectives. Then I met Dave Bluett, a computer analyst who lived outside the nation's capitol. We became close friends. Thereafter, with an open invitation to stay at his house, I made consistent week-long pilgrimages to the District of Columbia instead of sporadic daily assaults.

### Shipwrecks Farther Afield

I was willing to travel anywhere to dive on wrecks that were not fully explored, and to search for wrecks that my research told me lay as yet undiscovered. I first met Bluett when he put together a trip to a newly found wreck off Ocracoke Island, North Carolina. I found out about it through the friends and connections I had made throughout the years, and I signed up at once. This put me in the right place at the right time to make the first dive - with Bill Nagle at my side - on a previously unknown submarine. The story of the *Tarpon*, and the discovery and identification of other wrecks nearby, I wrote for magazine publication and later included in *Wreck Diving Adventures*, along with other experiences and discoveries of the decade.

Bluett was instrumental in opening the Ocracoke area to wreck diving. No dive shop or dive boat operated anywhere around, so it was necessary to charter boats from distant waters, and to bring extra tanks or a compressor. We had many great trips on the *Mary Catherine*, Captain Al Wadsworth, which operated out of Morehead City - a six hour boat ride. Later, and for years to follow, I chartered the *Gekos* out of Ocean City, Maryland, and helped Captain Larry Keen steer it three hundred miles to Hatteras and Ocracoke for extended expeditions to this wreck diver's tropical mecca.

## Halifax Harbor Shipwrecks

During this expansion of my wreck diving horizons I led trips throughout New England and as far north as Nova Scotia. Halifax has been a focal point of commercial marine traffic since colonial times. The more ships, the more shipwrecks. The harbor approaches are paved with wood planks and steel plates from unfortunate vessels that lost their way and ran aground on the rocky coast, or that had their bottoms ripped open by submerged geologic outcrops whose upthrust granite spears lay scattered invisibly across the harbor.

In those days there were few dive boats in the Halifax vicinity, and none that were reliable from one year to the next. Consequently, the only wrecks we could reach dependably were those along the shore. If this sounds like a bargain because there were no charter fees to pay, it wasn't. The lonely strands on which the wrecks came to rest were far from the nearest roads. We had to hump our tanks and dive gear as long as half a mile over rock-strewn hills, through uneven swamp, and down cliff faces that would have daunted a mountaineer. Those with weak constitutions found themselves exhausted by the end of the day - or crippled.

Nor were these dives necessarily shallow. Because of the uneven geological nature of the coast, the bow of a wreck might lie barely submerged at the water's edge while the stern lay beyond a drop-off at 140 feet or deeper. And the water temperature seldom rose more than a few degrees above freezing, even in summer.

Returning from one dive after a sudden squall, I found myself riding six-foot seas while attempting to climb the rocks. I charged for shore on the crest of a wave and got my feet on a narrow ledge, only to have the trough overtake me before I clambered out. Suddenly I was left high and dry, clinging to the wall like a limpet. Then gravity took its course, and the weight of my tanks flipped me backward into the crest of the following wave - which helped to break my fall instead of my neck and my camera rig, which I instinctively held at arms length to keep from smashing against the rocks. It required strength and timing to get out of the water that day.

(As a point of interest, Hulburt remembers this incident differently; or, he allows, I had two similar incidents. He recalls that I was entering the water when the wave fell out from under me and that I fell face down. He observed, "For a horrible moment I was sure you would be killed on the bare rocks. Had you held on one-half second less, you would have been. My laughter as your fins went over your head turned to horror and then relief. But the horror was never more than partially relieved - I still feel it today.")

Every year I tramped the docks seeking boats to take me and my group to the offshore wrecks - or even to the beach sites with less physical strain. The boats I did manage to charter were decrepit wooden fishing boats that had long outlived their useful life. They floated more by the grace of god than by sturdiness of construction. None had Coast Guard certification as passenger carrying vessels. All of them should have been condemned before the invention of scuba.

One boat's engine seized while we were lost in a fog. As we drifted helplessly in a contrary current, we spotted the buoy that marked the entrance to the inlet. Jim Murtha swam a sisal line to the buoy and tied it off. The rest of us pulled in the boat. After several dreary hours a sailboat emerged from the inlet, and the owner volunteered to tow us in to port.

Another boat ran out of fuel in the harbor, forcing the captain to motor the inflatable lifeboat to the dock for a couple of cans of fuel.

The only boat I could count on with any degree of consistency was the *Danny & Rickie*. Captain Harry Bartlett was a wizened old goat with a face that was wrinkled and weathered from a lifetime spent at sea. His boat had the same appearance and was obviously twice his age. He fished and trapped for lobster during the season assigned to him by the Canadian fisheries bureau. Otherwise he drank.

The planks on the deck were so old and rotten that one time my foot broke through a board and I ended up in the bilge. The bilge pump was a boxlike hand-driven affair that looked like an antique butter churner. Since it was situated in the middle of the deck, Bartlett had to set up a trough to get the water over the side. If the boat had ever been painted, no scrap of evidence remained to establish proof.

The wheel house looked like a homesteader's shack, or perhaps a farmer's tool shed that had seen better days. The roof was poorly fitted so that dirt accumulated in the many cracks and crevices. Grass grew there to a height of a couple of inches, moving me to ask the captain sardonically, "How often do you mow your boat?" He was innocently nonplused, but the group broke out in song to the strain of "Mow, mow, mow your boat, gently down the stream."

The ladder was fabricated from the trunks of two saplings that were covered with shredded bark. The rungs consisted of sawed-off floorboards that were nailed to the upright boles. This makeshift ladder was held in place by lashing the upper rung to the top of the wheel house with a loose length of rope. The ladder pivoted on the well-worn rub rail. When a diver stepped on the bottom rung to climb up out of the water, his weight pressed the bottom of the ladder against the crumbling hull - trapping his fin tips - and swung the top of the ladder outboard. Not only did the diver have to support his weight on the narrow tongue or groove, but he had to climb a ladder whose top overhung the bottom.

We considered this arrangement an improvement over other boats that had no ladder at all. In those cases we hooked an arm through a tire that was lashed to the side with rope, removed our tanks and weight belts in the water, then scooted over the gunwale with help from above.

Old Harry Bartlett was quite a character. On the wharf outside his dock house hung a smoked fish on a hook. It looked like a side of decomposed beef. The odor was nauseating. Whenever we came back from a trip he brushed the flies off the carcass and cut thin slices of meat which he ate on the spot. In between bites he took nips from the bottle.

About the only wreck he could take us to was the *Atlantic*, a passenger steamer that crashed into Mars Rock in 1873, with tremendous loss of life. Mars Rock was only a ten minute ride from the dock. We donned drysuits before putting off, then rigged regulators on tanks along the way. We were ready to dive as soon as Bartlett snagged the permanent moor. One time a strong onshore swell was running. It threatened to dash the boat against the rocks as it had done to the *Atlantic* more than a century before. Bartlett idled the boat nearby in obvious consternation.

Also on board was a group of single-tank Canadian divers who didn't like the looks of things. Bartlett kept shaking his head as he maneuvered the boat for position. After a while he called me into the wheel house by crooking his finger. I strode to his side in full regalia, complete with double tanks. I was eager to dive.

In his quaint Nova Scotian brogue, he said tremulously, "Gary, I don't know if I kin keep 'er off the rocks. What'll we do?"

I knew that he was on the spot and I was in the spotlight. It was the perfect opportunity to make use of the phrase that Mike de Camp made famous. I hammed it up dramatically, passed a knowing look at John Moyer and Gene Peterson, and intoned with deliberate overacted machismo, "If you can hook it, we can dive it."

He glanced at me, he stared at the rock, and he watched the waves crashing against the shore. He slipped the hip flask out of his pocket, unscrewed the cap, sneaked a drink, screwed on the cap, and replaced the bottle. He wiped the back of his hand over his mouth and across the stubble on his face. Then he repeated the entire process. Finally, with a look of utter despair, he declared in a cracked voice, "Gary, please, I kinna hook it."

The forlorn expression on his face made me want to retract my words. He felt bad because he couldn't put us on the wreck, and I felt worse for making him feel that way. I've never regretted making that statement more than I did at that moment.

Despite the strenuousness of diving from the beach and the hazards of waterborne transportation, I've made many rewarding dives off Halifax throughout the years, discovered wrecks that were previously unknown, and built a rapport with local divers that continues to the present. Halifax hospitality is exemplified by an incident in which Gene Peterson scrawled "Yankee go home" in the dust on the back window of a fellow diver's car, as a joke upon ourselves. Imagine our surprise when we came back from shopping to find the glass wiped clean! I surmised that the words were scrubbed out by a local citizen who was aghast at the unfriendly gesture made against American tourists. There are cities in the States where we might have found the words erased with a cinder block.

## KEY WEST AND THE *WILKES-BARRE*

At the southern end of the continent lies the sleepy town of Key West, an island of perpetual heat where the water is always warm. During a trip to Florida's westernmost key, de Camp met Billy Deans. Deans graduated college with a degree in chemistry, but he forsook the laboratory for the underwater world and earned an adequate livelihood by spearing fish commercially. When de Camp met him, Deans was working for Reef Raiders Dive Shop as a boat captain and dive guide.

Deans had just located a wreck called the *Wilkes-Barre*, a 610-foot-long light cruiser that was sunk in an ordnance test in 1973, in 250 feet of water. Deans was enthralled by the spectacular sight of a fully armed warship riding tall in crystal clear water, and was looking for people to introduce to the site. Deep wreck-divers were hard to find in Florida, where the emphasis was placed on diving shallow tropical reefs or Spanish treasure wrecks at snorkeling depth.

De Camp was no longer interested in diving deep wrecks, but he knew that I was, so he told Deans about me and said that he would call me when he got home. I was intrigued by de Camp's description of the wreck, and by the prospect of diving an intact warship under conditions that seemed ideal. I wasted no time in arranging a charter, so spring of 1981 found me headed for the Sunshine state. Accompanying me on the 1,300-mile drive were Bill Nagle and Jon Hulburt.

Deans was methodical in his approach to diving. He was used to taking tourists to shallow reefs and leading them by the hand so they didn't get into trouble. But he had no experience with dedicated wreck-divers such as the three of us, and I'm afraid that we made an

## 42 - The *Lusitania* Controversies

impression on him that wasn't what he expected. During the forty-five minute ride to the site we listened carefully as he described the condition of the wreck and outlined the procedure for getting down to it.

Explosives used to sink the *Wilkes-Barre* blew the hull in two amidships, into two equal parts. Because the bow section rolled over on its starboard side it presented a lower profile, with its port hull rising to 190 feet. One hundred fifty feet away, the stern section settled upright and gave the appearance of a badly encrusted ship sitting in dry-dock. Sheer vertical walls rose up from the bottom and stopped at the main deck at 210 feet. From there, superstructure deck levels were spaced at ten-foot intervals to a depth of 180 feet. Above that rose a pair of deck houses whose overheads reached 170 feet. The highest point of the wreck was the top of the smokestack, at 155 feet.

For ease of access Deans had established a permanent mooring system. An air-filled 30-gallon plastic carboy was suspended at fifty feet by a thick nylon rope that was secured to the top of the stack. A buoy on the surface would have been run down by merchant ships that plied the coastal route, if it wasn't first cut free and stolen by local anglers. The clarity of the water was so good that the buoy could be seen from the surface if one knew precisely where to look; or, if the water was somewhat turbid that day, someone got in the water with a mask. Deans had the loran numbers for the spot.

Acting as mate on our first trip to the wreck was Deans's closest friend and former spearfishing companion, computer programmer John Ormsby. After spotting the submerged buoy, Ormsby went down wearing a single tank and snapped the boat's anchor rope to the downline by means of a carabiner. Then Nagle, Hulburt, and I donned our drysuits. Deans and Ormsby made fun of us for diving dry in 71° water. I explained that we weren't wearing drysuits because of the warmth they provided, but because they were part of the system we were used to employing. Familiarity with one's equipment is an essential safeguard in deep wreck-diving. It isn't smart to switch gear configurations before venturing into unplumbed depths.

Deans briefed us on safety procedures without condescension, but in a fashion similar to the over-protective coaching method he might apply to a group of novices who knew nothing about the hazards of a reef. We went down as a foursome with Ormsby leading the way. Deans's plan called for Ormsby to show us around the wreck, escorting us much as a mother mallard would conduct a string of ducklings. That plan was dashed to pieces as soon as we reached the stack.

The three of us shot off in different directions, leaving Ormsby forsaken in shocked consternation. This had never happened before, and he didn't know what to do. As we disappeared in the distance at opposing points of the compass, he slowly resigned himself to hanging around the anchor line to wait for our return. Twenty minutes later and right on time, we all converged on Ormsby as if on cue. For us this was routine wreck-diving procedure, but for Deans and Ormsby it was an eye-opening introduction to the independent ways of hard-core northeast wreck-divers.

The Gulf Stream can flow over the *Wilkes-Barre* with a current of several knots, making it impossible to dive. Even when the current doesn't run as strong, the dive can be a challenge and decompression can be fatiguing. Deans had a way of dealing with this. He lowered a pair of heavy weights to a depth of fifty feet. Secured to the ropes that held the weights to the boat was a length of PVC tubing, positioned horizontally at twenty feet.

After ascending to a point above the marker buoy, we let go of the downline and let the current carry us back to the ropes beneath the boat. Ormsby then released the carabiner and the boat went adrift, carrying us along. Since we were drifting with the current there was no sensation of movement. We got more elbow room at the twenty-foot stop, where we spaced ourselves along the tubing like strap-hangers on a subway train. Regulators on long hoses connected to cylinders on the boat offered emergency air. Deans came down to check on us. When we indicated that we were ready to move up to the ten-foot stop, he got back in the boat and raised the tubing to the appropriate depth.

Compared to the Mud Hole and the *Andrea Doria*, a dive on the *Wilkes-Barre* was a walk in the park. The water was bright and visibility superb, the superstructure was uncluttered by monofilament, and the one or two nets that were snagged on the hull were easy to see and avoid. The wreck was also majestic and visually spectacular - one of the most photogenic wrecks I have ever seen.

On our second trip to the *Wilkes-Barre* a couple of days later, Ormsby stayed topside and Deans went with us on the dive. In the interest of efficiency we started getting dressed as soon as we arrived on site, now that we knew the procedure. There was a little delay in getting the boat secured to the buoy, and since there was no canopy or protection from the sun on the twenty-five-foot open runabout, I was soon roasting inside my drysuit despite the scanty underwear. I hadn't yet gotten on all my gear when I couldn't take any more heat. I rolled overboard without my fins and weight belt, then put them on in the water. A drysuit doesn't offer instant relief the way a wetsuit does. There's no cool water to wash over the skin. So I was still distracted by heat when we began our descent.

I felt cooler below the thermocline. For some reason, though, I commenced to have trouble breathing as I reached the upper platform. I thought I must be kicking too hard despite the lack of current. I let myself float down to the deck at 180 feet. My breathing became more labored. I signaled for Deans to come closer while I checked for the cause of the problem. Not being conversant with wreck-diver's hand signals he wasn't sure what I wanted, nor did he properly interpret my antics.

It took but a moment for me to ascertain that I was running out of air. This didn't make sense because my tanks had been full at the surface. Instinctively I reached for my pony bottle regulator - then discovered that I already had it in my mouth. My primary regulator was not held in place by a neck strap; it was dangling somewhere out of reach. I jabbed my thumb over my shoulder as another signal to Deans, then I rolled to the right in order to get the hose to swing away from my tanks and in front where I could reach it. I swept my arm behind me but didn't catch the regulator. My situation was desperate. I drew my finger across my throat - the classic out-of-air sign. Deans kept his distance from me.

During the next several seconds I repeated these frantic gyrations, alternating between signaling my problem to Deans and swinging sideways and reaching for the mouthpiece. Deans didn't offer to help, but at least he didn't swim away; he watched. Even though I was starved for air I didn't become distraught; I still managed to think the problem through, and realized that I wasn't communicating effectively. Instead of lunging for Deans and his regulator, I made a last-ditch attempt to convey my need with tact.

With my dying breath I spit out my regulator and grinned with composure that belied my great distress. Deans's eyes widened as comprehension dawned on him. He charged forward, yanked his regulator out of his mouth and shoved it into mine. I took two deep breaths and - yet unsatisfied - handed the mouthpiece back. He took two breaths and gave me the mouthpiece again. I took two breaths, then rolled over onto my side - and my regulator fell right into my hand. I inhaled long and deep until I caught my breath.

We exchanged "okay" signs without misinterpretation. The rest of the dive went smoothly. I stuck so close to Deans that I was practically wearing his wetsuit. With my pony bottle empty, he was my only back-up. I had more than enough air for the remainder of the dive and the fifty minute decompression.

I had some other anxious moments during deco. Jon Hulburt was nowhere in sight. I thought he must be dead, and already I was thinking about what to tell his wife Judy. I was considerably relieved when he finally appeared - much later - ascending the line as if nothing was wrong. Later he explained that his study of decompression theory led him to suspect that a slower ascent rate and deeper stops reduced the risk of sustaining decompression injury. He was right, but it wasn't until years later that this became the accepted norm. In this way and many others Hulburt was ahead of his time.

As soon as we got out of the water we began debriefing the dive. My mistakes were twofold: I assumed that Deans's in-water experience was similar to mine, and that he would respond to situations the same as a wreck-diver would at home. And I didn't concentrate on a single objective during the emergency, but divided my attention between communication with Deans and resolution of the problem.

Deans observed my frenzied gestures from a perspective based upon his background as a dive guide. As I danced on the bottom like a half-crazed puppet whose strings had gone into spasm, he interpreted my gesticulations as those of a tourist freaking out. Furthermore, my choreographed pantomime failed to describe the crisis because signals that were second nature to me were meaningless to him. It was a simple matter of the two of us signing in different languages.

"But the out-of-air sign is universal," I protested.

Deans replied defensively, "I thought you were joking."

I was quite serious when I said, "That's something we never joke about."

The incident was a learning experience for both of us.

The *Wilkes-Barre* left a lasting impression on me. The wreck was too large and awesome to be absorbed in a couple of dives. It served as the focal point for many more winter excursions.

## Bureaucratic Stupidity - An Oxymoronic Statement

A detour on the way home added both humor and frustration to a break in the long drive. When we pulled into Florida's Blue Spring State Park, to make a dive in the boil that created the headwater of the creek, the ranger at the gate demanded that we pressure gauge our tanks in his presence. There was a rule that diving wasn't permitted with tanks filled to less than three-quarters of their rated capacity. The purpose was to ensure that divers didn't run out of air in the underground shaft. The tanks that we loaded last in Nagle's pickup were only half full, but, I was quick to point out, we weren't using single 72's - with which it was permissible to dive - but double 80's, and a half-full pair of 80-cubic-foot tanks held more air than a single 72.

Logic wasn't the rule's strong point or in any way relevant. Before he would issue a permit to dive, we had to unload the truck completely and drag out tanks that were full - although we didn't have to use them once we were out of the ranger's sight.

The spring was warmer than the ocean off Key West. Hulburt thought of a novel way to stay cool in his drysuit: he crouched in the creek with the zipper open and let the suit fill up with water. When I pulled the zipper shut across his shoulders, he was wearing in effect an inflatable wetsuit.

Clear blue water boiled out of the ground from a circular shaft that plunged almost vertically to an impassable restriction at 120 feet.

When I swam over the restriction to peer into the shaft that extended beyond, the venturi effect suddenly catapulted me some twenty feet into a ceiling of solid rock. I did two somersaults on the way. It was like being caught in the curl of a storm wave beating upon the shore. We decompressed by clinging to the rock wall under swimmers in the open spring.

Hulburt's idea backfired after he swam downstream to the elevated platform on level with the parking lot. As he climbed up the wooden ladder from the creek, the water inside his suit settled lower and lower. Due to the weight of water pooling at his feet, each rung became a monstrous obstacle to overcome. By the time he gained the platform, nearly exhausted, his drysuit legs had swelled to gargantuan proportions. He waddled past a throng of gaping picnickers like a two-legged circus elephant. Then he lay down on his back on the macadam, and I opened his zipper to let the water drain out. Throughout the years, Hulburt has made numerous equipment modifications that achieved successful purpose; this wasn't one of them.

On our way out of the park we stopped at the entrance gate because I insisted on arguing logic with the ranger. "Can't you see the stupidity of basing permission on tank pressure instead of the amount of air?"

"I know, but that's the rule."

"But it's stupid not to let someone dive when he's carrying nearly twice as much air as someone else who's allowed. That doesn't make any sense."

"I don't make the rules, I just carry them out."

I didn't bother to mention that the rules said nothing about pony bottles or alternate air supply. My retort was a simple geometry axiom: "Then if you uphold a stupid rule, that make's you stupid."

He didn't like what I had to say. But I felt better for saying it.

## The *Seeker* and her Captain

Bill Nagle was doing well financially. He had forsaken auto mechanics for selling Snap-On Tools, and quickly rose to become their highest selling salesperson. In 1984 he bought a boat: a 35-foot Maine Coaster which he named the *Seeker*. He originally wanted to call the boat the *Rape 'N Pillage*, but decided against it because of the way it would sound over the radio should he require Coast Guard assistance. This turned out to be a smart choice, for he nearly ran the Coast Guard ragged throughout the years. He selected Brielle, New Jersey as the place to dock the *Seeker*, not far away from the *Sea Lion*.

With his own boat and a captain's license to go with it, Nagle entertained dreams of salvaging from shipwrecks those items that were greatly cherished by wreck-diving enthusiasts - although, I readily admit, the relics we recover have little or no value or significance outside the wreck-diving realm. Their value lies more in their uncommonness and in the difficulty of finding and recovering them. Non-divers look at my collection of artifacts and see nothing but junk. One person's junk, as they say . . .

Nagle's goals were lofty if strictly self-seeking. It had taken him two years to get over Dudas's death and to get back into deep wreck-diving on a consistent basis. Now he accepted the challenge with a vengeance. I helped him take the boat on a shakedown cruise to Virginia Beach, where we visited with our longtime friends Trueman and Nike Seamans. The next year we began diving in earnest. About the same time, however, Nagle's disposition - which had been degenerating over the years - took an accelerating turn for the worse. His wealth went to his head and brought out his basic character flaws.

As his closest friend, I was in a position to observe his personality change from one of subtle contempt to one of outright arrogance. His nature was becoming more and more abusive. His treatment of people - including his wife and kids - was intolerable to one with my temperament and regard for human sanctity. My arguments for common courtesy created a rift between us that continued to broaden.

Perhaps complicating the problem was his addictive personality. He was drinking heavily and soon got into drugs. Although I didn't know it at the time, he later told me that he was supporting a cocaine habit that cost a thousand dollars a week. He stopped sniffing the white powder only after it burnt a hole through his nose.

## Deep Salvage

The first big salvage job that Nagle wanted to tackle was the recovery of the auxiliary steering helm from the stern of the *Ioannis P. Goulandris*. We had looked it over on occasion throughout the years. It was a daunting proposition that couldn't be accomplished by conventional means. The steering station consisted of two wooden helms, each more than four feet across, connected to the rudder by means of a single steel shaft that was half a foot in diameter. One helm of the pair was mounted on the overhanging end of the shaft, the other one was mounted about four feet away on the inside of a supporting A-frame. Hacksawing manually through six inches of steel was hardly possible with the bottom time available at 175 feet. Nor could the job be done alone unless many multiple dives were planned. So Nagle purchased a Broco torch to burn through the shaft, and put out the call for help.

Four of us formed a pact. John Moyer, Artie Kirchner, and I agreed to help Nagle remove one helm from the *Goulandris*, then we would all work together on the second helm and on the double auxiliary

## 48 - The *Lusitania* Controversies

helms on the stern of the *Ayuruoca*, so that eventually we would each possess a helm. Thus we set out at the end of June 1985 on the first phase of a dedicated mission.

Jon Hulburt and Tom Packer weren't part of the pact but they volunteered to help by setting the hook. The Mud Hole was dark, visibility was poor, and the wreck was covered with fishing line, so their assignment wasn't a simple one. Yet they pulled it off admirably. Nagle dropped the grapnel just forward of the stern well deck. Hulburt oriented himself on the high side of the wreck, then, with Packer, dragged the grapnel aft and tied it to a spot less than ten feet from the helm station.

The Broco torch operates by means of oxygen and electricity, which combine to burn a sacrificial rod at an extremely high temperature. The oxygen is delivered to the rod holder through a low-pressure hose that is married to a thick electrical cable by means of rubber tape and cable ties. A couple hundred feet of the completed assembly weighed a considerable amount, even under water. Kirchner, the biggest and the strongest of the group, dragged the torch and cable/hose assembly down the taut anchor line to the work area.

Both Nagle and Moyer had practiced with the torch by burning through scrap metal in Nagle's back yard, so it fell upon them to do the actual cutting under water. My principal responsibility was to record the operation photographically, which I regretted not doing in 1973 during the recovery of the bridge helm.

Nagle, Moyer, and I went down as a threesome. Nagle began cutting the shaft of the overhanging helm: the easiest one to remove. A Broco torch is ignited by inserting a twelve-inch combustible rod into a holder, then striking a spark by arcing the tip of the rod on a metal plate to initiate the burn. The rod disintegrates at a prodigious rate as it generates heat that is hot enough to melt steel. Nagle had a laborious and frustrating time making an arc that would catch, so the burning did not proceed efficiently. Furthermore, the rod dissolved in a matter of seconds. The stub then had to be extracted and a new rod emplaced and arced. The hot stubs kept sticking in the holder. While I swam around snapping pictures, Moyer helped Nagle by removing the stubs from the holder and inserting new rods. Moyer also secured a liftbag to the top of the helm and inflated it part way. The purpose was to prevent the helm from falling hard onto the deck when the shaft burnt through: a drop that might break off the spokes or loosen the joints.

There was a possibility that oxygen would collect in a pocket burned into the metal and explode, with concussion enough to render the torcher unconscious. We experienced a few "pops," but none with enough force to have any harmful effect.

The first dive ended after twenty-seven minutes with the shaft only partially cut through. We decompressed for an hour. After several hours of surface interval we went down again, this time in succession. Working alone, Nagle kept cutting and created a big pile of slag. Then after fifteen minutes Moyer and I arrived and took over: Moyer burning while I took more pictures. Moyer completed the cut. The shaft sagged, the wheel broke free and dropped softly to the anemone covered deck, then slowly tipped over and dragged down the partly inflated liftbag. I tied a safety line to the wheel, Moyer attached another liftbag, then I supplied the air while he held onto the spool as the line unreeled and the helm slowly climbed for the surface.

We still had some bottom time remaining, so Moyer went to work on the shaft of the second helm. He made a few divots before our allotted time was up; we left at the end of eighteen minutes. By the time we got out of the water an hour later, the helm had been hoisted onto the boat by topside personnel (Drew Maser among them) and was hanging from the davit. Packer went down to retrieve the torch assembly by sending it up on a liftbag, and to untie the hook.

## Accident at Sea

By the time everyone was back on board and the lines were out of the water, it was dark. Moyer was so overcome by fatigue that he needed help in getting undressed. Nagle then started to maneuver the boat in order to retrieve the grapnel. Moyer's condition worsened. Mild pain in his shoulders spread to his elbows, then to his wrists and hands: a sure sign of the bends. His legs quickly followed suit.

I ran up to the flying bridge and relayed the news to Nagle, and asked where the oxygen system was stowed. Nagle was distracted by the operation of the boat and didn't have time to deal with insignificancies that didn't affect him personally. I was gruffly persistent, and eventually he suggested offhandedly that I try the storage in the V-berth.

I tore the mattresses off the bunks and looked in the spaces under the lids. By the time I finished groping through the collected junk and debris - without finding the system - the incipience of Moyer's decompression sickness had deepened to partial paralysis. With his condition deteriorating rapidly, he needed immediate treatment and evacuation to a recompression chamber. Yet I couldn't get Nagle to understand the seriousness of the matter. He was more concerned with his marker buoy than with Moyer's indisposition and potential paralysis for life.

Finally I climbed to the flying bridge and physically dragged Nagle out of the seat. "Forget the goddamn bleach bottle and find the oxygen system!" He grudgingly followed me down the ladder, cursing all the way. After searching for several minutes, he located the oxygen delivery system in a cubbyhole in which I wouldn't have thought to look - and not even close to where he had told me to search. "Now get on the radio and call the Coast Guard, and tell them we have to get John to a chamber right away."

I connected the hose to the oxygen bottle, gave the mask to Moyer, and adjusted the flow of gas. He responded quickly to initial treatment. The progress of the bends was temporarily arrested, he felt the paralysis receding, but he still required recompression under the supervision of a hyperbaric doctor. The Coast Guard recommended that Nagle steer the boat straight for shore - due west, not southwest toward the inlet - in order to meet the helicopter they were sending to pick up the patient.

A night evacuation! I couldn't believe the Coast Guard pilot's audacity. I thought they'd send a cutter. We prepared the after deck for the pickup by making space for the stretcher and by securing all the loose and lightweight equipment so it wouldn't get blown overboard by the downwash from the blades. As the helicopter approached it was aglitter with colored navigational lights and searchlight beams that played along the waves. It reminded me of an alien spacecraft from *Close Encounters of the Third Kind*.

The helicopter matched course and speed with the *Seeker* for a pickup on the fly. The stretcher was lowered to within reach, but we didn't touch it till it grounded on the rail. Once the static charge was dissipated, we pulled in the stretcher and disconnected it from the winch cable. The oxygen worked wonders, and Moyer was able to walk, albeit shakily, from the bunk to the stretcher. We strapped him in and sent him on his way, wondering if he would make it to the chamber or if he would end up in the mother ship that was orbiting the Earth. The Coast Guard officer who piloted the helicopter earned his pay and our respect that night with a fine display of flying under dark and difficult conditions.

After two and a half hours of recompression Moyer regained full movement of his limbs and digits. All tingling sensations were gone. Nor did he suffer any residual effects. The doctor told him that what contributed the most toward his complete recovery was the immediate administration of oxygen and the timely evacuation to a recompression facility. He was released the next day from the hospital.

It is of historical interest to note that the major technological advancement in wreck-diving during the 1980's was the recognition of the therapeutic value of oxygen. By mid-decade most professional dive boats carried an emergency oxygen medical kit on board. In its absence, clubs purchased their own kits to be taken on club-sponsored dives. Coming into vogue on organized deep dives was the use of oxygen for in-water decompression, as an added safety factor in the avoidance of the bends.

## 1980's: Down to Deeper Depths - 51

### *Andrea Doria* Bell Recovery

The downside of the above incident was the doctor's advice not to dive again for a month. This meant that Moyer couldn't dive on the *Doria* trip scheduled for the following week. But he went along anyway as topside crew member, thus relieving the rest of us from some of the boat duties we would otherwise have shared. The full story of that trip I related in *Andrea Doria: Dive to an Era*, so I needn't do more than touch the highlights here.

The specific purpose of the trip was to find and recover the ship's bell. Since Peter Gimbel had filmed the brass letters on the bow, only forty feet from the bell davit, I asked him if the bell was in place. He told me he never thought to look. As far as I knew, no one had ever bothered to look. This fit in perfectly with Nagle's new initiative, so he offered the use of his boat if we could assemble a group willing to donate their time to work on the project. Six of us agreed. Besides Nagle, Moyer, and myself, the other divers were Mike Boring, Kenny Gascon (a transmission shop owner), Art Kirchner, and Tom Packer, a sheet metal worker who was my dive buddy for the trip.

Because this was a team effort we dived on a rotation basis. Each team returned to the boat for debriefing before the next team entered the water. Due to the extreme depths at which we had to work, our plan of operations called for no repetitive diving - a safety concession. During a sand search, Packer and I touched down at 248 feet. I knew the depth precisely because of a new gadget that had just come on the market: an electronic decompression computer that was worn on the wrist. It was called the Deco-Brain.

This was no gas bag like the SOS meter of the 1960's and 1970's, but a fully programmed microprocessor that calculated decompression requirements based on an algorithm developed by Dr. A.A. Buhlmann. The Buhlmann tables were more conservative than the Navy Tables, and the computer gave more accurate readings because it went on the dive with the diver and therefore tracked the dive in actuality rather than by approximation. The integral timer and depth gauge updated the computer constantly with real-time data on the diver's accumulated exposure. All information was displayed digitally, including the ceiling depth and total decompression penalty.

If there was a major turning point in the evolution of deep wreck-diving, I'd say the Deco-Brain was the missing link. It introduced space-age technology to the underwater world. No longer did I have to rely on my crude extrapolations of the Navy Tables to make repetitive deep dives. Now there was a machine to do the work for me, and with greater precision and reliability. The dive computer was a distinct improvement over my Rube Goldberg technique. I've been using them ever since, although, because of my experience with lymphatic edema,

I still had a hang-up about wearing anything on my wrist, so I strapped the heavy device to my console.

Without repeating a story written elsewhere - and at the risk of underplaying an achievement of great significance in the world of wreck-diving - we succeeded in recovering the *Andrea Doria's* bell. Not the bell from the forward davit because that bell wasn't there. Packer and I discovered a bell in the stern, hanging from a davit above the auxiliary steering station, at a depth of 210 feet.

According to a pre-arranged agreement - written by me and signed by all - ownership of the bell was shared by all participants, including John Moyer who didn't dive but who was an asset nevertheless. Nagle never honored our other agreement, however, to recover the other three helms from the *Ayuruoca* and *Ioannis P. Goulandris*. He lost interest in the four-helm project once he had one helm in his possession, and when it would not benefit him to recover the remaining helms. He refused to take us out to complete the project. For this and similar false-hearted deeds he came to be known as Bill fi-Nagle.

## Another *Doria* Tragedy

Sal Arena tried to woo me back to the *Sea Hunter* because he no longer had a crew member who could locate the china hole - which was now more than ever a desirable objective that attracted divers who wanted a souvenir from the ultimate shipwreck. I demurred, and told him that because of our past differences I didn't think that we could work together. I suggested that he call Gatto and Packer. They had developed respectable reputations as experienced deep divers, and they knew the layout of the wreck. Arena took my advice, and they agreed to crew for him. Arena's treatment of his crew improved because Gatto and Packer wouldn't tolerate his insolence, and he couldn't afford to alienate them.

I still managed to make three more trips that year to the *Doria*, all aboard the *Wahoo*. These trips were run so close together that we established a permanent mooring next to the china hole and left a marker buoy on the surface, thus obviating the need to tie in anew on successive trips. The first two trips were noteworthy only in the quantity of china and jewelry that was recovered. I brought up enough jewelry to fill a one-gallon plastic milk jug. I also guided Steve Bielenda to the gift shop so he could collect a few pieces for himself.

The third trip was chartered by Spencer Slate, a dive shop owner from the Florida keys. Something came up at the last minute that prevented him from making the trip, but the boat left as scheduled with the rest of his group looking forward to their first *Andrea Doria* experience. Among the customers were my *Wilkes-Barre* friends Billy Deans and John Ormsby, and airline pilot Lou DeLotto. These three dived together on a familiarization tour of the hull and promenade deck.

No one will ever know why Ormsby didn't stick to the plan. He separated from the others and went his own way. When Deans and DeLotto realized he was missing, they turned back toward the anchor line and went looking for him. They didn't see him, but from the tie-in spot they could see bubbles rising from the china hole only eight feet away. Deans went to investigate. He followed the bubbles down into the darkness. There was Ormsby at 205 feet, at the end of the corridor that led aft to the dining room or down to the gift shop. He was tangled in cables worse than a fly in a spider's web.

Deans lunged in to help, but before he could begin to unsnarl the mass of cables, Ormsby grabbed him in a wild panic. Deans fought hard to break away in order to escape with his life. He barely managed to get free. Then he had to make the most agonizing decision of his life - to leave his best friend to die. Sacrificing his own life wouldn't help Ormsby a bit - but perhaps he could get help.

He raced up the anchor line with no respect at all for the safe speed of ascent. He stopped short of the surface, however, and hurriedly wrote on his slate that a diver was trapped inside the wreck. He showed the slate to someone at the ten foot stop - shallower than Deans should have ascended with his decompression penalty - and that person (I forget who) broke decompression to relay the message in a shout, then went back down before he got bent.

I was standing safety watch on the after deck, assisting customers with their gear, and waiting to desaturate before making another dive. My tanks were already rigged. The call for help provoked an instantaneous response. I jumped into my drysuit without taking the time to don my insulated underwear. Alternate captain Janet Bieser handed me gear as fast as I could put it on, and forced my mitts over my hands. Bielenda shoved a spare tank and regulator into my arms as I leaped from the deck. Not more than two or three minutes passed between shout and splash.

I kicked to the anchor line with straps undone and hoses flapping, and with no spare knife on my leg - just the back-up on my console. The shouted message hadn't mentioned where the trapped diver was located, but there no doubt in my mind that he had to be in the foyer deck. Fortunately I am blessed with quick clearing ears. I plunged down the anchor line as fast as a fireman sliding down a fire pole. When the wreck came into view I saw Deans hovering over the china hole. He had gone all the way back down! He looked up at me, indicated the cavernous maw, then plummeted into the darkness without waiting to see if I was following.

My eyes had no time to adjust to the blackness within. I dropped straight down toward a pinpoint of light that flashed momentarily in the dark internal abyss. My drysuit was squeezed tight against my body. For the first time I pressed the inflator valve in order to check my fall. From above I saw Ormsby's body convulse as Deans backed away from clutching arms. This gave me the impression that Ormsby was still alive.

I dropped in between Deans and Ormsby. Ormsby was hanging limp and upside down, and his regulator dangled loose below his head: a sure sign that he was either unconscious or dead. I shoved the spare regulator into his mouth and pushed the purge. Air burped out around the mouthpiece without forcing inhalation. Ormsby remained unnaturally flaccid and nonresponsive. I continued to hold the regulator in his mouth while forming a seal around the mouthpiece with my hands.

All my ministrations came to no avail. Reluctantly, I accepted the sad fact that John Ormsby was beyond reviving. The convulsion I observed was a counteraction imparted by Deans's hand against Ormsby's chest as he impelled himself away. It was not the movement of life.

Once the emergency was over I took careful stock of my surroundings. I had been to that location many times before, had in fact parted the curtain of suspended cables in order to slip into the gift shop another fifteen feet below. Now I realized that I was hovering amidst those very same cables that ensnared Ormsby so fatally. I backed away with caution. I flashed my light around the pitch black room and noted that Deans was no longer there - he had left after indicating Ormsby's situation. There was nothing more he could have done, and it would have been foolhardy for him to remain in light of his dwindling air supply.

If I couldn't save Ormsby's life, then at least I could recover his body. I rose up out of the opening in the hull and wedged the spare tank in a crevice by the anchor line. Then I descended into the gloom alone to begin the gruesome task of cutting the body out of the cables.

It was undoubtedly fortuitous that Ormsby was already dead when I reached him: there was no way I could have saved him from his predicament. I could only have prolonged his agony. Cables as fat as my thumb were wrapped around his right leg like the stripes on a barber pole, and so tightly wound that they were imbedded in the material of his suit. The sharp edge of my knife blade was largely ineffectual against the thick insulation. I gave up after ten minutes of vigorous slicing, with next to nothing accomplished except a careful observation of the details.

Ormsby wore a commercial weight system called a Miller harness. Straps across the shoulders were sewn permanently to the waist strap in the fashion of suspenders on trousers - with no quick release buckles. In this manner the weight of the lead was borne by the shoulders instead of by the waist or hips. Because the tanks were donned on top of the harness, the weight system couldn't be removed in an emergency. Ranged around the belt were D-rings and snap hooks to which he had clipped a liftbag and a variety of tools such as wrenches and a hammer. He had every appearance of an underwater construction worker. I unrolled his liftbag but left it clipped to his belt, then inflated it, hoping to lift the body clear of the mass of cables, but the 100-pound capacity wasn't enough to hoist it more than a few inches.

After twenty minutes of hard work Ormsby's body was just as firmly entangled as it had been upon my arrival. I had burned through quite a lot of air during my exertions. I needed what little remained for decompression, so I forced myself to quit my self-imposed responsibility before I found myself in trouble. I was hanging at forty feet when one of the customers came down the line with a slate on which was written a question about what to do. I wrote, "Try hacksaws." He read my words in shock, then wrote, "Is he alive?" Only then did I realize the difference in our perceptions: he didn't know that Ormsby was dead and thought that I was still working toward his rescue, whereas I alone knew the truth. I shook my head. He carried the bad news to the surface.

Now I felt the lack of adequate thermal protection and began shivering. Deans was conducting a protracted decompression due to his exceptionally long exposure. He covered his anguish with anger. Under water by mime and later in words he rebuked Ormsby severely for not sticking to the plan, and for doing something stupid that resulted in his death.

## Heart-Throbbing Recovery

The next morning I went down with Rick Jaszyn, co-owner of a sheet metal fabrication firm. We were armed with hacksaws and wire cutters. We also had a nylon rope that was seventy-five feet in length, to each end of which was secured a brass gate hook. We snapped one end of the rope to the base of the anchor line, the other end to a D-ring on Ormsby's weight belt. We added a 200-pound liftbag to the one already emplaced, and inflated them both. This raised not just the body but the weblike mass of cables in which it was wrapped, while leaving the unconnected cables behind and giving us a clear view of the extent of the job. Cables stretched taut were easier to cut.

The cables were so many and so twisted that we hardly knew where to begin. We sawed and severed furiously for several minutes. The work was slow going and arduous. We had streamlined our equipment more than usual so as to reduce the chance of entanglement. Despite this precaution, one cable slipped under my pony bottle valve and prevented me from turning or backing away. I tried to free myself by reaching over my shoulder and pulling on the cable, but when my first effort failed I did not continue to fight. I signaled Jaszyn with my light and pointed behind me with my thumb. Then I lay still and let him do what he had to do. He squeezed my upper arm to let me know when it was all right to move.

It was amazing to me that Ormsby could have become so entangled. We sawed and clipped through one thick cable after another. The body rose higher with each succeeding cut - due to the buoyancy of the liftbags - yet still it wouldn't come loose. Finally, when a sufficient number of cables were cleared out of the way, I saw the causative cable that was holding the body in place: it had slipped through the gate of a clip on the back of Ormsby's belt. In a terrible flash of insight I imagined the awful scene.

I visualized Ormsby brushing past the veil of cables. He rotated slowly to glance at his surroundings, to orient himself before striking off for the dining room. Unknowingly he must have backed into a cable. When he tried to swim off he found himself held from behind. He couldn't see or feel the cable in the clip, and even had he ascertained the problem he couldn't have pressed open the gate in order to work the cable out - not with his hands behind his back, working with thickly gloved fingers and by feel alone. As he spun around frantically to see what was clutching him from behind, the cable twisted around his leg, binding him even worse. He kept twisting and fighting until he was trussed up securely, irrevocably. He struggled, he fought, he gasped. Air was delivered in restricted quantity. Fear set in, then panic. He couldn't get a full breath. He contorted violently. A half breath. A quarter breath . . .

Often worse than the pain of death is the abject fear of impending death.

John Ormsby suffered horribly in his final moments.

A person trained in the ways of wreck-diving wouldn't have worn the rig that Ormsby wore: neither the non-releasable weight belt nor a hook with a pivoting gate. An experienced wreck-diver is aware of the hazards that cables and monofilament present. I don't fault Ormsby for his ignorance because he didn't know any better. A dive on the *Wilkes-Barre* is not sufficient training for a dive on the *Andrea Doria*, any more than a hike on a steep hill is suitable preparation for a climb up a vertical cliff. The *Wilkes-Barre* has depth and current and long penetration, but the wreck is clean of entanglement. It had been down only eight years when Deans and Ormsby first dived it.

Ormsby was not a wreck-diver, he was a person who dived on a wreck. The difference is enormous.

I hacksawed through the last ensnaring cable. The body didn't skyrocket upward after the sudden release, but rose slowly, trailing a conglomeration of dangling cables that were still fouled in the wreckage. As the liftbags floated upward, Jaszyn and I tugged vehemently on the body in order to position it beneath the opening in the upper side of the hull. We couldn't do it. The drag and the friction of many cables snagging debris created too great a resistance for us to overcome in the time allowed. The liftbags scraped by the side of a bulkhead that extended six feet down from the hull: the side of the bulkhead opposite to the entranceway. (Ironically, this was the same protruding bulkhead that had caused me such trepidation in 1974.)

By this time I was overbreathing my regulator and was close to losing control. I shook my head at Jaszyn, and he agreed. We ascended together through the opening. I lay down on the mammoth hull with only one thought on my mind: to get my breathing rate back to a comfortable level. I was gasping like a fish out of water. Jaszyn made a calming motion with his hand that signified "let's slow down." I nodded in agreement. For perhaps two minutes we hugged the hull and concentrated on passing air into and out of our lungs.

I had been on the point of calling it quits, but getting my breath back restored my confidence. We left our tools on the outside of the hull then returned to complete the job. We both knew what had to be done. Wreck-diving is often like a mathematical formula or an exercise in logic. There's a certain way of doing things - a method of approach - which wreck-divers understand: not through intuition but through common sense based on experience. Thus no direct communication was necessary between us. We each chose a necessary job and did it.

I ducked under the barrier and went up the interior side to the overhead. My light enabled me to locate the purge valve on the larger liftbag. As I let out air and the bag began to sink, Jaszyn reached through a large rust hole in the bulkhead and pushed down on the body. Our movements were coordinated by a shared background. We didn't want the body to drop all the way to the bottom of the corridor, only enough to duck it under the bulkhead. I let out just enough air so that, coupled with Jaszyn's pulling, the top of the liftbag cleared the bottom of the bulkhead. Together we towed the body toward a position under the opening, then added air to the liftbag.

The body floated up out of the foyer deck and toward the surface, still secured to the safety rope that was clipped to the anchor line. The loop of rope rose off the deck, and after the body was lost to our sight and the safety rope went taut, the clip slid up the anchor line exactly as planned.

Jaszyn and I gathered the tools and made our ascent. It was good that we had used a safety rope. The liftbags burped out all their air on the surface and the body settled back into the depths. On the way up we passed the safety clip on the anchor line and saw the rope leading down, but we had neither the air nor the bottom time to deal with this new situation, so we continued our ascent. One of the customers came down and hauled the body up from that point on the anchor line, reinflated the liftbags, and sent the body up for the second and final time.

Because Ormsby had been Deans's closest friend, Deans suffered his loss with dire despair. Further turmoil ensued after his return to Key West, when he recounted to Ormsby's parents the circumstances of his death. No longer did diving to unplumbed depths hold the fascination for him that it once had. His momentum was broken by a personal tragedy that only one who has been there can feel. For several years afterward he cut back on diving deep, even on the *Wilkes-Barre*, the wreck on which nearly all his experience was based.

There is clearly a difference between diving deep, and diving deep on aged shipwrecks. Nor is this intended as an indictment. The greatest hindrance to wreck-diving education at the time was the dearth of instructional materials on the subject. There was none.

## The Mind of an Author

I suppose I could make the claim that Ormsby's death inspired me to correct the situation, but it wouldn't be true. I had already begun the series of articles that I later compiled in book form as the seminal work on wreck-diving techniques.

My writing career - my dream - was coming about slowly. I was neither a fast nor facile writer. I had to work long and hard in order to achieve the imagery I wanted to evoke. The result was that I produced precious little copy until I organized my thinking pattern into the proper mode. This process of rewiring my mind took years. The writing of a novel is a quantum leap above the drafting of a letter.

Finally, something clicked in my brain and my productivity increased dramatically. At breakneck speed I wrote the rough draft of a science fiction novel about time traveling dinosaurs. The second and third drafts went slower because I was so picky about the selection of words. *A Time for Dragons* was the first book I sold, although, due to the publisher's delays it was not the first to appear in print. After that, writing came more naturally and somewhat less of a struggle.

Every science fiction author in the history of the world has been asked the same old question, "Where do you get your ideas?" I was no exception. Whenever someone put that query to me I gave my standard reply, "I don't *get* ideas, I *create* them." Most people don't grasp my meaning and I can't say that I blame them. Creativity is the stuff of imagination. Ideas are not something you go out to the store and buy: they are generated within by some unknown mechanism of the mind. For a science-fiction author it helps to have a broad base of knowledge in the sciences, so that a story can be told with relative verisimilitude and with a reasonable suspension of disbelief. A person with no imagination cannot understand the concept of creativity any more than a person who is color blind can comprehend indigo.

Creativity is a gift.

It can also be a curse. If you were to ask a thousand authors how their thought processes worked, you'd get a thousand answers. Mine works constantly: while I'm eating, while I'm reading, while I'm walking, while I'm talking, while I'm listening, and - most inconvenient of all - while I'm drifting off to sleep. Thoughts, notions, snatches of sentences may pop into my head unannounced at practically any moment. It's distracting and often quite annoying.

I'm forever halting an activity in order to put something down on paper before I forget it. A tape recorder next to my bed and in my vehicle have gone a long way toward making life easier at times when my mind relaxes and soars aloft.

I miss parts of conversation because my mind drifts uncontrollably. Then I "wake up" and - quite embarrassed - have to ask the person to repeat what he just said. People must think that I'm not interested in what they have to say, or that I'm not paying attention. But the same thing happens while I'm reading a book - my eyes continue to scan a page long after I've entered some imaginary world. I have to re-read whole sections. If a youngster exhibited these symptoms in school he'd be classified ADD: attention deficit disorder. But I was never this way as a child. I became this way as an adult. I leave it to the psychologists to decipher my mental aberrations.

While on one hand my creativity is unruly, unordered, ambiguous, and mischievous, on the other hand I can "work" it: that is, I can encourage the flow of ideas toward a specific objective. I can't speak for others, but that's how creativity operates in me. It's certainly not for everyone, but I've gotten used to it.

## *Advanced Wreck-Diving Guide*

As a sideline, and practically as an afterthought, I decided to try my hand at magazine publication. I knew a great deal about wreck-diving that I thought needed to be shared. And I had a large portfolio of photographs that I wanted to showcase beyond the presentations I gave at clubs and conferences.

To one used to writing in the length of books, an article came as a kind of relaxation. The short form allowed me to complete a writing project in days rather than years. I began producing magazine pieces and illustrating them with my underwater photos. After breaking through this conceptual barrier, I formulated my ideas for a wreck-diving guide.

I outlined the subjects that I wanted to cover and that had never been seen in print. My purpose was to introduce to up-and-coming divers the proper approach to wreck-diving. I had learned the hard way by acquiring information piecemeal and through the inefficient mediation of trial and effort. By passing along techniques that I had picked up or developed, I could save others from the drawn-out process and enable them to avoid the terror-filled moments that had attended my own education.

Each topic heading began as a self-contained article that I planned to incorporate as a chapter in the book. My grand design was threatened at first because much of what I had to say was forbidden in the dive rags. The mention of deep diving and decompression was considered heresy among the established periodicals, and I was branded a heretic. I got around this resistance to novel ideas by seeking out newer or lesser known journals that were willing to accept controversial material as a way to attract subscribers who were dissatisfied with the pabulum being fed to them by the older magazines that were entrenched in mediocrity. Eventually I found what I was looking for.

After completing the series of articles - and before all of them found their way in print - I proceeded to search for a book publisher. Jane Butler, my literary agent, dealt primarily with fiction. She wasn't familiar with the non-fiction market. I did the legwork by polling the various publishing houses that issued diving books. She made submissions in my name.

They wouldn't touch it. Once again, the subject matter was too controversial. So I went outside the field and found a willing publisher in Cornell Maritime Press, a house that published technical books and how-to guides on marine and nautical subjects. The company had never published anything on diving and knew nothing at all about it. This was a valuable asset to me because the editorial staff held no preconceptions about my ideas - they didn't suspect that there was anything heretical about them. And I certainly wasn't going to enlighten them on the matter.

Butler worked out the contract. I assembled and revised the articles, and the *Advanced Wreck-Diving Guide* became a reality. In the years since, the specialized skills, techniques, and gear configurations that were introduced in the book have grown to become common knowledge among wreck-divers worldwide.

Later writers who quoted or misquoted passages, or who imitated my work in book or magazine form, often demonstrated how little they truly understood. Some of them couldn't even copy accurately. One case in point is the jonline. The jonline is a length of rope with a loop at either end. It is choked around the anchor line so a diver can decompress comfortably while holding onto the trailing end, away from the sickening up-and-down motion that an anchor line assumes in rough seas. Jon Hulburt invented the concept but I created the word. I immortalized him in the book by giving him credit for the invention. Inattentive mimickers consistently misspell the word as "Jon line," "John line," "johnline" or some other variation that is equally un-eponymous.

## THE POPULAR DIVE GUIDE SERIES

While casting for a publisher for the *Advanced Wreck-Diving Guide*, I happened to come across Sea Sports Publications. This was a small press with only a handful of titles. It was privately owned by Bob Bachand. He wanted me to write a book on New Jersey shipwrecks.

I must admit that the thought never occurred to me. I regarded wreck research as a hobby, not a profession. I traced the histories of shipwrecks to satisfy my own curiosity and to assuage my thirst for knowledge. My files contained enough material for a number of books on shipwrecks. But I wanted to write fiction about adventure and flights of fancy, not documentaries. Yet there was unquestionably a need for such a book, or a series of books.

No original shipwreck guides were available. Those that existed were either long out of print or horribly childlike and inaccurate, based as they were on secondary source materials that were poorly researched to begin with. Copycats merely accumulated more errors and destroyed all semblance to truth, creating grave misconceptions among readers about the histories of the vessels under discussion and the circumstances of their sinkings.

I took Bachand's initial idea and expanded it. The humble first volume was *Shipwrecks of New Jersey*, which I subtitled "A Popular Dive Guide." During the wind-up research and writing I prepared to tackle a project that would take years to complete - indeed, I'm still working on it: a Popular Dive Guide Series that would eventually cover every major shipwreck along the entire east coast.

Original in the first book of the series - and continued throughout succeeding volumes - was a list of hundreds of shipwreck positions given in loran coordinates. This enabled people to plot wrecks on the charts and to locate them in the ocean. I compiled this list in order to make the wrecks more accessible, to make the book more useful.

Some people saw the list as a threat. Those who did believed that only they possessed the loran numbers that I shared so readily with the public. In actuality, none of the numbers was secret, else I wouldn't have been able to get them in the first place. All those numbers existed in various notebooks, scratch pads, scraps of paper, backs of envelopes, and the like, in total disarray. I merely assembled them in one location, and arranged them so they could be found with ease. For every enemy I made by publication, I gained a thousand friends.

## THE END OF A FRIENDSHIP

Life means growth, and growth inspires change; the process is inevitable. The development of interpersonal relationships is part of the process of human activity. Most relationships reach a stable plateau and remain there, some become stronger through common interests and shared interaction, a few languish or even decline because of philosophical differences. Coming to a close was one of my longest diving friendships.

Bill Nagle was becoming insufferable to me because of our contrasting temperaments. I perceived in him a steadily increasing arrogance, a lack of respect for his fellow man, a disinclination to treat other folks the way he expected to be treated, and an abiding indifference to his family and friends. I suppose one could make a case for the effects of drugs and alcohol, or excuse his behavior on the grounds of his addictive personality, but I'm not that forgiving. I think people have a choice in how they wish to conduct their lives. Some of them choose poorly.

Exacerbating Nagle's inherent sense of superiority were the roots of all evil: money and power. His downfall began long before he inherited more than half a million dollars from his grandfather. Sudden opulence merely afforded him the means to exercise his base inclinations.

His disdain was brought to the fore on the *Ocean Venture*, a freighter torpedoed by a German U-boat during World War Two. I helped him drive the boat three hundred miles from Brielle to Virginia Beach, where we again enjoyed the hospitality of Trueman and Nike Seamans. We met the rest of the group at their house, then left for an overnighter.

Nagle found a telegraph near the end of his first dive. He planned to retrieve it that afternoon. I saw his good fortune as a photographic opportunity. But for some reason that only he could fathom, he wouldn't reveal the location. I argued passionately, citing the diver's code. In order for him to invoke the code he had to describe the artifact's position and placement in such a way that someone else coming upon it would recognize it and know to leave it alone. Else it might be a different telegraph, one that was up for grabs.

He didn't want me to photograph it or want anyone else to see it. He remained vague and intentionally misled me by giving the wrong direction from the anchor line, and by stating that the telegraph lay under an overhang so low that it couldn't be photographed. That proved his undoing.

On the next dive, Jon Hulburt recovered a telegraph that we later determined - after Nagle depicted its whereabouts in exacting detail - was the one that Nagle found originally. Hulburt contended that the diver's code was invalidated by Nagle's false lead, since the telegraph that Hulburt discovered lay on the opposite side of the wreck and not under a low-hanging hull plate. Nagle then changed his description, and after listening carefully Hulburt allowed that had he come upon the telegraph from the angle that Nagle had taken, it might have appeared the way Nagle now described it. Hulburt had approached the telegraph from another direction, from which it presented a completely different aspect due to the nature of the hull collapse. But that didn't mean that Hulburt was willing to give it up. He found it fair and square and without any help.

If you're going to cheat by flipping a two-headed coin, you need to make sure you'll be the one to call the side. Nagle's attempt to play the game both ways against the middle didn't work. He thought he was secure in a position of domination, that he could exercise subjection with impunity. He believed that people with money were better than people who were without, that money ordains the privilege to treat the have-nots with contempt, as if they were inferior beings. As though affluence were the sole criterion of a person's worth, he believed that he was above the law and beyond the accepted codes of ethics and morality. Virtue was for the poor. He learned otherwise.

Hulburt was bold enough to call Nagle's bluff. Conversely, had Nagle been inbued by more faith in his friends, the troubles he fomented wouldn't have arisen.

Back at the Seamans house we took a vote on ownership in order to settle the dispute. Nike read the sealed ballots that we each put into a box. I don't know how the others cast, but I voted for Hulburt. Nagle lost by a margin of five to two. Hulburt maintained ownership. Nagle accepted the ruling with surprising resignation. Nor did the disagreement cause as much of a rift as one might expect. Despite his air of superiority, Nagle desperately needed the approval of his peers.

However, as time went on, Nagle forgot - or perhaps never learned - the moral of the lesson. His domineering manner intensified. He began judging people harshly by their ability to dive, disparaging all those who were less skilled than he, in some twisted belief that underwater performance alone and no other competence or quality bore relevance to a person's merit. He was withdrawing into a solipsistic world in which the only attributes that counted were the ones which he possessed.

Because of his demeanor and subsequent actions he began losing friends like a tree dropping leaves in autumn. I was the first to go.

The choice was mine and mine alone. When I ran out of tactful ways to point out his overbearing, I laid it on the line in no uncertain terms. "I don't like the way you talk to people, I don't like the way you treat people, and I don't like the way you treat your wife and kids." His wife Ashley had been my friend since before her marriage to Bill, and I had attended their wedding. I steeled myself for what in good conscience I could no longer avoid declaring. I held to a higher standard than he was willing to maintain. It was as difficult for me to pronounce as it was for him to hear. "You can't be my friend if you keep acting like this. Either you change your ways and start treating people with respect, or our friendship is over."

Nagle had the strength of ego to deliberate what I considered to be his shortcomings. We had a healthy exchange of ideas. But it all came to naught. He desperately wanted my friendship and respect, but not at the cost I quoted. For months afterward he called me on the phone late at night, reaching out for help but unwilling to mend his ways. Always in the background of these conversations I heard the tinkle of ice in a glass of whiskey and the hissing inhale of marijuana: the drugs that obliterated his mind and which protected him from reality.

Here, too, our philosophies were antithetical. The total quantity of alcohol I was likely to consume in a year was the sherry in a bowl of snapper soup or the rum in a commercial cake. After breaking my back in a skiing accident, a doctor prescribed codeine for pain. I threw out the leftover pills when I felt I could do without them. I was too high on life to require artificial oblivion.

Our relationship degenerated to the point where I disconnected my phone after eleven o'clock at night: the time when he usually called and wanted to talk for a couple of hours. He had no respect for anyone's convenience. When I protested that I needed to go to sleep, he refused to stop talking. More than once I hung up without his consent. When he couldn't get through to me he called someone else at random, with the result that many of us shared his dissipation and wasting away.

But none of us could cure him because he didn't want to be cured.

## MORE SURGERY

My Vietnam legacy continued to plague me. One fine spring morning I stopped on my way to the shore for a cup of coffee and a doughnut. The doughnut shop was on the opposite side of a divided highway. I jumped over the low cement median and came down wrong on my bad ankle. An audible "snap" was followed immediately by a sharp, stabbing pain. In the moments it took to get back to my vehicle with my breakfast in a bag, my ankle was swollen to twice its normal size.

I was in so much pain by the time I reached the shore that I could barely limp to the dock. I needed help loading my gear onto the boat. I made the dive, but the surface of the fin acted as a large lever, so the slightest movement of water against it twisted the ankle with agonizing pain. I feathered my left foot throughout the dive and decompression. Hulburt had to remove my fin before I could climb up the ladder.

## 1980's: Down to Deeper Depths - 65

The swelling went down during the week but the instability remained. I didn't need a doctor to tell me that the reconstruction had broken: that the transplanted tendon in my ankle had torn apart. Throughout the rest of the summer I suffered continuous falls as my foot folded under at the slightest provocation. I knew there was only one cure for the problem: another operation.

It's difficult to describe the sense of overpowering dread I felt at having to admit myself to a military hospital for examination and treatment. The thought evoked memories of that long-ago year of pain and isolation. But I knew what had to be done. After the diving season was over I contacted the Veterans Administration and made an appointment at the hospital in downtown Philadelphia. The consulting doctor quickly confirmed my initial diagnosis, and recommended another transplant operation. I wanted the surgery scheduled so I could recuperate over the winter, in order not to impose upon the following year's dive agenda. In mid-December I took a bus and a train to the VA hospital and checked myself in.

Despite my anticipated apprehension, once the formalities were out of the way I felt confident and unafraid. After preliminary x-rays were taken I spent a relatively peaceful night reading in bed. The next morning I was wheeled into the preparation room. The anesthesiologist explained the procedure and warned me about the potential side effects of general anesthesia. I waved away the prep shot that would put me out before entering the operating room. I quipped, "I don't need that; I'm falling asleep already. I stayed up all night reading *Coma*." He couldn't help but smile and comment about my uncommonly good spirits.

"Wake me when it's over," cried the Cowardly Lion. That's my philosophy, too. Sodium pentathol carried me through the surgical procedure unaware. The first night was a blur of pain and freezing cold despite injected analgesics and four thick blankets. The next morning the surgeon stopped in to see me. I complained jokingly that I came in for a trim and instead got a shave - both legs were completely smooth. He explained that I was running out of spare parts.

Prior to cutting me open he wasn't sure that the selected tendon on the inside of my left calf was thick enough to do the job - that tendon is vestigial and inadequate in some people. Since doctors prefer to use a tendon from the same leg - in order not to have both of a patient's legs incapacitated - he had my upper leg shaved as well. But the damage there was so great from the gunshot wound that he doubted he could find a salvageable tendon, so he had my right leg shaved just in case.

I settled down for a solitary weekend with the books I had brought to pass the time. Sunday night I had a surprise visitor: Bill Nagle. On his way home from a dive he drove considerably out of his way in order to personally check on my health. It seemed strange that the only visitor I had during my stay was the person I least expected to see. It made me realize how important our friendship was to him. I thanked him for stopping by. Nothing came of it, though, because he still refused to alter his adopted behavior. My sense of loss was nearly as great as his, yet I couldn't reconcile myself to accept the person he chose to become. He continued to grow away from the path of reason.

At first it seemed as if I might have to spend Christmas in the hospital. But my progress was so good that the doctor released me a few days before the holiday. Drew Maser picked me up and drove me home, then made sure that I didn't pass the festive season alone. I spent time with him, his wife Linda, and daughters Laura and Marie. Laura was my godchild.

Full recuperation took six months. For the first two months my lower leg was immobilized in a non-walking plaster cast. At no time was I permitted to put any weight on the leg, so I had to get around on crutches. I alleviated much of the pain by propping my leg on my desk while I worked at my computer. (I had replaced my typewriter with a word processor - a costly investment that nearly put me in the red, but which was worth every hard-earned penny it cost. I paid for it with savings from lecture fees and magazine publications.) My productivity suffered only slightly.

Getting around a two-story house wasn't easy. My bedroom and study were upstairs, the kitchen and dining room downstairs. After I got over the wobbles I could negotiate the stairs with caution. I cooked while leaning on my crutches. (I had laid in a supply of canned and frozen foods prior to the operation.) My chief handicap was climbing the stairs with a cup of coffee and not spilling it. I had to crawl up each step on my knees while placing the cup on the step above. After a while I got a bucket that I could hold by the bail in my fingers below the crutch handle. This became my purse and primary method of transporting items around the house.

During this phase of recuperation I was scheduled to deliver a talk and slide presentation in Washington, DC. Maser acted as my chauffeur for the weekend.

When the cast came off and I saw my spindly, sticklike leg, I thought I'd never walk again. The muscle had atrophied to the diameter of a toothpick. I was fitted with a stainless steel brace that was built into a shoe. A flat, vertical support rose on each side of my leg to a broad leather band below the knee. A Velcro strap held the leather band tight. The device allowed my foot to pivot up and down but not to bend from side to side.

Not only did I find that I could put weight on the leg without collapsing, soon I was walking almost naturally, and then I resumed my jogging - wearing the brace and dress shoes! Except for sleeping and

## 1980's: Down to Deeper Depths - 67

showering, I wore the brace continuously for the next two months. Then I was given an inflatable brace that fit tight around my ankle. I wore that for two months, even inside my drysuit. For six months or a year afterward I wore the inflatable brace whenever I jogged or contemplated walking on uneven ground, and especially when playing racquetball.

### Unexpected Blessing

My ex-wife prevented me from seeing my son shortly after our divorce. The whole sordid saga would take a book of its own to recapitulate. Here's the short version.

Philadelphia's family court assigned visitation rights so that Michael and I could spend time together. We went on hikes through the woods, canoed local creeks, had picnics with friends who had children close in age, and visited his paternal grandparents. The event he looked forward to the most was the Mummer's parade on New Year's Day - not as an onlooker but as a participant. He was only twenty months old when I first carried him along Broad Street in costume. My father had been strutting with the comic division since his teens.

After my ex remarried she renounced all association with me and with my side of the family. Subsequently, when I went to her new home to pick up Michael at appointed times, no one was there. If I hung around they never appeared. At other times she was home but refused to open the door. My only legal recourse was to take her to court. It took several months to schedule a hearing.

Invariably the judge would admonish her for not letting me see my son, state for the record that she had to permit visitation rights in accordance with the dictates of the court, and dismiss the case. The very next week she did the same thing again, and the process would start anew. If she didn't show up for the hearing, the judge issued a warrant which was served on her by the sheriff. Then she might show up for the next hearing several months later - or she might not.

This went on for years. She wouldn't even let Michael march in the parade.

The only punishment she ever received from the court was stern language. To me the message was clear: the court would enforce my responsibilities but would not protect my rights. Divorced mothers could scoff the law with impunity.

The court required that Michael - the object of the case - appear at each hearing. Those were the only times I saw him. His life was in constant turmoil. He wanted to be with his father and I wanted to be with him, but his mother would never allow that to happen, and the court would not compel her compliance. My ex used him as a cat's-paw to create stress and anguish for both of us.

Whenever I managed to see him on the sly he was depressed afterward for days, sometimes weeks, because he never knew when he would see me again. His mother punished him for crying about wanting to be with me. Thus my sporadic visitations and follow-up absences did more harm in the long run than good.

Eventually I perceived that my motivations were less than pure. I wanted my visitation rights enforced because of what Michael meant to me, not necessarily because it was beneficial to him. I rationalized that a stable life - even without my influence - was more important for his development than a relationship with his natural father, however loving and needful that relationship might be.

If I refused to let his mother use him as a weapon, if he could put my existence out of his mind, that part of his emotional trauma could be cauterized. The decision to let him go was perhaps the most difficult decision of my life. I hoped it was for the best.

I saw him occasionally after that but he didn't see me. I swung past his house and the nearby playground, and sometimes I parked in the dark, waiting for a glimpse of him. He never knew that I watched him secretly through the playground fence.

Michael's approaching eighteenth birthday inaugurated change: he was graduating from high school and, because mandatory child support was coming to an end, his mother would no longer receive her weekly checks. Unexpectedly, my ex showed up at my door one day with Michael in tow. She was getting divorced, was moving away, and wanted Michael to live with me. I grasped the gift from heaven with an open heart.

Finally I had the opportunity to show Michael who his father really was - as opposed to any description he might have heard from his mother. By this time he was no longer the child of my dreams and memory, but a young man, so I treated him as an adult. There were house rules of course, but I exercised no other authority over him. If he didn't like what I cooked for dinner, he was free to not eat it and to cook something that he liked. We had and continue to have a warm and rewarding relationship. My lasting regret was that I could not be a part of his formative years.

We converted the basement recreation room into a bedroom. It wasn't totally reclusive because I had to walk past his bed to get to the laundry room. But the back door enabled him and his friends to come and go in privacy.

My ex was concerned that I might not let her in the house to see her son. She was more afraid that she wouldn't be able to see her other son, who went to live with his father. I said, "Are you afraid that he will do to you what you did to me?" She nodded tearfully. "Paybacks are tough," I said. I never prevented her from visiting. I even let her come inside when Michael wasn't home.

Once, while waiting for him to come home after work, she said, "I'm really glad that after all these years we could become friends."

I said, "We're not friends. I hate you. I tolerate your presence because you're Michael's mother."

She never understood how different we were. She expected me to behave the way she had behaved when our situations were reversed. Her refusal to let me share my life with my son was as much a part of her character as my permitting her to visit was a part of mine.

## THE *MONITOR* DEBACLE

During this time and for years afterward I was deeply involved in a legal wrangle with the government over access to the Civil War ironclad *Monitor*, which foundered off Cape Hatteras, North Carolina in 1862. Since that vicious, protracted saga is covered in depth in my book *Ironclad Legacy: Battles of the USS Monitor*, I won't repeat my travails between the present covers. But the *Monitor* as a shipwreck and the lawsuit that resulted from bureaucratic obstructionism have relevance to the present narrative and a deeper meaning for wreck-divers than might at first glance be supposed.

One of the first things I learned in my blind pursuit for justice in the case of the *Monitor* was that people in general are overwhelmed with apathy. For a couple of years I fought alone against the machinations of petty bureaucrats, until I met Peter Hess, an attorney and wreck-diver who volunteered his services to take the case to court, and without whose assistance the case never would have been won. During the six-year struggle to open up the wreck for divers, not one person offered support - either moral or financial - even those who wanted to see the wreck for themselves.

It is ironic that this public cause was subsidized by one with so little wherewithal. The reason that destitution didn't dissuade me from my course is because of an inherent weakness: I don't have the strength to shake deep-seated convictions no matter what the cost in terms of money or emotional turmoil. I would be a happier person if I could learn to live with injustice the way other people do. Instead I am plagued by an irrational and self-defeating sense of righteousness.

The *Monitor* case cost thousands of dollars that I didn't have, and tipped the delicate balance of my economic position into debt. I was forced to borrow money from the bank in order to pursue the cause, and then pay interest on the loan. I hated the insecurity of living on credit, but no other recourse was available to me. I wouldn't drop the case.

## The Abandoned Shipwreck Act

A far different cause concerning wreck-diving was being waged at the time. A minuscule minority of marine archaeologists and petty bureaucrats were trying to force a bill upon an unwilling public that would take from American citizens every shipwreck within the territorial waters of the United States and turn them over to the adjacent States. This was done in the wake of the fantastic trove of treasure found on the *Atocha* by Mel Fisher and his investors after sixteen years of searching for the Spanish galleon's mother lode. Instead of locating shipwrecks on their own - and at their own expense - marine archaeologists, estimated at less than fifty in number, sought to usurp the tremendous efforts of treasure salvors by bureaucratic fiat. Their philosophy was: why earn something when we can take it by passing laws? That might make theft legal but it doesn't make it right.

The story is one of epic proportions, with enough material for a full-length book. Perhaps someday I'll write it. Peter Hess has suggested the title: *The Treasure Wars*. Here is a condensed version of the plot.

Under the guise of "historic preservation," marine archaeologists lobbied for the nationalization of American shipwrecks as a way to ensure future employment among their ranks. Quick to jump on the bandwagon were opportunistic bureaucrats who came to perceive shipwrecks as a valuable political ploy. Historic preservation was enjoying enormous popularity, and popularity equates to votes. Thus there were economic incentives to such an endorsement, not only for the politicians' continued occupation of office, but for the States they represented - in the form of tax impositions, licensing fees, tourist attractions, and so on. The proposed legislation was unequivocally anti-democratic, at the expense of private enterprise, and against every principle on which the Constitution was founded. This didn't bother some lawmakers, however.

The bill that was framed was called the Abandoned Shipwreck Act, a semantically appropriate title. "Abandoned" was intended to refer to shipwrecks without traceable ownership and those that were derelict or had otherwise been deserted by their owners: in other words, every wreck on which there was no current claim of commercial salvage - and this included unknown and as yet undiscovered wrecks. The dictionary also defines "abandoned" as "shameful" and "immoral" and "corrupt," terms which come closer to the truth as descriptions of the Act and the purposes for which its passage was sought, especially considering the manner in which it was ultimately passed.

The underlying theme of the Abandoned Shipwreck Act was control and the lust for power. In their shotgun approach to shipwreck "preservation," archaeologists sought to possess totalitarian authority over all wreck sites in the country, aggregating to some tens of thousands of wrecks of which only a trivially small fraction were historically significant. The majority of wrecks had already been destroyed by time and the elements. Nearly all were the remains of commonplace vessels such as barges, ordinary sailing ships, and twentieth century steamships. It is patently absurd to spend taxpayers' money to "preserve" a wreck when ships just like it still ply the seven seas. It's even more absurd to suffer unfounded beliefs that a wreck under water is preserved from the awful forces of nature. One might just as well "preserve" rare works of art by exposing them on windy, snow-covered mountain tops.

The archaeological community was a juggernaut gone wild on bad faith.

Founded to contest the ASA was the Atlantic Alliance for Maritime Heritage Conservation, a nonprofit organization led by Charles McKinney, an archaeologist with the federal government and whose full-time job was to investigate proposals for historic landmarks and places; and Duncan Mathewson, the outspoken and volatile consulting archaeologist who worked with Mel Fisher on the *Atocha*. Primary financial support for the Atlantic Alliance was provided by Mel Fisher and other treasure salvors seeking to protect centuries-old Admiralty jurisdiction and to preserve private enterprise. Fisher had already made his millions so he had little to gain by getting into the brawl, yet not only did he travel to Washington, DC to protest the ASA, he brought with him his brilliant oratorical attorney, Dave Horan.

I began supporting the Atlantic Alliance as soon as I found out about it, and joined some of its members in testifying before the House of Representatives in opposition to the ASA, citing my tribulations with NOAA over denial of access to the *Monitor* as an example of bad faith on the part of the government, but by that time the fracas was in its final stages of inequity and my testimony was purely incidental and completely ineffective, particularly in light of the opening statement made by Representative Bruce Minto, who was the head of the subcommittee to which I presented my evidence. He announced haughtily that the hearing was a mere formality and would not affect the outcome of the passage of the bill. Bureaucratic minds were already made up.

By 1987 the Abandoned Shipwreck Act had been proposed for five years straight, and had been defeated every time. It would seem that the people didn't want it. But what the majority wanted didn't matter to the infinitesimal minority whose personal ambitions denied accountability to the American public they were supposed to represent.

Communism may have its politburos, but democracy has polit.

burros and donkeys of a shadier kind. Chief ass in this case was Senator Bill Bradley of New Jersey. Since he couldn't get the ASA passed honestly, he resorted to chicanery and deceit in the bill's ultimate session. On the surface it appears that he presented a forceful oral argument to a full and attentive Senate, pleading that the august body should overlook five years of undesirable legislative proposals and the strong opposition they engendered, and should this time pass the bill. His speech was published in the Federal Register, which purportedly records all Senatorial dialogue as well as preliminary discussions on proposed bills and the votes themselves. The Federal Register duly noted that on December 18, 1987 a vote was taken and the Abandoned Shipwreck Act was passed by majority consent.

Like a merchant ship, a proposed bill requires two passages in order to complete a voyage. The bill was required to pass in Congress before it became law. Congress generally rubber-stamped bills that were already passed in the Senate, and vice versa, on the premise that an unpopular or controversial bill wouldn't have gotten passed by the majority in the opposing house. Representatives don't like to appear antagonistic toward their counterparts in the legislative assembly unless there is a compelling reason to do so. Otherwise, one house might find itself opposed by the other when the situation was reversed and some other pet bill came up for vote. Furthermore, representatives don't generally like to go against the tide. If they do, they get a reputation as obstructers, they lose the favor of their constituency, and they find themselves without backing in their own favorite causes. Since nearly half the States in the Union had no coast line, and therefore were not affected by the Act, those state representatives had no vested interest in the bill and no self-serving reason to oppose it.

The Abandoned Shipwreck Act of 1987 became law. It was the greatest land grab since the Louisiana Purchase. The submerged territory added untold thousands of square miles to State ownership and control.

Yet, as I discovered, the ASA was based upon a fraudulent premise and passed by Machiavellian machinations that, although technically legal under the present system of government, were not condoned by the spirit of the law under true democratic principles.

The Atlantic Alliance for Maritime Heritage Conservation has been unfairly criticized by some for failing to achieve its objective to overthrow the Abandoned Shipwreck Act, partly because of in-fighting among the competing salvage outfits that were its sponsors, and partly because it did not coordinate the voices of recreational divers who were adversely affected by the shrapnel of archaeologists' shell fire aimed primarily at the treasure salvors. This isn't true. Salvors may have had their differences, but they offered a united front, else the Alliance wouldn't have existed. Regional directors such as Joyce

## 1980's: Down to Deeper Depths - 73

Hayward and Pam Warner expended considerable personal energy toward galvanizing sport divers for the cause - and were greeted largely with yawns and inertia. Recreational divers were content to let someone else fight the battle. They weren't willing to be inconvenienced. They languished indifferently and let their shipwrecks be legislated away from them.

The opposition camp, on the other hand, was not only strongly organized but it was infinitely well subsidized - by taxpayers' money. And whereas Atlantic Alliance members were volunteers, all the archaeologists and politicians who framed and supported the bill were paid full-time to do so. That was part of their job. Despite these obstacles to freedom, the Abandoned Shipwreck Act was overthrown four times in four consecutive years.

What ultimately enabled the bill to get passed was trickery and conspiracy: the hoodwinking of the American public through political opportunism. McKinney told me that only a handful of senators were in attendance when the bill was put to vote. I didn't believe him - didn't *want* to believe him - and spent years tracking down the truth. The truth, when I blew the lid off it, was worse than I could possibly have imagined.

I got suspicious right away when Bill Bradley refused to talk to me or to reply in writing to any of my dozen or so letters. A couple of times his underlings called on the phone to stress that there was no record of how many senators were in attendance the day the ASA was passed, or how any of them voted, because a roll call vote was taken. This means that instead of requesting written ballots, the chairman simply called for a show of hands and asked for "ayes" and "nays." In such a case, the Federal Register simply showed whether or not a bill was passed, not by what percentage. I claimed that this didn't matter. Since it was Bradley's prize bill and he was there to promote it, he would certainly remember the details. Undoubtedly he did, but he wouldn't talk to me because - as I finally discovered - in order to answer my questions he would have had to admit his culpability.

I eventually uncovered the squalid and unvarnished truth after years of persistent effort and research. All Senatorial and Congressional proceedings are now videotaped, and these tapes - after a certain period of time - can be viewed at and purchased from the Library of Congress. I managed to have the appropriate tapes retrieved from the long-term storage facility and made available to me in the viewing room at the Library of Congress in Washington, DC. This is what I saw.

December 18, 1987 was the Saturday before Christmas. Nearly all the senators had already gone home to spend the holidays with their families. Only a handful remained in Washington, and of those even fewer stayed in session. Those who had bills to propose or statements

## 74 - The *Lusitania* Controversies

to make did so in the morning, then left. By late afternoon only two senators were present in the chamber: Robert Byrd and Alan Simpson. Brock Adams resided as Chairperson. Bill Bradley didn't appear.

Byrd and Simpson extended their stay in Washington for one reason only: to pass unfavorable legislation that would otherwise have been vetoed by those senators not in attendance. Byrd and Simpson had their own precious bills to pass, and they were in cahoots with Bradley to pass his bill. Byrd submitted a copy of Bradley's speech and asked that it be appended to the record *as if it had been read*. Thus, one perusing the Federal Register would be deceived into believing that Bradley was present that day and delivered his pleading in person. The Federal Register doesn't distinguish between written and oral statements, and surrounding dialogue is deleted so the truth cannot be determined.

After the false submission, Byrd asked that a vote on the bill be taken. Adams waived the reading of the bill and went through the pre-voting formality as if the Senate were in full session. "All in favor?" The camera focused on Byrd and Simpson who stood side by side. Both stated "Aye." Adams then asked "Opposed?" He could clearly see that behind Byrd and Simpson the chamber was totally empty, and he knew that they had already voted in favor. Who was he trying to kid with this sham? After a moment of silence, he said, "The ayes have it."

That was how the Abandoned Shipwreck Act got passed. And it was legal. But that's not the end of the story. Byrd went on to propose another half a dozen bills, and Simpson parroted his agreement. The pre-voting litany was repeated for each bill, the chairman called for a vote each time, and these additional bills were passed into law. Still, the worst was yet to come.

The session closed. The Senate chamber dissolved from view and was replaced by a title screen which stated simply that the Senate was not in session. I was not able to *see* that the chamber was empty; I was led to believe that it was. There was no reason for me to expect the session to resume, but just to be thorough I fast-forwarded through eight hours of blank tape. Incredibly, just before the official midnight closing, two senators sneaked into the chamber and reopened the session.

Byrd stood alone on the floor. Assisting him in his nefarious deeds by acting as chairperson was John Glenn, the space hero turned senator. Like an automaton, Glenn pronounced the procedural litany as if the full assembly were gathered. Byrd proposed a bill by name and number only, Glenn accepted the proposal as if it had been read in its entirety, then called for a vote. Byrd alone voted aye, whereupon Glenn asked the empty chamber for all opposed, and, hearing no dissenting voices, said "Passed. Next bill." In this manner Byrd forced through another half a dozen unpopular bills, all by himself and with no one there to oppose him. When Byrd passed these bills he did not represent the majority of the nation's voters, nor did he represent the majority of their representatives. Yet his whim, his personal ambition, and his single vote constituted law.

Thus were passed into law that day more than a dozen bills that were binding upon some 240 million American citizens, because one or two people wanted it so. Thus were the American people raped by Senatorial fiat.

I must have been asleep in class the day my history teacher covered the legislative process and a legislator's responsibilities to the people he is supposed to represent. I don't remember being told that such shenanigans could be conducted in the highest office of the legislative process. Legal it may be, ethical it is not. Beware the system that allows circumvention of the people's will, beware the leaders whose only interest is their own. Liberty may be short-lived and freedom a hollow word.

## Great Lakes Shipwrecks

Through the Atlantic Alliance I met Joyce Hayward, a grade school teacher from Ohio. In 1988, she offered to introduce me to the wrecks of the Great Lakes. I am forever grateful. I quickly came to learn that submerged in the Great Lakes are the greatest wrecks on the continent, because they have been so well preserved by the cold, fresh water. Schooners and square-riggers that would have been lumps of worm-holed wood in the ocean exist in the Great Lakes as museum showpieces, despite having sunk more than a century ago. Steamships that might appear as collapsed hull plates rusted from immersion in sea water look like they're docked at a Great Lakes wharf. The Great Lakes became an annual pilgrimage. I've been going back ever since.

I also came to learn that many Great Lakes wreck-divers were highly skilled and experienced. In fact, there was an entire community of deep wreck explorers who rival in numbers those who dive along the eastern seaboard. Perhaps because deep water lies so close to shore and within comparatively easy reach they had more opportunity to explore the deep range than ocean divers, who must go many miles to sea to find an equivalent depth. Yet there was little or no communication between these geographically distinct groups whose interests corresponded so closely.

I was fortunate enough to be escorted by such Great Lake explorers as Paul Ehorn, Tom Farnquist, Ryan LeBlanc, Gary Shumbarger, Emmett Moneyhun, Dave Trotter, and others. I hope that someday someone writes *their* story, for a rich story it is.

Gary Shumbarger and I had more in common than a given name.

Shumbarger dived with one fin because he had only one foot. One leg was amputated below the knee. He lost his leg in Vietnam. After being sent back to the States he was treated at Valley Forge General Hospital. Despite my long convalescence there I didn't meet him because he was discharged prior to my arrival. Coincidentally, we were both operated on by the same orthopedic surgeon, Dr. Sargent.

## Isle Royale Escapade

Of the many exciting adventures I've had while diving shipwrecks in the Great Lakes, I will relate one incident of note. Joyce Hayward and I met Don Edwards in Thunder Bay, Ontario. Edwards owned a small boat which he wanted to take to Isle Royale in order to dive the wrecks. But he couldn't interest anyone locally to go with him. Hayward and I were delighted to offer our company. We dived by day and camped at night, enjoying the wrecks and the camaraderie. During the return to Thunder Bay we experienced mechanical difficulties with the outboard motor. Bad fuel kept clogging the line and shutting down the engine. Finally the engine sputtered and quit altogether. No amount of effort would restart it.

Edwards called for help on the radio. The northern side of Lake Superior is patrolled by the Canadian Coast Guard, which responded immediately. They had a cutter in the vicinity and after hearing our plight, dispatched it to our assistance. It wasn't out of the cutter's way to tow us to Thunder Bay as that was the cutter's station.

We drifted idly in the afternoon chill, awaiting rescue. The cutter called to ask our precise position. Edwards checked the loran and relayed the coordinates. A couple of minutes later the Coast Guard came back with apologies: they wouldn't be able to assist us because we were on the American side of the border by about two miles. They were very nice about it, but no amount of logic or persuasion would convince them to cross the line. If we could reach the political boundary, however, the cutter would send its inflatable to meet us there.

Edwards' boat was equipped with a "kicker" motor for emergencies. It was about the size of a lawnmower engine. Edwards pulled the cord and got it started. For the next hour we took turns steering with the motor handle, creeping along only slightly faster than we could have rowed with oars. In the lakes that divide the United States from Canada there is nothing to mark the border other than treaty and electronics. The inflatable was waiting for us on the opposite side of the manmade but otherwise invisible barrier. It towed us to the cutter, which then took us aboard and towed Edwards' boat to Thunder Bay. The crew gave us hot coffee to take away the chill.

I was grateful for the rescue, yet I couldn't help but feel annoyed at mankind's foolish political schisms. Where does absurdity end?

## The Wilkes-Barre Re-enters the Loop

In the autumn of 1988, I had the opportunity to resume my affair with the *Wilkes-Barre* after a four year hiatus. Traumatized by John Ormsby's death on the *Andrea Doria*, Billy Deans imposed a personal moratorium on deep diving. This is not to say that he gave it up entirely, but that he ceased promoting the *Wilkes-Barre* while he reassessed the inherent dangers of deep wreck-diving with newfound respect. Diving on collapsed and crumbling shipwrecks was not the same as diving in open water. Nor, for that matter, the same as diving on or in a wreck that had been recently scuttled and which had yet to reach the stage of breakdown that made wreck penetration hazardous. Ormsby learned that lesson the hard way by presuming that expertise in one field of diving automatically conferred expertise in an altogether dissimilar field, where different rules applied. Deans was now more cautious.

He was looking farsightedly toward the future. He got back into deep diving with aggressive resolution and dynamic enterprise - so much so that he bought out the Stock Island facility of Reef Raiders Dive Shop and named it Key West Divers, with the immediate intention of expanding his deep diving operations.

## Shipwreck Research Interlude

My reestablished southward migration proved rewarding in many ways, and added several concrete building blocks to the foundation of projected far-flung ventures that formulated continuously in my mind. This helped turn some of my most fanciful dreams into reality. During the years when Deans was immersed in his deep diving doldrums, we maintained a correspondence and kept in touch by phone. Since retirement gave me more free time for pure research, on an avocational basis, I haunted the archival facilities in Washington, DC, and made some surprising finds, all of which I carefully annotated, photocopied, and filed for later retrieval.

As a result, I accumulated extensive documentation on many shipwrecks that no one else ever heard of. I also had a great deal of information on wrecks that were known but hadn't been located. It was my policy to share this knowledge with anyone who professed an interest, in the hope that it would inspire others to search for unlocated wrecks. It was more important to me that these wrecks be found than that I be there at the time of discovery - although, as an explorer, I naturally felt the latter case more fulfilling.

## The S-5 Swindle

As an example, this kind of encouragement eventually led to the discovery of the submarine *S-5*. I had been looking for the wreck since 1975 when I chartered the *Sea Lion* for the task. Ever since then, I had been disseminating thirty-page packets of material to anyone who showed the least bit of interest in locating the wreck. These efforts paid off after ten years, but not in the way I hoped. Milt Herschenrider and a handful of friends (George Hughes, Joe Milligan, Steve Sokoloff) worked diligently toward finding the *S-5*, largely by towing Sokoloff's side-scan sonar unit in overlapping lanes where the sub was reportedly lost. I accompanied them on some of these search trips.

Purely by happenstance, Herschenrider obtained a hang number just outside the perimeter of our search area. I was out of the country at the time. He and his group checked out the hang during the next good weather window and, to their surprise, found the long-lost and sought-for submarine. Herschenrider called and told me the good news as soon as I returned, then took me to the site when the first opportunity was presented. He allowed me to make one dive on the wreck. Then, instead of sharing the location with the diving community, he and his conspirators decided to keep the position to themselves. I was incensed by their opportunistic selfishness.

Later, Herschenrider admitted to me why he refused to share the submarine's location: he and his group lacked the necessary skills to compete against more experienced wreck-divers in the recovery of artifacts. The only way to ensure that he and his group could hope to obtain the prime artifacts was to exclude the competition. His reasoning was valid.

## Planting Ideas of a German U-boat

Notwithstanding the above, I refused to let such an example of bad faith sour my outlook on life. I sent to Billy Deans a package of information about a German U-boat that was sunk off the Dry Tortugas in a rocket test after the war. The latitude and longitude placed the wreck in the 200-foot range. I suggested that he poll his boating acquaintances about hang numbers in the area. Deans put the *U-2513* on the back burner. I kept sending him reminders, but he was so busy developing his new business that he had no time for a project that was more important to me than to him. Then, after pressing him in person, he promised to look into it more closely.

## The *Wilkes-Barre* Again

The *Wilkes-Barre* was just as visually spectacular as I remembered it. I made a refresher dive with Deans, then otherwise went my own way, except for a dive I made with one of his cave diving friends, Jim King, who owned a manufacturing plant. That day we descended a new mooring line that Deans had established on the forward section of the wreck. The bow was separated from the stern by about one hundred fifty feet, and it lay on its starboard side rather than upright. The highest point of relief was 190 feet.

After alighting on the port hull and exchanging okay signs, King and I dropped over the edge to about midpoint. From there we could see two triple-gun turrets lying in the sand at 250 feet. One had toppled out of its mount, the other had sheered off at the top of the barbette. Unlike the *Doria*, which was dark, dismal, and dangerous, the Bear was open and uncluttered, without nets and monofilament, and sitting in water whose clarity was astounding: visibility often exceeded one hundred feet.

We swam into the open barbette. Shining my light along the horizontal interior was like looking the wrong way through a spyglass: the inside diameter narrowed the same as a telescoping tube. We swam thirty or forty feet along the barbette to the entrance of a long steel cylinder. It reminded me of a round sewer pipe with successively diminishing diameters. At the base of the barbette, deep within the bowels of the ship, a hatchway led to the handling room, through which shells and powder bags were at one time passed. Inexplicably, the "floor" of the closet-sized room was littered with military helmets called "steel pots." When I looked back out from this compartment toward the tiny circle of faintly bluish light, I felt like a paramecium on a tinted glass slide, staring up at some Cyclopean eye that was pressed against a one hundred diopter lens.

From there King and I faced forward. We squeezed through a doorway into an even smaller room from which two other doorways exited. At this point, according to our plan, I maintained a grip on the sill of the doorway we had come through, while King disappeared from sight into one of the offshoot corridors. I waited alone in darkness for him to return. With the beam of my light I examined the "floor": actually, since the wreck lay on its side, a bulkhead that was horizontal. I spent my time rummaging through debris, finding more helmets, but was careful not to stir the silt unduly. After a few minutes King returned. We exchanged nods, then I led the way out the proper doorway. If I had chosen wrong, we could have been in big trouble.

Outside the wreck we finished our remaining bottom time by angling downward toward the bow. We settled on the sand under the shadow of the stem. I pointed out a six-foot shark that was snoozing on the bottom, then disturbed its nap by grabbing its tail. The shark erupted off the sea bed like a bucking bronco in a rodeo show, and charged off in a swirl of sand and shells. I turned to King and grinned broadly. He thought I was crazy. We rose up to the point of the stem, then followed the curvature of the hull back to the mooring line.

## Inception of the Ostfriesland Project

Dives like the one just described are challenging and exciting and serve to maintain skills. Of equal value to me moreover are the people I meet and the friendships that are forged. On the boat that day were three folks from the Washington, DC area. Although I didn't dive with them I saw them on the bottom. They dived as a threesome: John Terry (a stone mason), Jeff Addis (an airline pilot), and Ken Clayton. I remember Clayton in particular because I saw him scoop a sea shell off the bottom. After the dive he offered it as proof of his deepest dive.

A few months later, in early 1989, I ran into Clayton again. I was scheduled to give an evening slide presentation to the Capitol Divers. In order to make efficient use of the travel time - three hours from Philadelphia - I turned the week into a research trip. I stayed with Dave Bluett as usual. Clayton attended the club meeting and cornered me at once.

He handed me a photograph of the German battleship *Ostfriesland*, and asked if I knew anything about it. By a coincidence so strange that it gave me chills, I had been researching the *Ostfriesland* that very day for a volume of my Popular Dive Guide Series. I was shocked to learn that there was another person in the world who knew of its existence, much less about its loss off the Virginia coast. It was a highly esoteric piece of knowledge.

After the presentation, Clayton drew me aside for a private consultation. He was a civilian computer analyst with top secret security clearance who specialized in government and military installations. He'd been diving for only three years, but with a hard regimen of progressive depth attainment he had felt comfortable on the bottom at the depth of the *Wilkes-Barre*. And he had the sea shell to prove it!

He held a special fascination for the *Ostfriesland* that went all the way back to his teens, when he saw a picture in a book that showed the giant lumbering battleship under aerial bombardment, then rolling over and sinking. He dreamt about the ship for the next three decades while he graduated from high school and went to college, got married and raised a family, later got divorced, then embarked on other adventures such as whitewater rafting. It never occurred to him that he could see the wreck for himself, but after he got certified and discovered the challenge of diving deep, the long-ago image that haunted his past resurfaced. He did some research and located a book - not the same one from his childhood - that contained a similar photograph. Then he began to wonder: was the *Ostfriesland's* position known, and could it be dived?

I gave him the bad news quickly. I had first researched the *Ostfriesland* in the mid-1970's, at which time I learned that, according to official documents, the wreck had been sunk intentionally "beyond the fifty fathom curve" in order to conform with the terms of the Naval Arms Limitation Treaty. The reason for the specification of depth was to preclude the possibility of salvage. That meant that the German battleship lay more than 300 feet deep.

Clayton took it well. Suppose, he rationalized after opining for a moment, that the wreck had settled upright like the stern of the *Wilkes-Barre*. Then it might stick up high enough to reach on conventional scuba. I agreed that it was possible.

We didn't talk long that night, but we communicated on the same wavelength with a depth that was intellectually stimulating for both of us. We shared a common interest and a rare meeting of minds. I promised to send him my complete file on the *Ostfriesland* and the other German warships that were sunk in the same test. We also discussed another battleship that was scuttled a few miles away: the USS *Washington*. I was researching that one as well for my forthcoming book, *Shipwrecks of Virginia*. I knew it was deep, but I didn't know how deep; I suspected it was shallower than the *Ostfriesland*. I had only an approximate position from the historical records, but indications were that the *Washington* showed more potential as a possible dive site. We both committed ourselves to looking into it further.

When we parted it was with a strong alignment of purpose. Thus was formed a bond that bore us farther and deeper than any wreck-diver had ever explored.

## GGP

At that time I was fully committed to another venture. I had sold seven books with moderate success, but the publishing industry was collapsing like a house of cards that were not yet all on the table. The global economy was cyclical. Embedded in the panoramic view of cycles were thousands of markets that have their individual ups and downs - the whole taking on the aspect of sine wave caricatures within caricatures. The interaction and interdependence of worldwide commerce created rampant fluctuations in the consumer index, producing a system of such enormous complexity and variability that, like the weather, an overall pattern may be vaguely discernible but localized effects are impossible to predict with any useful degree of accuracy. A meteorologist may forecast a fifty percent chance of precipitation, but no one can say for sure whether it will rain on your particular house. The book business was undergoing such a temporary aberration - a transitory decrease in readership. Or was it plummeting down an irretrievable slide? No one knew.

The indications, however, were not good.

Publishing houses were going bankrupt left and right, or were merging into conglomerates that hoped to benefit by reducing the competition in a diminishing marketplace, or were being gobbled up in corporate takeovers and tolerated as barely profitable subsidiaries on the theory that the whole would emerge greater than its failing parts.

Blatant manifestations of the general trend were the proliferation of videotape rental stores and book remainder outlets, together furnishing evidence that the public was viewing more and reading less. Large publishing houses spread the propaganda that remainders were the result of printing overruns. In truth, print runs were the same but fewer copies were being sold, resulting in large returns that cluttered the warehouses.

Publishers might claim that alternative forms of recreation such as television and movies would never replace an industry that went back five hundred years, but such a declaration was self-indulgent and counterproductive - automobiles replaced horses against the same kind of logic. Horses are still around, but not as a primary means of transportation.

More subtle omens were visible to those of us in the profession. I had several unsold manuscripts that represented years of work, and publishers weren't buying. Mass market paperback houses were overstocked with manuscripts they'd already bought but were slow to publish because they'd had to retard their production schedule on account of the number of books that were already on the shelves and not moving. Publishers initiated a policy of canceling agreements with authors who failed to meet contract obligations on time, even if a manuscript was submitted only a few days late; failure of an author to uphold his side of the bargain meant not only that the book went unpublished, but that the author had to return the advance.

The handwriting - or the word-processing - was on the wall, but no one wanted to admit it. Despite the hopeful Presidential decree to the contrary, the country was in recession. That coupled with the regressing rate of literacy, also contested in political circles, spelled an uncertain future for many publishing houses - not because they couldn't sell books, but because they couldn't sell them in sufficient quantities to earn a profit after production costs were subtracted. Mass market paperbacks retailed for so little because they were mass produced. Take away mass production and the price would have to be raised, often beyond what the majority of the market was willing to bear. It's a self-defeating cycle of decline.

It has often been said that authors *need* to write, meaning that they have a deep-seated desire to put down words on paper. This is only part of the equation. I want more than to write for my own satisfaction - I also want to be read. Take away the readership and writing becomes an exercise in futility: there's more to creativity than burying manuscripts in a filing cabinet.

Many full-time authors who relied solely on advances and ever-shrinking royalties to live on were driven out of the field, usually forever. They had to seek real jobs that provided some sense of financial security, however small. The true state of affairs was made evident to me when my literary agent began dropping clients and took a part-time job as a waitress in a small-town diner as a way to make ends meet. I viewed that final indictment as the apocalyptic prediction of literature's doomsday.

If these dire prognostications on the future feasibility of the book publishing industry sound prescient in retrospect, I submit that any impartial and rationalistic approach would have reached the same conclusion. Accepting my own analysis, as I look back on my precarious financial position, I wonder how I ever had the confidence to "go for broke" a second time, when I had more at stake to lose.

With my eyes open to bitter reality I started my own publishing business.

I know this doesn't sound like a reasonable formula for success, but it depends upon the measure. There was no doubt that this was not a good time to go into the publishing business - or any business, for that matter. The difference between me and other small presses with targeted markets was that I didn't need to earn a profit in order to support myself, I just had to break even.

The primary reason for starting my own business was my dissatisfaction with the job that my publishers did with my work. My diving books were printed in black and white on cheap rag paper, the artistic design and layout were woefully bland and amateurish, and the photographic selection and reproduction were poor. The resulting products were cheap looking and shoddy in appearance. The books sold because of their content, not because of their visual appeal. Furthermore, anticipated publication schedules fell far short of my pace of productivity, which I expected to increase as I cut back or stopped writing fiction altogether. In short, I thought I could do a better job myself.

I had done a little stock trading throughout the decade in order to manage my portfolio with greater efficiency and to increase dividend yield during market fluctuations. Now I took an investment risk that was larger than the one I took ten years earlier, when I retired from the electrical field. I reconstructed my portfolio by selling off poorer-paying utility stocks and purchasing high-yield mutual funds, whose broad diversification gave me a greater sense of security in addition. In the process, I withheld a large portion of the funds I needed to capitalize the business.

This shrewd market strategy left me with an income deficit: fewer stocks and bonds equated to lower monthly dividends. I was used to living beneath my means, but I couldn't live without means. So I had to purchase debt.

The concept of borrowing money would have been more devastating to my ideal of independence had I not already resorted to such measures to pursue my case against NOAA over access to the *Monitor*. Thus I had a psychological defense already prepared. Anyhow, all the money I borrowed - for the *Monitor* case and for the business venture - was fully backed by my own assets: I put up my stocks and bonds as collateral, and their value exceeded the amount of the loan.

Now not only did I have less income, I had interest to pay.

By borrowing against my own resources I had no outside investors in the business. I was the sole proprietor, and I worked without a salary. It is usual to create a fictitious name for a business even if it is owned exclusively by a single individual. This makes it sound like a legitimate company or large corporation of long standing. The owner may even call himself the president or CEO, in order to lend the impression that there are a large number of employees. I entertained no such pretensions.

Every part of the production process I implemented myself: the research, the writing, the editing, the photography, the artistic design, the layout, the paste up, and so on. I had no research consultants, no editorial staff, no art department, no secretary, no assistants of any kind; also no overhead or salaries to pay. I alone was responsible for the choice of every word, the placement of every comma, the selection of every photograph. I had total artistic control over the product. So naturally I called the business Gary Gentile Productions, or GGP.

I still laugh when I get calls or letters from book stores and distributors asking for the ordering department or billing department or some other such corporate division. Callers are sometimes astonished to find the "president" answering his own phone instead of a receptionist. In this sense, GGP is perceived as people imagine it must be: a publishing firm with plush suites occupying several floors of an office building or skyscraper. Little do they suspect that I produce every book and run the entire business from a single desk in a room that is six feet wide and eleven feet long - smaller than many walk-in closets. The reality also makes me wonder how "real" publishing houses could possibly require so much space and so many employees, and how extravagantly and inefficiently they must operate.

The only part of the business that I don't do myself is the accounting, which is largely tax work. For that I paid Drew Maser, my long-time friend and accountant, now for Temple University Hospital. He also handles the orders and shipping when I'm away.

In the spring I brought out my first book, *Andrea Doria: Dive to an Era*.

## WRECK-DIVING REACHES NEW HIGHS AND LOWS

In more ways than one, 1989 was a crucial turning point in the history of wreck-diving, witnessing moments that were the lowest, the deepest, the most exhilarating, and ultimately the most victorious. Apply the adjectives as you see fit.

Steve Bielenda's routine was to run two *Doria* trips in July. This year he decided to move one of the trips to June. The reason, I learned after the fact, was to beat the *Seeker* to the wreck. And the reason for *that* was to take advantage of a what the *Seeker* patrons had accomplished the year before, in near secrecy, but had not yet had the opportunity to capitalize on. The primary patrons on the *Wahoo's* June trip were members of a Long Island dive club called the Atlantic Wreck Divers.

Since Nagle and I had gone our separate ways he had purchased a newer and larger boat with his inheritance. He kept the name *Seeker*. He quit selling Snap-On Tools in order to run the boat full time. The fledgling business soon ran aground, initially because of the belligerence he displayed toward the passengers: people paying charter fees don't like being taunted and ridiculed by the captain. Worse than that, however, was his drinking problem. Too many times he pulled out of the slip drunk and with a hangover. By his own admission his alcoholic consumption was a fifth of whiskey a day. Several times he was so inebriated by the end of a trip that he fell off his own boat while stepping onto the dock, and had to be hauled out of the oil-laden water in the marina. The long-standing joke was that he had hired three new mates: Jim, John, and Jack (for Jim Beam, Johnnie Walker, and Jack Daniels).

Dive shops stopped chartering the *Seeker*, not just because of complaints made by returning customers, but because of the liability. If a shop were ever sued by a passenger's surviving relative, there could be no defense against the argument that the charter arrangement had been made in full knowledge of the captain's alcohol abuse. Dive clubs adopted a similar attitude.

Despite the negative aspects of Nagle's character, he developed a clientele of deep wreck-divers who were experienced and extremely competent and who were pushing the limits of deep wreck exploration - in the Mud Hole as well as on the *Andrea Doria*.

On the *Wahoo's* June *Doria* trip my job, as usual, was to tie in. My long-time tie-in buddy was again Gary Gilligan, a house builder and general contractor. Bielenda showed us on the plans where he wanted the anchor line shackled: in the stern, just abaft the enclosed promenade deck. Gilligan and I weren't privy to any ulterior motives for such a placement - we were innocent dupes - but I noted right away that the spot was near a number of places that I had "worked" in the past, and near others that I suspected had the potential to yield artifacts, but that I hadn't yet explored. (The wreck is enormous. I still haven't seen it all.) I also knew that the spot was near where the *Seeker's* divers had forced a route to the third class kitchen the previous year.

The grapnel snagged about fifty feet abaft the optimum tie-in position, and two decks lower. We knew exactly where we were. We dragged the downline only a few feet against an energetic current before acknowledging that nature was stronger than we were. We slung the chain around a stanchion in order to take off the strain. Then I unscrewed the bolt from the shackle - and dropped it!

I felt like such a fumble thumb. The bolt fell fifteen feet and vanished into the silt on the horizontal bulkhead of the walk-around deck. I left Gilligan holding the chain while I descended slowly into the open corridor. Luck was with me that day: there was a perfect imprint of the bolt in the top layer of silt. I stuck my hand into the thick mud and wrapped my fingers around the bolt. I carried it up triumphantly to Gilligan. Now we wouldn't have to tie the chain in a square knot. We released a couple of Styrofoam cups to let the topside crew know that the tie-in was successful, so they could slack off the anchor line and let the boat ride on the downline.

We had some bottom time remaining, so we went exploring. Together we dropped into the walk-around and looked forward into the alcove - and that was when someone's bizarre sense of humor nearly caused us to drown. To explain, I must first recapitulate relevant preceding events.

Beyond the alcove in question lay storage cabinets, a kitchen, and a dining area, all clearly shown on the plans but with no easy way for a diver to get there from outside. A large metal grating used for ventilation was a tease, for from the alcove one could peer into the beckoning darkness within but couldn't squeeze between the steel bars in scuba gear. I once contemplated passing horizontally along two flights of stairs from the promenade deck, but was quickly convinced to the contrary, for the steps spiraled disorientingly around a staircase that was partially collapsed and choked with timber and debris. I was terrified before I even began the penetration, and gave it up.

Then Pete Manchee conceived a bold and novel approach. Manchee was a long-time correspondent who first wrote to me in the late 1970's about restoring a gun sight he had recovered from the *St. Cathan*, a British armed trawler sunk in collision off South Carolina during the war. I responded, and we maintained a pen-pal relationship afterward. Several years later he called to ask how to get on the *Doria*. I smoothed the way for him, and we met for the first time in 1987, on his first *Doria* trip. Because by then we were working the deep side of the wreck, I showed him how to extrapolate the Navy

Tables in order to make repetitive dives beyond 200 feet. He was fascinated by the wreck, undaunted by the depth, and signed up for future trips.

Manchee started diving as a teenager in 1964, when his friend received a complete scuba outfit as a present from his father. His friend shared the equipment and let Manchee dive in his pool. He was hooked at once. He lived in New Jersey at the time, but moved to Columbia to attend the University of South Carolina. He graduated with a degree in biology and with a minor in chemistry. He spent some time at sea on the research vessel *Eastwind*, but jobs were seldom available in his field so he drifted away. For five or six years he managed health spas. In 1974 he moved to California for two years where he worked as a commercial diver. Here he met Bernie Campoli and Jack McKenney, who fired him up about the *Andrea Doria*. He moved back to Columbia because his wife took off with his best friend while he was working at sea, and it was the only way he could have custody over his son. He then sold insurance for five years before returning to the health spa business as a chief executive with access to his own Leer jet. He oversaw the construction and the operation of new facilities. He and several partners then started their own company, Health Designs International. When business slowed down he went back to selling insurance for a living. The common thread throughout these changes in occupation was wreck diving, primarily off South Carolina. Manchee loved to dive. Manchee *lived* to dive. He also owned a dive boat and was a licensed ocean operator.

His scheme to reach the kitchen and storage cabinets was to remove his doubles in the alcove, slip between the bars, then explore the area beyond with a single tank that a support diver would pass to him through the grille. I lent some help by carrying a single tank down to the wreck, but Manchee's actual support diver was Ed Suarez, an accountant for the Social Security Department. Suarez stayed by the grille and held a light to mark the way out while Manchee went exploring. The operation was a success. Not only did Manchee make it out alive, he brought back souvenirs of his exploit in the form of dinnerware.

Nagle supported a plan that was just as bold but was considerably less risky. Afterward he bragged about how he had pulled one over on the *Wahoo* and her patrons, but in actuality he had little to do with the plan or with its implementation other than to charter the *Seeker* to those who did. He received full payment for the charter. The people who deserve the real credit were John Chatterton and Glen Plokhoy.

Chatterton did a replay of Manchee's maneuver. He squeezed between the bars while Plokhoy and Bernie Saccarro stayed outside the grille to act as safeties. Chatterton brought back some mementoes: china, silverware, and, in his words, "the conviction that a larger hole was needed." He proposed to burn off the grille's slender bars with Nagle's Broco torch. He and Plokhoy organized a venture dedicated to doing just that. The job was every bit as demanding as cutting off the shaft of the auxiliary steering helm on the *Ioannis P. Goulandris*, and deserves high kudos for conception, execution, and effectuation.

Jon Hulburt, who was part of the operation, characterized it this way: "This group effort was possibly the largest among sporting wreck divers up to the time, comparable in effort to that expended in retrieving the *Doria's* bell." He further noted that "cutting open the bars was at least a small step forward in demonstrating the capabilities of *Doria* divers," and "the effort to cut open the stern was not trivial."

Hulburt described the operation to me: "The task of cutting the opening took more than 3 dives from everyone involved. Some of us reset the tie-in and placed the cutting hoses and cables. Divers were needed to feed the rods to Chatterton and Nagle. Some of us documented the action with film and video. It was an impressive and well-orchestrated effort. Despite the talent and skill I observed applied, it took nearly the entire trip to set up and cut through both sides of the thick angle iron blocking the entrance to be. Then cables, hoses and other equipment had to be brought to the surface and stowed."

The removal of the grille left a square opening some three feet across, which provided easy access to the compartments inside. But so much time had been spent in clearing the way that there was little leftover to work the area. Only a few trinkets were recovered before foul weather cut the trip short. The people did not have a chance to "clean up," although the word that got out was that they did. This happened in August 1988. It was the last opportunity of the season to dive the *Doria*.

But the story doesn't end there. Once the way was open, Nagle became distressed by the thought that divers from other boats - the "enemy" - might enter the compartment and recover some of the artifacts that "his" group left behind. His notion was fully justified, for that was precisely why Bielenda moved up the date for his first *Doria* trip the following year - and he sold out the charter on that basis.

Little did Bielenda know that his claim-jumping tactic was to be pre-empted. The diver's grapevine is more like a multi-media broadcast than a backyard gossip channel. Just as Bielenda and the Atlantic Wreck Divers heard about the torching of the grille and the opening of the way to the third class kitchen, Nagle heard about Bielenda's stratagem to clean out the cabinets before the *Seeker* divers had the opportunity to reap the rewards of their own efforts.

Anticipating such a trick, Chatterton and Plokhoy decided to prove a point by reclosing the opening in such a fashion that only those who went to the *Doria* on the *Seeker* would have access to the rooms beyond. They did this by having a sophisticated stainless steel

## 1980's: Down to Deeper Depths - 89

gate constructed. The measurements were calculated from Plokhoy's video footage of the opening. A welding fabricator built the gate to their specifications. Crosshatched bar stock and flat iron created a mechanism that looked like a miniature jail cell door, complete with sliding dowels and matched holes for a lock. The gate was so heavy that it had to be lowered to the wreck on a liftbag and maneuvered into place.

After the vertical slides were extended inside the opening, the gate was secured with a chain and lock mechanism which was brainstormed by Hulburt and Nagle. Hulburt: "The end result was a square linked specially hardened chain, which could not be cut by either bolt cutters or by metal saws, and fastened by an eccentric recessed bolt. The head of the bolt had a peculiar Allen-wrench-like head that took an equally peculiar Allen wrench to turn." Then a well-oiled brass lock was clamped in place.

The gate was installed during a "sneak" trip to the wreck only a week before the *Wahoo* was set to sail. The entire operation was such a closely guarded secret that even the Russians didn't know about it. That in itself was a monumental achievement, considering how much divers like to prattle about their exploits.

To Chatterton and Plokhoy, the issue at stake was not the recovery of third class china - or the prevention of its recovery by others - but the principle of the matter. They wanted to send a clear message to the diving community in general and to the opportunists in particular that certain patterns of behavior were not to be tolerated. In order to ensure that the message was clearly received, they cable-tied a white plastic sign to the gate. It read:

CLOSED FOR INVENTORY
PLEASE USE ALTERNATE
ENTRANCE
THANK YOU
CREW AND PATRONS OF
## SEEKER

It was this sign that nearly caused me to drown, for I went into such paroxysms of laughter that I spit out my regulator, and even after I found the mouthpiece and put it back between my teeth I couldn't stop laughing. Gilligan suffered just as badly. After recovering from our laughing fit at 190 feet, we broke off the sign and carried it up with us. All during the decompression we kept exchanging looks and breaking out into guffaws that again and again nearly resulted in our demise.

What Gilligan and I treated as comic relief met with shock and indignation aboard the *Wahoo*, as though it were Grand Guignol. If we hadn't brought up the plastic sign we wouldn't have been believed. No one else saw the irony or the exquisite black comedy of the plotters being outfoxed by their intended victims. Perhaps instead of dogging the footsteps of others they should have been making their own path. Be that as it may, it didn't make for a happy lot who signed up for easy pickings and wound up empty-handed. The charter was overcome with gloom and anger.

To me the sign presented a challenge. I studied the plans with the idea of trying one of the alternative routes that circumvented the gate. On our next dive, Gilligan and I swam to the aft end of the foyer deck and entered a doorway into a tiny cubicle at a depth of 200 feet. The corridor leading forward to the rear wall of the dining area was blocked, but a transverse passage going down appeared to be free of obstruction. Because this was a service area not intended for passengers, the passage that connected the two parallel lengthwise corridors was narrow: only slightly wider than a person's shoulders. It was also dark and forboding.

Gilligan maintained station while I descended into the shaft. It was like falling into a surrealistic and incredibly deep coffin. My elbows scraped both sides. When I reached the end of the vertical passageway I found myself at the junction of a horizontal corridor that extended fore and aft like an inverted tee. The way forward appeared to be open if no wider. Before continuing out of Gilligan's line of sight, I rolled over and shone a steady beam upward - our pre-arranged signal that all was well - then checked my gauges perfunctorily. The depth was 220 feet.

I surveyed my surroundings carefully for snags and possible entanglements. The corridor was clear. With only three feet of clearance, anti-silting techniques were impossible to implement. I swam little more than a body-length when the "floor" beneath me - actually a horizontal partition - fell away to reveal a closet. The closet was empty, but the next one forward gleamed of shimmering white porcelain through an irregular rusted opening: from side to side and top to bottom it was filled with cups and saucers. I stared in disbelief. All thoughts of going any farther were abandoned.

I backtracked. Rolling over to look up, I found myself staring at receding steps in an enclosed stairway. After my heart started beating again I realized that I hadn't gone far enough. A few more feet brought me to the vertical entrance corridor, where Gilligan's light shone down like a friendly beacon. I shook my light as a signal for him to come down (it could also have meant that I was in trouble). He was by my side in an instant. I waved for him to follow me. A few feet away I aimed my light into the closet below. His reaction mirrored my own.

He and I did not need to communicate further. We had dived

together often enough to know exactly what roles to take. I squirmed down into the closet while he unleashed his goodie bag and held it over the lip. He illuminated the rim of the bag with his light. I reached into the muck, grabbed a handful of china, and dropped the pieces into the bag. After a couple of grabs I was blinded by agitated sediment, but it didn't matter: wherever I put my hand I touched unbroken cups and saucers. I didn't need to see. Time and again I rose up out of the black obscuring cloud, spotted the dim glow of Gilligan's light on the mesh bag's metal hoop, made my deposit, then plunged back down for more. In short order we accumulated forty-two cups and saucers. There was no end in sight - or in feel. Gilligan kept an eye on his watch and let me know when it was time to go.

We retreated in reverse order. Passing under the stairwell, I waited at the bottom of the vertical shaft until Gilligan reached the top before making my ascent. We exited the cubicle without difficulty, then made our way to the anchor line. Gilligan clipped the mesh bag to his harness, and for an hour he had to bear the weight of our haul. Our return to the boat was met with jubilation. The somber atmosphere dissolved at once and an air of hope prevailed.

Our entire collection was laid out on the deck and photographed by the now-exuberant customers. Excelsior and packing straw, surviving these many years, and the lack of shelving in the closet, suggested that the ware was packed in boxes to replace breakage. Gilligan and I posed grinning with our booty. We held between us the *Seeker's* closed-for-inventory sign as an additional note of sarcasm. High-spirited comments of derision ensued from those who felt they'd been cheated by the *Seeker's* patrons' successful counterplot.

No sooner had we doffed our gear than the deck plans were thrust beneath our noses. The customers wanted to know where we had gone. I am partially to blame for this presumptuous display of arrogance. Ever since I had shown the way to the first class dining room, Belienda and the *Wahoo* patrons had selfishly come to expect that I was obligated to reveal my secrets. I didn't feel such an obligation, although admittedly I enjoyed the veneration.

Gilligan and I pointed out the compartment that was the storage closet. More important than its location was the convoluted way of getting there and the extreme working depth of 223 feet. Without exception, none of them had ever been that deep before. This posed serious problems with respect to individual tolerances to nitrogen narcosis: an unknown quantity whose effects would likely be exacerbated by confinement and by length of penetration. The consequences of these psychological factors are not to be pooh-poohed. I was careful not to underrate the difficulty of the dive, and described in exacting detail every point of reference along the way, in particular the overhead stairwell that initially might go unnoticed. Upon return, a narked diver who turned upward too soon and entered the stairwell instead of the adjacent shaft could not be expected to overcome the resulting disorientation.

In the event, most people couldn't find the doorway to the cubicle, while others found a doorway whose interior layout didn't match our description. After post-dive debriefings, I determined that they had gone to the wrong deck: an easy error to make under low-light conditions where the decks looked so much alike, and where lack of familiarity with the wreck and the plans led to confusion. Later a few were more successful, some choosing to dive in teams of three, with one safety stationed at the top of the shaft, another at the bottom, while the third moved laterally to reach the closet. A guideline down the shaft would only have created an entanglement.

We returned to the same spot on the next trip several weeks later. This time performance increased so dramatically that repeat customers recovered several hundred cups and saucers. Even first-timers went home with souvenirs that were gratuitously donated by those who were more fortunate. While the customers were busy scavenging the "cup hole" - as it came to be called - Gilligan and I moved on, to explore unknown territory and to find other items that were rarer and more unusual. We were keeping ahead of the pack, as it were.

This episode is not without its irony. The *Seeker* arrived on site about a week after the *Wahoo's* departure. Chatterton unlocked the gate to enable the crew and customers to work the kitchen. When the *Seeker* departed, the gate was left lying on the bulkhead so that entry was unopposed. Yet on the *Wahoo's* second trip, although we moored within a few feet of the opening, *no one went there.*

## THE BATTLESHIP *WASHINGTON*

Meanwhile, Clayton and I were working diligently on the primary phase of our extreme depth project. Our first target was the U.S. battleship *Washington*. The ship was three-quarters completed when construction was halted in compliance with the terms of the Naval Arms Limitation Treaty, in 1922. Two years later the hulk was sunk in an explosives test some fifty miles off the approaches to the Chesapeake Bay. The ship was slightly shorter than the *Doria* but had a wider beam.

We acquired a set of hang numbers from Roger Huffman, a commercial fisher and dive boat captain. These numbers coincided closely with the historical location I obtained from archival sources. The description of the depth recorder readings compared favorably with what we expected for a battleship. Local dragger captains called the hang the 44 Fathom Wreck or the British Aircraft Carrier. They believed that the flat-topped profile was the flight deck of the

*Illustrious*, which was brought into Newport News for repairs during World War Two. Since the *Illustrious* didn't sink off the coast of the United States, I was more inclined to think that the long level readout was the side or the keel of the battleship in question. If the wreck lay on its side at a depth of 264 feet, the 97-foot beam would produce an impressively high relief.

Our problem was getting there to check it out. In scouting for a boat that could go the distance and accommodate us overnight we discounted several possibilities before finally chartering the *Sea Hunter*. This was the same boat on which I made so many trips to the *Oregon*, *San Diego*, and *Andrea Doria*. Sal Arena sold it to Mike Boring, who moved it to Virginia Beach. That spring and summer we were hounded by bad weather that compelled the cancellation of four consecutive charters. The weather wasn't marginal for three of them. It blew hard enough and early enough that Boring called them off before anyone left home. Once I made the six hour drive only the have the trip canceled at the dock.

On trip number five we got out of the marina, but made it only half way to the site when the weather turned snotty and we had to abort. We made two dives on the *Lillian Luckenbach*, a freighter sunk in a collision during World War Two, then hightailed it for port. We blew off the second day.

On trip number six, not only was the weather perfect but the forecast was ideal. Barely a ripple marred the sea as we motored through the inlet at midnight, full of anticipation in finally getting underway toward our goal with no impediments in sight. Only an hour from the dock, however, ominous clanking sounds threatened the accomplishment of our mission. Boring idled the engine and made a cursory examination. Something mechanical was amiss but he didn't know what. We were barely able to continue at partial speed. Boring had no choice but to turn the boat around and ride the gentle swells back through the inlet before a serious breakdown occurred and left us stranded on the high seas.

The next day dawned still and radiant, with not a breath of wind to blow away the clinging, torrid air. Sweat poured off my brow even in the shade. Boring spent all day patiently disassembling, repairing, and reassembling the transmission. Not until that night did he announce that we would give it another try. The mid-July weather was holding fine, so once again we made a midnight departure that was timed to put us on site the following morning - and once again the transmission acted up about an hour from the marina. We turned back, docked for the night, then went home in the morning, frustrated at having wasted so much time and a prime weather opportunity.

Boring laid up the *Sea Hunter* for repairs, but promised to have the boat ready for another attempt next month. In light of the distance

## 94 - The *Lusitania* Controversies

from shore that we had to travel, Clayton and I lost faith in the serviceability of a boat evidencing such mechanical unreliability. I explained this to Boring, who was my friend, and he understood. About that time, Larry Keen accepted delivery of a brand new boat to replace the *Gekos* - a boat that I had chartered throughout the years. He named the successor after the old one. The interior work on new *Gekos* was not yet complete - wires dangled from the overhead, lights were not installed, the bulkheads were unpaneled, and so on - but Keen was willing to take us out on a shake-down cruise if we didn't mind the inconveniences. He assured us that the boat was mechanically sound. He quoted a price, Clayton and I accepted, and we scheduled a three-day trip for August.

If ever there was a trip that went like clockwork, this was the one. The boat operated flawlessly, the coordinates were right on target, and despite a couple of misses, grappling the wreck proved simple. The depth recorder drew a picture of the hull exactly as the fishermen described it, although it was deeper than we anticipated: 275 feet to the bottom, 240 to the top.

Clayton and I were the first pair down the line. The blue-tinted water was exceptionally clear and warm - almost tropical. The Gulf Stream brought in surface water with a temperature of 78°. Even below the thermocline it didn't get colder than 56°. A barely perceptible current kept the anchor line taut. We had an easy descent.

The hull coalesced before our eyes like a mist forming imaginary patterns. So huge was the wreck that at first I mistook it for the sea bed. The illusion was caused by the lack of distinct edges on a rounded hull that was similar in color to the sand on which it rested. The wreck lay upside down. It appeared like a gigantic, over-arched turtle shell whose breadth and one end I could clearly see. In the opposite direction, the wreck faded into the incomprehensible distance.

As I drew closer and the wreck loomed larger I had the sensation that I was shrinking: a trick of perception evoked by the change in comparative size between my body and the massive hull. This impression was instigated by ambient light visibility that exceeded one hundred feet, thus permitting me to envision the wreck nearly in its entirety. I understood how a flea might feel if it were to parachute onto an elephant's gray back.

Settling down on the hull was like sliding across home plate and scoring the winning run. The depth was 250 feet to the highest part of the inverted hull. The chain was draped over the starboard outboard shaft strut. (The propellers and shafts were not in place when the *Washington* was sunk.) I followed the chain over the edge of the wreck in order to inspect the grapnel. Two tines were firmly embedded in the overlap of a hull plate. The grapnel was secure as long as the anchor line stayed taut.

The hull appeared to be perfectly intact and was only thinly encrusted. The veneer consisted of indiscriminate marine fouling organisms such as barnacles, whose calcareous shells and exuded adhesive substrate clung to the steel plating like a carpet with a hard, sharpened pile. Slender hydroids and fine, close-cropped algae resembled patches of weeds on a lawn that needed mowing. Everywhere lay black bulbous purses half the size of a cigarette pack, from the pinched corners of which extended thick tapering filaments. These were elasmobranch egg cases such as those found washed up on the beach and usually attributed to skates. But the purses of other cartilaginous fishes such as sharks and rays were similar in appearance and difficult to distinguish except by educated scrutiny. Perhaps they were the egg cases of the chain dogfish sharks that inhabited the wreck in the thousands. These deep-water sharks were only fifteen inches in length. They didn't swim about, but lay still and allowed themselves to be touched without reacting. I estimated that each square foot of the *Washington's* hull contained three of these unidentified egg cases.

I clearly saw the other three shaft struts and, at the extreme stern, the multi-ton rudder that rose ten feet above the shallowest part of the hull. I exchanged okay signs with Clayton. We both felt good, so we dropped off the starboard side of the hull toward the bright reflective sand, seduced by uncommon clarity into going deeper than we had planned.

In wreck-diving parlance, a plan is not a detailed program of inviolate restrictions, but a basic outline or framework that serves as the minimum threshold from which alterations are made, contingent upon the circumstances encountered and conditions in the water. We headed straight for the bottom.

We alit on the sand at 283 feet. Because the rays of the sun angled down from the other side of the wreck, the proximate portion of the hull lay in shade. I played my light along the undamaged hull, which towered above my head like the wall of a monstrous building. It pained my neck to look up. We were each mesmerized by personal thoughts and observations. I moved away from Clayton toward a washout in the sand, adjacent to the hull. No superstructure was visible, the weight of the hull and the suction of the sand having crushed and buried the upper works long ago.

In the washout I saw a row of vertical beams from which the steel plates had rusted through. I swam down the slope to where I could look under and into the hull. The openings were not large enough to fit through, so I contented myself with shining my light into the vast cavernous interior.

All my equipment was functioning flawlessly. I kept a constant check on my gauges, pausing frequently to monitor my air, depth, and decompression penalty. The computer mounted on my console displayed the current depth: 289 feet - an unsatisfactory maximum for one with a relish for round figures. I lay horizontal in the lowest part of the washout, shoved my console down beneath me, and watched with satisfaction as the digital readout flipped to the next highest number. Then I backed out of the hole.

I rejoined Clayton who floated a few feet away. We agreed to ascend. Once again I checked the grapnel and noted that it still held fast. A capsized hull as smooth and as unbroken as this presented few places where a grapnel could snag securely. After witnessing the prime condition of the wreck, I realized how fortuitously the grapnel had caught.

Air and bottom time remained, so we swam across the broad beam of the hull and dropped down to the sand on the sunlit side of the wreck. Here the brightness was so pronounced that my light didn't cast a visible beam. I thought the filament must have burned out until I looked directly into the bulb and saw the dim pinpoint of light.

Despite the luminosity I began to feel the effects of nitrogen narcosis. I didn't hand my regulator to a passing fish - the absurd fictional description that has been so widely overstated. I knew who and where I was and what I was about. Instead, I experienced a loss in mental acuity that manifested itself in visual disturbances. The wreck appeared blurred and I had difficulty focusing my eyes on my gauges. There was no distortion. Rather it looked like the snow on a television screen when the set is tuned to a channel that is broadcast from beyond the range of reception. With intense concentration I managed to decipher my gauge readings, but the snowstorm wouldn't go away from the wreck. My brain couldn't process all the incoming data.

I signaled to Clayton to ascend. He acknowledged. We reached the anchor chain fifteen minutes into the dive. From there we made our ascent at the rate of 30 feet per minute, as prescribed by my computer. By this time our eyes were fully adjusted to the lower light parameters at depth. Usually, a wreck disappears from view within seconds after leaving the bottom. In this instance, due to the bright ambient light, we were able to observe the wreck for the next five minutes as we crawled up the line. Since the mild current flowed from the stern to the bow of the *Washington*, the anchor line lay directly over the capsized hull. This enabled us to see more than two-thirds of the battleship's length. I couldn't take my eyes off it. It was an experience that I didn't want to end.

Clayton and I decompressed according to different timetables. He used the Navy Tables, which somehow worked for his particular body form and circulatory system, while I relied on a computer whose calculations were more conservative. My ceiling was 80 feet. I hung there for a couple of minutes until the computer cleared me to ascend to the next stop ten feet higher. Clayton did the deep stops with me until his Table kicked in at fifty feet, then he surged ahead with shorter stops.

The next two down were Steve Gatto and Tom Packer. I waved my computer for them to see so they would be prepared for a depth that exceeded our predictions. It was well that I did, for although they carried slates on which they wrote contingency Navy Tables to 300 feet, they carried only analog gauges whose maximum notation was 240 feet. When they touched down in the washout in order to peer under the hull, each gauge needle swung into unmarked space then lodged against the zero detent on the second sweep around. The stress stretched the spring out of calibration, so the needle no longer indicated the depth correctly.

That sixth sense that comes with experience made them distrust the readings. What they were not able to determine, though, was how much the gauges were in error. This meant that they couldn't ascertain the appropriate decompression depths. Eyeballing the surface the way Mike de Camp did in the 1960's - when the deepest decompression stop was twenty feet and the penalties averaged fifteen or twenty minutes - was an expedient whose time had passed. Packer ascended above his ceiling long enough to compare his depth gauge with my digital computer. From the difference he calculated the true depth. When Gatto and Packer emerged from the water, their gauges still registered a depth of thirty feet. The gauges were not repairable and had to be replaced.

Jon Hulburt and Greg Masi, the last pair down the line, relied on Hulburt's computer for depth reading and decompression. By then, due to a subtle shift in the wind or the direction of the current, the anchor line was chafing where it passed over the strut. Hulburt moved the line but it eventually chafed through anyway. Hulburt videotaped the entire dive. His video light, rated to 140 feet, was permanently distorted on the bottom yet continued to function.

My computer required 101 minutes to decompress. Clayton beat me to the surface by more than half an hour. But if I had relied on Navy Tables I would have run the risk of inviting adverse side effects. What works for one person doesn't necessarily work for another. Perhaps my twenty years of diving compared to Clayton's four made some physiological difference.

Clayton stared at my computer in stunned amazement. "How did you get a maximum depth of 290 feet when I only got 283?"

I looked at his gauge and shrugged. I told him about the washout. He hadn't noticed it, nor had he seen me peer under the hull. To me the point was of no consequence. But to him it mattered a great deal that I had gotten a deeper reading than he got. Thus began a friendly competition that frustrated him for years.

## Second Discovery

We were ecstatic over the achievement of our primary purpose, but the trip was far from over. We anchored that night over the *Merida*, a passenger liner lost by collision in 1911, and resting in 210 feet of water. The wreck was a disappointment because it had been demolished through the years by sundry commercial salvors in their quest for a non-existent treasure. The site lay far off the beaten path, and few wreck-divers had ever visited it.

After diving the *Merida* we set our course for home. I had some hang numbers that I wanted to reconnoiter. One set lay right across our path. Keen barely revved down the engine when a 40-foot spike jumped up on the recorder from 180 feet. Gatto, Packer, and Masi rolled over the side minutes later for a short reconnaissance dive. The wreck we found that day we didn't identify until we returned a couple of weeks later. It was the freighter *Ethel C.*, which foundered in 1960.

## The Pledge to Dive the *Ostfriesland*

The trip proved satisfying beyond our expectations. Yet it was only a scouting expedition: the trial phase for deeper explorations that Clayton and I were planning. Perhaps more important than what we accomplished was what we learned.

The depth recorders with which boats were equipped were inaccurate for our purposes. Time and again I've heard captains claim that the readings were based on the distance between the sea bed and the transducer, which was mounted in the hull below the waterline. I find it difficult to believe that no manufacturer has considered compensating for the difference. Compounding the error was the tendency of shipwrecks to create an all-around depression in the bottom as they settle into the sea bed. This combination produced actual depths that were ten to fifteen feet deeper than the surrounding sea bed. We needed to take this difference into account when diving unexplored sites.

We were exceeding the design limitations of standard recreational equipment. Our tanks did not hold enough air to provide a reasonable margin of safety. Gear configurations and back-up systems were inadequate for our needs. Surface support was lacking. Training was non-existent. When all was said and done, the exploration of unknown territory demands the development of new ideas and techniques. Lest this come across too much like "winging it," I submit that Clayton and I held long, in-depth discussions about our future goals and ambitions, and the assistance we would require to carry them off.

Divers had gone deeper on air than 290 feet. But those were stunt dives with momentary bottom times - a mere touch and go - that accomplished no useful work. In order to achieve the objectives we set for ourselves, we had to remain on the bottom long enough to explore our surroundings and to remember what we saw. A bounce dive was not worth the effort. Viewing a wreck while stupefied by nitrogen narcosis was not only risky, it was nonproductive. None of these deterrents presented any physiological barrier. In the subconscious gestalt of the wreck-diving community the predominant obstacle was attitude: a mental condition which for some divers didn't exist.

We decided to go ahead with our search for the *Ostfriesland*, and to make the dive breathing mixed gas. Cave divers were already forging into unplumbed depths under the earth, breathing a variety of mixtures in which a percentage of the nitrogen was replaced with helium, which was non-narcotic. If they could reach 300 feet in caves, we reasoned, we could reach the same depth or deeper in the ocean. So we set about purchasing the gear we would need, designing open-water techniques, and studying the literature on theory and practice.

This was not something to be taken lightly. It needed full commitment.

## THE *U-2513*

A month later I went on another junket to Key West. I started out by making a couple of dives on the *Wilkes-Barre*, but only as a prelude to greater adventures. The primary objective was the *U-2513*. Deans had found someone who thought he knew where it was. His name was Don DeMaria.

DeMaria was a fisherman of a different sort: he didn't catch fish with hooks or trawls but with plastic bottles and a butterfly net. He collected tropicals alive for sale to pet stores and aquariums. Sometimes he had to dive deep for rarer species. He got to know Deans because Key West was a small community, and not that many divers needed double-tank fills. DeMaria had friends among the commercial draggers, and it was from them that he obtained hang numbers that purported to be shipwrecks. He even verified that one of them was a submarine.

The wrecks lay some twenty or thirty miles west of the Dry Tortugas, a remote island group that was located sixty or seventy miles west of Key West. DeMaria worked from his own boat, the *Misteriosa*, and although he had never run charters he was willing to give it a try. Also participating in the trip were three fellow U-boat enthusiasts: Hank Keatts, a college professor who taught marine sciences; Brian Skerry, a salesman for a cardboard box manufacturer, but who had a degree in film production and who was working his way into underwater pictures; and Frank Benoit, a mechanic.

The boat ride from Key West to the Dry Tortugas took all day. We anchored in the protected harbor in the shadow of Fort Jefferson, then left from there at dawn. DeMaria had no trouble locating the wreck.

## 100 – The *Lusitania* Controversies

It was the *U-2513*, just as he promised. This experimental U-boat presented a distinctive hull form that was designed to move underwater at more than twice the speed of conventional German U-boats. It also incorporated a host of other technological innovations. It has often been said that if the type XXI had reached full-scale production a year before the end of the war, it would have posed a serious threat to Allied supremacy of the seas.

The first day I went on a tour with Brian Skerry. The depth was 210 feet, visibility around forty. Rockets littered the sand around the wreck like confetti after a parade. The U-boat was sunk in a test rocket attack, and these were the duds that failed to detonate on impact, perhaps because they struck at an oblique angle. This meant that the warheads were very much alive. Each missile measured some four feet in length. The four stabilizing vanes on the tail were encircled by a shroud.

We returned to the protection of the Dry Tortugas that night, then went back out to the wreck in the morning. I went in alone this time, and had one of the scariest dives of my career. The seas were a bit bouncy - just enough to make the boat ride up and down the waves, possibly enough to jerk the grapnel out of the wreck. I checked the grapnel when I reached the bottom - as did the others - and it appeared secure. One might expect that squeezing through a jagged damage hole and penetrating alone some fifty feet past hanging cables and debris to the forward torpedo room would be the feat that caused my fright - but one would be wrong.

At the end of my dive I noticed that the grapnel had slipped from its original emplacement and was dragging along the port side of the hull. No one else noticed because they completed their dive on the starboard side and converged with the anchor line above the wreck. With a long decompression to face, I thought I'd better rehook the grapnel so the boat wouldn't go adrift and wallow sideways in the troughs. Imagine my horror when I beheld the true state of affairs: the trailing grapnel had snatched up a rocket by the tail and two of the tines were hooked firmly in the shroud. Worse yet, the rise and fall of growing seas yanked the anchor line up and down. When the boat rode up on top of a crest, the rocket was lifted clear off the bottom. When the boat fell into the following trough, the nose cone slammed down hard against the sand. The action was that of a pile driver with three feet of ram travel.

I was stricken with panic. If the warhead exploded, the concussion would kill everyone in the water, including those who were decompressing a couple hundred feet away. They would never know what hit them. Neither would I, for that matter, although I might see the flash a millisecond before I felt the concussion. Death would be instantaneous.

## 1980's: Down to Deeper Depths - 101

I knelt on the sand next to the wreck in a posture strangely reminiscent of prayer. On the rocket's next downward plunge I gripped the cylindrical body with the idea of twisting it out of the grapnel's clutches. So powerful was the upward thrust of the anchor line that the rocket was pulled right out of my hands. When the nose cone plunked into the sand again I made another grab, better prepared for the violent boost, and held on fast. This time I was lifted bodily with the rocket some three feet off the bottom. My wetsuit-clad body was then scraped up and down the barnacle-encrusted hull as if I were a speck of hardened grease clinging to the bristles of a bottle brush. I held on for several swipes before I was ripped off, battered and bruised by the jerking rocket, and flung aside, thoroughly thrashed.

Without hesitation I leaped back into the fracas. I wrapped the fingers of my left hand around the sharpened metal edge of the shroud. With my other hand I seized the shank of the grapnel. For an instant I tried to pull the rocket and grapnel together and separate the tines from the shroud as one might free a fishhook from a snag. But the tines were not simply curled under the shroud, they were wedged between the vanes. I might as well have tried to bend an iron bar with my hands at either end. Then the instant passed and the rocket plunged to earth, crashed, was yanked up, attained burnout altitude, and began the cycle over again.

I held on fiercely. Each periodic burst of speed was attended by a spell of slack. During these transient moments I fought to maintain my grip and jiggle the rocket off its overhead launching pad. Between times, I was wrenched vertically to and fro like a piston rod on the cam shaft of a high speed engine. My arms were being jerked out of their sockets.

All this while the anchor line was sliding sideways along the hull, adding another vector of abrasion. To gain a stable position with respect to the rocket and the grapnel, I bent at the waist and brought my knees forward, then quickly wrapped my legs around the lurching metal cylinder. I rode the rocket like a rodeo cowboy on a maddened bucking bronco, or perhaps like some maniacal child on a pogo stick in a storm.

I hoped that straddling the rocket would enable me to apply sufficient leverage to pry the grapnel out of the shroud. No way. My wetsuit was grated like Parmesan cheese and I was thwacked on the head by the chain. Despite the terror of the moment I was surmounted by my sardonic sense of humor: I couldn't help but think of *Dr. Strangelove* and the scene with Slim Pickens riding bareback on an atomic bomb as it dropped from the bay of a jet.

I pulled and shook the shank to no avail. The tines might just as well have been welded to the vanes. And time was running out. Despite my best efforts I couldn't free the grapnel from the shroud. I had to leave. Dizzily, I pushed away from the rocket, watched helplessly as the nose cone continued to wallop the sand, then caught my breath as I made my slow ascent.

For the next hour and a half I agonized over that unguided missile trying to impact itself on the bottom. My friends were blissfully ignorant. On the boat I warned DeMaria to be careful pulling the hook. He tugged the anchor line a couple of times, then the rope parted, and we lost both the grapnel and the rocket. I was relieved.

### THE PARK SERVICE STING

As if that weren't enough adventure for one day, I nearly got arrested that afternoon. A slow, low-altitude drug interdiction plane flew overhead on the way to the Dry Tortugas. The pilot scrutinized the after deck, which was loaded with tanks and scuba equipment, then went on his way. We were not harassed. We docked at Fort Jefferson in order to take in the tour. (The Dry Tortugas is a National Park.) The water under the dock was littered with trash from more than a century of disposal. I struck up a conversation with a friendly park ranger, and asked if it was okay to pick up some of the discarded bottles that were mingled with the rubbish. He gave me permission to do so.

I was swimming alongside the dock when someone gave me the high sign. (It was one of our guys, but I can't remember who.) He made frantic motions for me to drop the bottles I had collected. I ducked under water, scattered most of the bottles on the bottom, but stashed a couple of displayable ones next to a pile that supported the dock. Later he told me that he came upon the ranger crouching behind a bush at the water's edge. Upon asking the ranger what he was doing there, the ranger, who evidently didn't suspect that he was with our group, explained scornfully that he was waiting to catch me looting cultural material from a protected site. Pepsi bottles are cultural material?

The ranger had set me up for a National Park Service sting operation.

It is satiric to note that had I tossed such a bottle in the water first I could have been fined for littering, then fined again for picking it up and placing it in a proper receptacle.

### MIXING AIRS – OR AIRING MIXES?

During these weeks in Key West and off the Dry Tortugas, I told Deans about the *Washington* dive and about our intention to dive the *Ostfriesland*. Coincidentally, he was already playing around with different gas mixtures and had experimented with them on the *Wilkes-Barre*. He invited me to try it. I jumped at the opportunity to make a

## 1980's: Down to Deeper Depths - 103

pair of comparison dives: one on air and one on mix. We went to a local commercial gas supplier and purchased the helium in cylinders. Using the filling station at the shop, Deans showed me how to mix helium with air by means of blending fractional components.

He also worked me through the decompression schedule, which utilized nitrox and oxygen to accelerate the process. I was familiar with the theoretical concepts, not just from reading the literature but from actually diving on nitrox with Deans. Nitrox is sometimes called enriched air because it contains a higher percentage of oxygen than ordinary air. More oxygen means less nitrogen, and consequently longer bottom time or shorter decompression. But a limitation on depth was imposed because oxygen becomes toxic under pressure.

Helium is an inert gas which, like the nitrogen in air, serves no metabolic function. If some or all the nitrogen is replaced by helium, that helium is inspired and absorbed by the tissues in the same fashion as nitrogen. But helium takes longer to eliminate - much longer. This means that decompression is longer, inordinately so. The way to accelerate decompression is to switch to other gases during ascent: various blends of nitrox and, above twenty feet, pure oxygen.

The schedule was provided by Bill Hamilton, an ex-fighter pilot who was now self-employed as a physiologist and computer programmer. He designed his own decompression model and wrote a program to run it. He called it DCAP, for Decompression Computation and Analysis Program. Hamilton was already well respected among cave divers who used DCAP for long deep cave penetrations in which they staged bottles along the route for emergencies and to extend their bottom times. If he wasn't already the guru of mixed-gas decompression schedules, he was clearly on his way. The DCAP profile included not only stops and times, but prescribed the gases to be breathed during ascent. Extra tanks must be taken in order to contain these alternative gases.

The *Wilkes-Barre* was an ideal training ground for mixed-gas wreck-diving. The site was a short jaunt from the dock, and the mooring system that Deans established offered a certain measure of control. Warm, clear water relieved much of the stress that derived from the cold and dismal conditions encountered farther north.

On the air dive I found an abandoned fish trap on the bottom. It was old and covered with growth but was still catching fish. It is my policy to break in the slats of wooden traps or open the doors of wire traps in order to end the senseless cycle that results from dead fish attracting other fish to their deaths, ad infinitum. I couldn't figure out how to release the catch. On the gas dive I understood the mechanism perfectly: a dramatic demonstration of the mind-clearing benefit of mixed gas. There was no snow on my mental screen as there had been on the *Washington*.

I asked Deans if he wanted to be part of the *Ostfriesland* dive team, but he declined. I then encouraged him to teach mixed-gas diving as a way of extending his business. He demurred, stating that he wanted only to do it himself and possibly show a few friends. But I suspect that the thought had already entered his mind, and hadn't yet quite gelled. He was simply biding his time and reckoning how best to proceed.

## MONITOR VICTORY

The year ended on a triumphant note for wreck-diving. After six years and at a cost of thousands of dollars, my suit against NOAA over access to the *Monitor* went to court. Based upon a full evidentiary hearing the judge overruled the agency's self-indulgent policy of denying American citizens the opportunity to see the historic wreck for themselves. Since I've already written about the precedent-setting case and its aftermath in *Ironclad Legacy: Battles of the USS Monitor*, I don't want to belabor the legal issues here. But I would be delinquent with respect to the ambition of the present volume if I did not mention in passing the significance of those features of the action that dealt directly with deep wreck-diving.

One of the judge's conclusions read: "The standards adopted for Agency use by NOAA and/or the United States Navy may not be imposed upon the public sector merely because the proposed activity is to be carried out within a Marine Sanctuary."

This meant that never again could any government agency limit the depth to which a diver may descend.

The judge also concluded: "The appellant and other staged decompression divers are not sport or novice divers. Their training, experience and certifications reflect a substantially greater proficiency."

This seemingly profound judicial finding reflects a point that I have reiterated throughout the years, and that I went to great lengths to establish on the witness stand. There are varying degrees of expertise in all forms of human endeavor. If everyone displayed the same scholastic aptitude there would be no reason for taking tests and giving grades. If everyone exhibited identical physical prowess there would be no Olympic competition. Diving is no different in this regard. Some divers are more skilled than others. To unite all divers under a single banner is equivalent to lumping high school grads with Ph.D.'s, joggers with marathon runners, foot doctors with brain surgeons.

Those who refuse to acknowledge individual ability, especially that which is greater than their own, do so because of their inherent insecurities. They create groups as fodder for their prejudice. They put people in their place by lodging them in artificial categories with convenient and often derogatory headings. This enables them to rationalize suspicion, disrespect, ridicule, or hatred: unreasonable attitudes that deserve repudiation.

In order to prove to the court's satisfaction that the *Monitor* should be open to the public, I had to prove that wreck-divers, called "staged decompression divers" in legal brief lingo, were capable of making such a dive. Thus those two salient sentences in the judge's ruling undertook a special significance, one that outweighed the recommendation to permit divers to dive the *Monitor*, which was the object of the suit. Now there existed a legal recognition of deep diving as a proficiency all its own.

The judge didn't state what "staged decompression divers" were, only what they were not.

A new definition was in the offing.

# Part 7
# Early 1990's - Mixed-Gas Revolution

## *Ostfriesland* Gestation

If 1989 instigated a transition in the concepts of deep wreck-diving, 1990 saw the implementation of those concepts.

During the winter months, Clayton and I worked on organizing a trip to dive the *Ostfriesland*. This was no ordinary dive, and no ordinary trip. The variables and complexities of such an enterprise were manifold. Nothing like it had ever been done before, so we had no guidelines to follow. To pull it off we had to plan for every foreseeable contingency, as well as for eventualities that were totally unanticipated - perhaps unimaginable. Most of what we had to do was either novel or untried.

Some of our prerequisites were the position of the wreck, a decompression schedule, mixed gas, a suitable dive boat, support divers, and equipment that could do the job. Plus we had to develop reliable in-water techniques for exploring the wreck, returning to the anchor line, and conducting a long and complicated decompression that required four different breathing gases, none of them air, plus argon for drysuit inflation.

My original research notes indicated the general area in which the German warships were sunk. Subsequent research narrowed the location to a specified locale some sixty miles off Cape Charles, Virginia. The records mentioned by name some of the vessels that either participated in the exercise or that served as observation platforms for a multitude of navy and army officers, elected heads of state, foreign dignitaries, and reporters and photographers. From the deck logs of these vessels I obtained anchored positions in latitude and longitude - from sextant readings - as well as notations about other vessels in sight.

After I gave this information to Clayton, he went to the National Archives and studied the deck logs of additional vessels. In some of these logs were recorded not only the position of the recording vessel, but the compass headings and either distances or running times to the target vessels. From this elaborate pattern of positions and directions, Clayton wove a fabric whose design approximated the mooring layout for the entire operation. I bought a couple of hydrographic charts that showed enlargements of the area in question - and that indicated the presence of sizable obstructions. We then combed hang logs for coinciding coordinates.

When all this was done we needed verification of the selected sites. Clayton went on an overnight trip on the *Gekos* when Keen was planning to pass through the area. They checked out the most likely targets with the depth recorder, and verified loran coordinates. One obstruction rose 60 feet off the bottom. It could be none other than the battleship *Ostfriesland*.

The wreck lay 380 feet deep. The highest point of relief was 320 feet.

To say that I was undaunted by the depth would be a lie. I felt great trepidation about the feasibility of such a dive: on one hand whether it was humanly possible, on the other hand whether I was up to the task. On an intellectual level I supposed it could be done, but on an emotional level I entertained serious doubts about my own capabilities. I kept on with the project nonetheless.

With a firm maximum depth established we could begin to think about the gas requirements and decompression schedule. Clayton obtained the promise of support in that regard from Bill Hamilton the previous autumn, after the *Washington* dive. Clayton attended a cave-diving conference in Florida and had a long talk with Hamilton about our mixed-gas diving objectives: the discovery and exploration of what we called the "Billy Mitchell wrecks."

Clayton told Hamilton how deep we wanted to go. Hamilton thought the dive was feasible. We placed complete trust in his expertise and in the accuracy of DCAP to provide us with the appropriate mixtures to breathe, the gas switches to make, and the decompression schedule to follow. This was a lot to put on faith because DCAP was not based on field experimentation, but was instead a theoretical mathematical model - more sophisticated of course than my Navy Table extrapolations of the 1980's, but a hypothetical construct none the less. The profile that DCAP computed for us had never been verified in practice. The algorithm yielded a calculated prediction based upon the current state of knowledge of decompression theory.

The charge to generate the schedules for a single depth with a spread of five alternative bottom times was $400.

Hamilton recommended two gas mixes for the dive: heliox-12 and heliox-16. Heliox-12 consists of 12% oxygen and 88% helium. Decreasing the percentage of oxygen in the mix reduced the risk of toxicity at extreme depth. But the percentage of oxygen was so low that he thought we might suffer mild hypoxia on or just below the surface, especially if we had to swim hard against a current. Therefore he recommended a travel mix with 16% oxygen for the initial descent and down to 320 feet, at which depth we would switch to bottom mix. We would also switch back to travel mix during the ascent (at 320 feet) and breathe it during the deep stages of decompression, which began at 160 feet. At the shallower stages of decompression, beginning at 130 feet, we would first breathe nitrox-36 (36% oxygen, 64% nitrogen), then switch to pure oxygen at 20 feet.

Once we knew the mixtures, we had to obtain the gas. Known to anyone who has had to fill floating party balloons, helium can be purchased from a commercial gas supplier. We soon found that Potomac Air Gas could not only supply the raw helium that we needed, but was willing to blend it with oxygen in the required proportions and deliver it already mixed. The company would deliver the gas in 300-cubic-foot storage cylinders, but it was up to us to have the gas transferred to our scuba tanks.

The final cost of heliox per single scuba tank came to $75. Since I planned on carrying four tanks of heliox, my cost for helium for a single dive was $300. This did not include the cost of decompression gases.

Due to circumstances that made the *Gekos* unreliable, we had to find a dependable boat that could carry all our equipment, our support divers, and our support divers' equipment; that could withstand rough seas in the open ocean and protect the passengers from the weather; and that could provide comfortable overnight sleeping accommodations. We got more than we bargained for.

The *Miss Lindsey* was a 50-foot aluminum boat especially fitted for diving. In form and layout it was reminiscent of the crew boats so prevalent in the Gulf of Mexico: those that transported oil-rig workers stationed on offshore drilling platforms. Mike Hillier, owner of the Lynnhaven Dive Center in Virginia Beach, designed the boat from the keel up with overnight trips in mind. Instead of a bus-like interior consisting of rows of padded seats, he built in bunks that converted to seats and tables for dining. Not only was there a galley with the amenities of a modern kitchen in miniature, but food and a cook were provided. A second captain and a diving deck hand rounded out the complement.

Tanks and gear boxes were stowed primarily on and under benches that lined the after deck. The solid canopy doubled as a full-width sun deck. Extra storage space was available topside under the forward extension of the sun deck and behind canvas curtains that checked spray. Cameras and clothing could be stowed below deck. And because the boat was equipped with a compressor, tanks could be refilled with air on board, thus reducing the number of doubles needed for two- and three-day trips.

More important, Hillier was willing to accept the potential liability of an admittedly risky undertaking.

A boat so large afforded lots of room, but it also cost more than we could afford. Charters are generally run on the premise of filling the boat with divers who pay an equal share of the cost. The problem we faced was convincing divers to sign up on a trip to a wreck that was too deep for them to dive. One way out of this dilemma was to make the trip an overnighter on which only one dive would be deep and the rest would be shallow. That way the support team would get something for their money. Also, as a sales pitch, we could bill the trip as a training exercise. There was no doubt that the support team would learn nearly as much about deep gas diving as we would - and at considerably less risk. By promoting the support role as an educational opportunity, we hoped to inspire supporters to carry the torch on future dives.

Clayton and I were starting at the top, in a sense, of a new diving hierarchy. There was no other way, since neither training programs nor apprenticeships existed. We had to develop a complete methodology just to make a single dive. And we had to get it right the first time, or suffer the consequences.

What I found most frightening about the proposed dive was that it violated one of my most basic wreck-diving principles: that of progressive exploration, whether it be progressive penetration into an intact wreck or progressively deeper descents. The step we were about to take was in no way incremental. It was a plunge. Yet, by its very nature, mixed-gas diving was no half-way measure: it was a quantum leap into the depths. And despite the freedom from narcosis, 380 feet was a long way to the surface.

Clayton advanced another plan for fleshing out the charter: advertising my participation as a way of attracting customers. Clayton was unknown in the diving community, but he thought people might sign up for the trip just to be on the same boat with me. I speculated that my reputation was of dubious value at best, but despite my vocal skepticism, that was chiefly how he filled the boat. I suppose that a long history of lecturing, giving workshops, and writing books and articles counted for something after all.

In the end, we had three categories of participants: gas divers, support divers, and recreational divers. The support divers were those who supported the gas divers in the water by setting up lines, by staging tanks of nitrox for decompression, and by meeting the gas divers during ascent to relieve them of side-slung bottles that were no longer needed, and to hang around in case some other assistance was required.

The recreational divers were not so designated derisively, despite the fact that they were selected less on the basis of ability than on the basis of availability, and that their primary function was to add money to the pot so the gas divers didn't have to pay so much for the trip. It is more correct and less disparaging to call them "topside support personnel." They performed duties and functions on the boat that relieved the rest of the team from overwork.

At the risk of sounding trite, in one respect we were all in the same boat: we wanted the mission to succeed. Only through cooperation could this be accomplished.

## SPREADING THE WORK LOAD

It is appropriate at this point to remark that with respect to the division of labor, Clayton handled more of the day-to-day details than I did. There were several reasons for this. He had been laid off and while unemployed he could devote full time to the project. He was geographically closer to the facilities that handled the gas delivery and transfer. He was an absolute stickler for detail, leaving no stone unturned in the quest for an answer to a question however insignificant. It is also my opinion that he enjoyed the titular role of "trip leader." I'd been running trips for years, so the charisma of the position had long since lost its veneer.

My time was more than fully occupied. In addition to running a publishing business and doing all the work of a fully staffed company, I was researching and writing books and articles, maintaining a lecture schedule, giving workshops, diving, and doing the photography to illustrate my numerous publications. I've often said that if a boss or employer drove me as hard as I drove myself, I'd have quit long ago. I was my own worst taskmaster.

My work was quite overwhelming. An author generally leaves fine finishing details such as continuity, sentence structure, word usage, and spelling to the editorial staff: the managing editors, copy editors, line editors, proof readers, secretaries, and so on. I had to do it all myself. So while I take credit for the placement for every period, I must also accept responsibility for every mistake (except for typos introduced in the typesetting process).

In addition, I had sole responsibility for organizing, implementing, and conducting a ten-day trip to the *Monitor*, which was fraught with its own complications, logistic as well as bureaucratic.

After two decades of diving and arranging trips, I was used to ironing out matters as they came along, without seeking hitches that didn't exist. I often relied on my experience to carry through a task or to overcome a difficulty. In this regard I was perhaps too complacent. Thus our personalities complemented each other in a working relationship that weighed my underwater experience against his fresh originality in which nothing was taken for granted.

I didn't ride herd on issues that Clayton wanted to address himself, and he passed all concerns by me for advice and approval. We sometimes disagreed on minor points, differed greatly in our philosophy of conduct and management, but always came to terms without loss of prestige between us. Out of dissent always emerged resolution before a final decision was made.

## Mixed-Gas Initiative

In order to spread the word on what I already believed would be the next major development in deep wreck-diving, and to promote mixed gas as a means of reaching deeper wrecks, I coordinated a mixed-gas diving initiative. It began with a meeting at Bart Malone's house. I chose his home instead of mine because he was more centrally located in New Jersey: closer to where most of the deep divers lived.

Malone had taken his own initiative with respect to artifact display. Frustrated at the disinclination of museums to exhibit relics of the nation's maritime past if those relics had been recovered and restored by recreational divers instead of by accredited marine archaeologists, he built his own museum in his back yard. The peaked-roof single-story building was the size of a large cottage. Inside were shelves, showcases, cabinets, and display racks filled with artifacts that he had brought up himself, plus those belonging to many local divers, myself included. He also maintained a library of books and reference materials and files on individual wrecks.

Malone was particularly interested in the concept of mixed-gas diving because of his low tolerance to narcosis. I should emphasize that an individual's threshold or response to nitrogen narcosis has nothing to do with ability or experience, any more than the way one is affected by, say, an analgesic. Codeine reduces the severity of a headache for some, and puts others out like a light. The depth at which narcosis becomes an impediment is determined by innate and so far unexplained physical phenomena. Only the psychological effects can be overcome by training.

The meeting was an informal gathering. We swapped stories and traded anecdotes, but my discourse on mixed gas went over like a balloon filled with lead instead of with helium. People listened but were unimpressed - or were scared off by the exceptional depth of the proposed dive to the *Ostfriesland*. I stressed the value of breathing mixed gas at intermediate depths, where replacing some (but not all) of the nitrogen with helium could reduce the debilitating narcotic effect. Instead of extending the depth range of working dives, mixed gas could permit all divers to perform at their full potential unaffected by narcosis. In that sense, helium could be viewed as an equalizing agent.

These ideas were too far ahead of their time. The seeds were planted, but the germinating process was slow. The time for conceptual breakthrough had not yet arrived.

## Drafting Cohorts

Meanwhile, the search went on for fellow divers and support personnel. First to share our enthusiasm for the venture into unplumbed depths was Jon Hulburt. As one of the prime innovators in wreck-diving techniques, he brought with him a wealth of deep diving experience as well as engineering knowledge and skill. He was in charge of Dupont's high-pressure experimental facility in Wilmington, Delaware. He worked "under pressure" every day. His thorough occupational understanding of pressure vessels, booster pumps, gas mixes, and chemistry made him a prime candidate for embracing new and unproved technologies.

Pete Manchee had often told me to call him if any exotic diving opportunities arose. I did, and his response was promptly positive.

No one else was willing to make the commitment. With four divers in the team we decided to dive in two pairs: Clayton and I, Hulburt and Manchee.

In April I went to South Carolina to dive with Manchee on his boat, the *Atlantic Lady*. I distinguished myself by puking my guts out in the rough seas, but still managed to get in the water and dive. Manchee not only dived, but manned the helm like a beardless Captain Nemo indifferent to the chilly spray and chop. During calmer moments ashore we talked about the *Ostfriesland*.

## Tortugas Adventure

In May I went to Key West to continue exploring the wrecks off the Dry Tortugas. Don DeMaria had more hang numbers to check out. The impetus for the trip was an iron-hulled sailing vessel that Deans and DeMaria had dived once and christened "the schooner." They hoped that I might establish the wreck's identity. It lay at a depth of 215 feet. Visibility exceeded one hundred feet. It took only a casual glance to determine that the "schooner" was in fact a brig. Schooners don't have crosstrees.

The wreck stood upright with fifteen feet of relief. Except for the port bow, where the anchor, hawse pipe, and surrounding plates were torn asunder and lay sprawled in the sand nearby, the hull was structurally intact. The stern presented a delicately rounded fantail atop an upright rudder. The foremast and mainmast were broken off at the upper deck and lay on the sand to starboard. At one time the ship had been fitted with a mizzenmast, but only a stub now remained. The mizzenmast had been sawed off during the ship's career, just below the weather deck. Incredibly, some of the wooden decking was extant, as were all the planks below deck, which sealed off the cargo compartment.

From these cursory observations I determined that the vessel had originally been rigged as a three-masted ship or bark, then reduced to a brig. The profusion of deadeyes and the absence of blocks to tighten the rigging, both standing and running, led me to conclude that the vessel's construction pre-dated the turn of the century by quite a number of years. After about 1900 most windjammers were fitted with blocks, or pulleys, instead of deadeyes. The prior decade or so was a period of transition, and saw a combination of deadeyes and blocks.

I saw no sign of a bell, but near the end of the dive I found a couple of portholes and a lantern lens. I stacked these items on the deck next to the bulwark, close to the anchor line. DeMaria did the same with some bottles and brass knickknacks.

Deans conducted a circumnavigation of the wreck on his Aquazepp, a torpedo-shaped diver propulsion vehicle, or DPV, known more familiarly as a scooter. I looked up whenever I heard the dopplering propeller sound, exchanged okay signs, then went about my business after he scootered by. He was conducting a photographic reconnaissance with what I called "roving video." He had secured a video camera to a swivel mount on the Aquazepp's nose cone. As he made sweeps around the hull he swiveled the camera so the wreck was always in view. I had convinced him to go on the *Monitor* trip, to shoot roving video, so he used the "schooner" as a proving ground to perfect his roving video technique.

Because Deans was a dealer and shop owner he had more toys and bona fide equipment than most divers own in a lifetime. On that dive he wore three computers, then violated their decompression requirements by surfacing ahead of the time they proposed. He did this purposely with no fear of getting bent because he decompressed on oxygen. Violated computers lock up, then don't clear for twenty-four hours. To prevent this from happening, he put the three computers in a mesh bag which he hung over the side of the boat so the computers could complete their decompression.

When we started to dress for the second dive, Deans discovered that the mesh bag had become unclipped. More than a thousand dollars worth of computers - to say nothing of a fine mesh bag - were gone. Deans wrote them off in disgust.

I argued persuasively that he had a good chance of finding the bag full of computers. Visibility was superb, and the live bottom consisted only of small, low patches of coral and grass scattered over predominantly white sand. As examples of success in similar situations I cited two recoveries that I had made: Mike de Camp's camera in the 1970's, and my own camera rig in the 1980's. I threw a buoy over the side despite his protestations.

DeMaria checked the compass and determined that the boat had swung at least forty-five degrees. This was not good. I suggested tying a line to the buoy line and making ever-increasing circles. Deans finally relented, but decided to make circles with the Aquazepp because he could cover more territory.

Deans went down alone. In the nineteenth minute of a planned twenty minute repetitive dive he found the bag and the grand of computers. He was decompressing on a line hung below the stern of the *Misteriosa* when I went down to the wreck with Steve Gamble. Gamble had been a carpenter with a company on the west coast for more than ten years when he'd been struck with wanderlust. He quit his job as foreman and went to see the country, with no particular destination in mind. He wound up in Key West low on funds, and took a position with Deans as dive guide and general shop assistant. He was the crew on this Dry Tortugas trip.

Gamble wanted to see the wreck for himself. I wanted him to have the opportunity to do so. There were no full doubles and we had no compressor, so he had to make the dive on a single tank. Our plan was short and simple: put all the artifacts into a mesh bag, clip the bag to the anchor line, send the bag to the surface on a liftbag, then pull the hook. I estimated that we could do everything in ten minutes tops. We didn't have to travel more than twenty feet from the anchor line. I would fill the liftbag so that Gamble could conserve his air.

The dive went like clockwork with an exploding spring. Things started to go wrong before we even reached the wreck. The grapnel was hooked in the wreckage of the mainmast on the sand. DeMaria was right about the boat having swung. Whereas before the anchor line paralleled the hull, now it lay across the bulwark: and rested not on the chain but on the rope. I automatically checked for chafing and saw that the strands were severely abraded. Instead of collecting the artifacts I headed straight for the grapnel. I wanted to release it and reset it temporarily on the bulwark or inside the wreck.

No sooner did I put my hands on the tines than the rope parted. The wind blew the boat away in a flash, leaving us stranded on the bottom without an ascent line. Not to worry: I had a decompression reel and an extra liftbag. I didn't want to lose DeMaria's grapnel, so I kicked and swam the heavy steel device from the sand up to the bulwark. Gamble then helped me hump it across the deck supports to where a section of planking offered a platform on which to set it down. I took my decompression reel off my back, pulled off some line, tied the end to the grapnel, clipped on a liftbag, sent the grapnel to the surface, and tied the line to a beam.

I checked my time and air. Gamble did the same. Then we set about doing the job we had gone down to do. I was shocked to find that where I had left two portholes there now lay three. The appearance of the third porthole was never explained, as neither Deans nor DeMaria took credit for placing it there. It's an unsolved mystery.

At the time, however, my primary concern was to get everything to the surface. Gamble and I swam the portholes, lantern lens, bottles, and knickknacks to the tie-off point and stuffed everything in a bag. Gamble stopped me to show me his pressure gauge: he was getting low on air. We had expended quite a bit of time and effort on fooling with the grapnel. I asked if he wanted to go up and let me finish, but he shook his head. By the time I sent up the second liftbag, sixteen minutes had passed. We had overstayed our planned bottom time by six minutes. Since this was a repetitive dive for me, after a short surface interval, my computer showed more than an hour of decompression.

Gamble was now really low on air - he had less than 500 psi. I was breathing pretty heavily after all my exertions. I got my respiration under control during the ascent. We were only halfway to the surface when Gamble showed me his gauge again. It now registered 200 psi. He wasn't going to make it on his own.

Because of the upcoming *Ostfriesland* dive I had updated my equipment. Gone were my old aluminum tanks that held only 80 cubic feet of air. Now I wore steel 100's whose manifold incorporated captured o-ring valves and twin regulator ports. I passed my spare regulator to Gamble. I had a pony bottle besides.

We continued up the line to the first stop shown on my computer. Gamble rode piggyback most of the time, but because my regulators vented to the side instead of to the bottom as in traditional regulators, the mouthpiece could be swiveled so we could hang side by side and look at each other for ease in communication. Gamble was calm. He breathed easily without a bit of tension. Usually, once I get settled down on the line - and if there is neither current nor bounding seas - I can get my breathing rate down to three breaths per minute: an eight second inhale followed by a twelve second exhale, with no skip breathing. I have on occasion of dire necessity gotten my breathing rate down to two breaths per minute, but the inhalations never satisfy my demand for air. Neither of us made unnecessary movements or exertions. We simply hung on as calmly as we could because we knew we were on our own. We couldn't expect help.

Deans had barely begun his decompression when we had made our descent. That meant that DeMaria couldn't move the boat until Deans was out of the water. By that time the boat would have been blown a considerable distance away. If I required more than an hour of decompression, Deans needed more.

Activity of the brain consumes oxygen, but I couldn't prevent my mind from calculating furiously. No matter how I juggled consumption rates with the quantity of air available (in my doubles, in my pony bottle, and with the little left remaining in Gamble's single), and taking into account that Gamble had a shorter penalty, I didn't see how we could make it. The extra time on the bottom had increased our decompression time significantly. According to my best estimate we'd be twenty minutes short. Longer, more than likely. We nodded and rolled our eyes at each other.

All we could do was settle down and breathe as slowly as we could. I was concentrating so hard on controlling my air intake that I didn't hear the whine of the approaching engine. Not until I felt a tap on my shoulder did I spin around and see Billy Deans's grinning face. Then I looked up and saw the silhouette of the *Misteriosa's* hull. We were saved!

DeMaria tossed a loop around the liftbags. Then he threw the surface supply hoses over the side. Gamble and I each grabbed a regulator and inhaled great lungfuls of oxygen, fed to us from large storage cylinders on deck. Only then did I feel the deficit of oxygen that I was laboring under.

How had it come about? DeMaria pulled in the miscellaneous dangling lines after the boat went adrift. He dived overboard, pulled himself down the weighted line to where Deans was decompressing, and signaled for him to hold on tight. Deans got the message. It was obvious from the boat's motion that it was no longer anchored. He expelled every bit of reserve buoyancy and descended to the weight at the bottom of the hang line. DeMaria climbed back aboard, started the engine, checked to make sure that Deans was deep and clear of the propeller, then gently eased the transmission into forward. He nudged the boat ahead with slow revolutions. Despite the slow speed and the weight on the line, Deans soared aloft as if he were straddling an underwater sled. He hung on with both hands while his mask threatened to leave his face. He was barely able to maintain ten feet of depth. He was probably as glad to see me - and the end of the ride - as I was to see him.

## *AquaCorps* Redefines Deep

Diving off the Dry Tortugas was always an adventure. Twice we were nearly run down by merchant vessels running on autopilot and apparently without lookouts or anyone at the wheel or radio.

Back in Key West, while marking time to return to the Dry Tortugas, I made a few more dives on the *Wilkes-Barre*. Visiting the Keys from California was a man who was soon to have a profound impact on the evolution of deep diving. His name was Michael Menduno. He had been a marketing consultant in Silicon Valley, making high technology investments for venture capitalists, but had abandoned corporate America to seek his future in the underwater world. Only a couple of months previous he had published the premier issue of a slim magazine called *AquaCorps*, which was subtitled "The Independent Journal for Experienced Divers." The magazine was the first to address the issues specific to the growing movement of diving deeper and staying longer.

The initial volume was launched with editorial comment on defining the activity and giving it a label that distinguished it from other forms of diving. Menduno proposed the term "professional sport" diving, but encouraged further discussion. Bill Hamilton, in his article, pointed out that neither word sufficed: "professional" because it referred to an occupation in which a diver was engaged for pay instead of for fun, and "sport" because it implied competition. The certifying agencies preferred "recreational" diving to "sport" diving.

Hamilton suggested "advanced" diving - the term I had coined in the *Advanced Wreck Diving Guide* - but he restricted the definition to divers who exceeded the no-decompression limits wearing wetsuits. Divers who wore drysuits, double tanks, and carried computers instead of decompression tables, he called "high tech," although he also allowed for refinement of the terms and for alternative designations as demanded by future considerations. He described both advanced and high-tech diving as non-traditional, in the sense that only a small minority of divers exceeded the basic qualification by going beyond accepted practice and procedure.

Perhaps the most prescient statement in the magazine was a prediction made by Walt Hendricks, Sr., a former training director for a national certifying agency. He thought that since training agency concepts were mired in the orthodox conventions of novice instruction, new training and certifying agencies would be created to cater to the needs of hard-core divers who refused to be bound by artificial limitations.

*AquaCorps* was billed as a quarterly, but because the magazine was at first a one-person show, its publication schedule was somewhat erratic - to the continual frustration of its subscribers. Each issue carried progressive views and up-to-date information that were available nowhere else. An expanding coterie of divers was drooling over every word.

Menduno had considerable journalistic ability but little in-water experience. He himself was not a "high-tech" diver. But he was willing to learn the ropes in order to report with verisimilitude. Hence the reason for making a junket of the east coast and the cave-diving mecca in Florida. Deans and I took Menduno down on the *Wilkes-Barre* for an introductory dive on air, all the way to the sand. The next day we repeated the dive on trimix-17/50 so he could experience the difference. (trimix-17/50 is shorthand for a blend consisting of 17% oxygen, 50% helium, and 33% nitrogen.)

He was ecstatic. Or should I say "more ecstatic?" Menduno was always ecstatic - that's the way he comported himself. The dive gave him something to write about from first-hand experience.

## The Araby Maid

A few days later I boarded the *Misteriosa* for another trip to the Dry Tortugas and the unidentified "schooner." I went down with Deans and DeMaria, and in true veteran-diver fashion our paths managed to cross a couple of times. Deans zoomed around on his scooter. DeMaria and I plodded in the slow lane using fins and leg muscles for propulsion. While I was off on my own I inspected the capstan and saw that it was topped with a brass cover. I scratched away some of the encrustation and uncovered some engraved lettering. I was excited. I didn't have tools with me, so removal of the cover would have to wait until the next dive.

On the boat I waxed enthusiastic over the find. In response I received a couple of blank stares. Finally Deans said, "What's a capstan cover?"

I explained that it was a decorative top that fitted on the capstan like a skullcap. Often the cover was imprinted with the name of the manufacturer, sometimes with the name of the vessel.

Deans went with me on the next dive. I worked on the hexagonal nut in the middle of the cover while he knocked the encrustation off the lip. I soon learned that the nut was an ornamental part of the casting, not a method of securing the cover to the capstan. I started banging under the lip with a hammer and chisel, and with additional leverage from a crowbar we popped the cover free.

On the way back to the Dry Tortugas for the night, Deans patiently tapped away the last vestiges of encrustation until the inscription was fully revealed: ARABY MAID. And underneath, the date: 1868. Later, after months of archival research and correspondence with several Norwegian museums, my observations of the wreck were borne out. The *Araby Maid* began life as a bark, then was converted to a brig. She sank after a collision in 1903 when the stem of a freighter rammed her port bow.

## The Rhein

The next day, Deans, DeMaria, and I dived together to 240 feet on a huge steel passenger-freighter that eventually proved to be the German motor vessel *Rhein*, which had been scuttled by her crew in order to avoid capture by British warships, in 1940. The wreck measured more than 400 feet in length and rose some forty feet off the bottom. Portholes lying loose on the deck bore mute testimony to the fact that the wreck had never been dived.

## Dissent on the Doria

I managed to squeeze in a trip to the *Doria* before the *Monitor* trip began. The *Wahoo* was chartered to a group of cave divers with Bernie Chowdhury in charge. As Gary Gilligan and I got dressed for the tie-in dive, Steve Bielenda quietly gave us instructions on where he wanted the downline shackled: on the stern near the "cup" hole that we had found the previous year. It made sense to me.

The *Doria* was a vast wreck with places all over that I wanted to explore. We discussed our options and made alternative plans depending upon how much time we had left after tying in. Despite the fact that the grapnel snagged on structure far aft of where we were supposed to shackle the downline, the current was negligible and the tie-in went smoothly.

We did our job and moved on to the auxiliary steering station where Tom Packer and I had found the bell five years earlier. There should have been a compass and binnacle there, but we had found only the protruding studs that bolted down the base. I reasoned that the nuts had rusted off and that the binnacle had fallen to the bottom. With Gilligan acting as safety, I ducked under a billowing trawler net and dropped down to the sand. My heart pounded with deep anxiety. It was dark as a pocket because the net was thickly overgrown with sea anemones that blocked all incoming light. I swam fore and aft, and away from the deck as far as the net would allow. There was no sign of the compass or the binnacle. They must have sunk into the sand over the years. No parts were exposed.

Back on the boat I pointed out on the plans exactly where we tied in. This was the first trip for the cave divers to the Grand Dame of the Sea. They were eager to get in the water and see the shipwreck of their dreams. Most of them had little or no background in wreck-diving, but I was mightily impressed by their incredible array of equipment, ingenious gear configurations, and extensive experience at extreme depths, lengthy penetrations, and long decompressions.

I must have seemed to them like a moonstruck schoolboy. I shot pictures of their rigs as they were getting dressed and going overboard. I made a nuisance of myself by continually asking questions about what appeared to the uninitiated to be a confusing superabundance of paraphernalia. But they were all affable and more than willing to share their knowledge. Most of them had read my book on the *Doria*.

However, a rift soon developed between the customers and the crew (Gilligan and I excluded). When Chowdhury originally set up the charter he stipulated that the boat be moored at the forward end of the promenade deck, near the first class china hole. The group had specific objectives in mind: they mapped out routes to where they wanted to go and made plans to penetrate difficult-to-reach areas using advanced cave-diving techniques. Bielenda agreed, and accepted their money based on that understanding. Chowdhury didn't discuss the matter with me or Gilligan. Ostensibly there was no need.

When the group first learned where the downline led they grew reserved. Anticipation overcame initial disappointment. But after the dive they turned glum. I overheard some of their low-spoken grumbling and asked what was wrong. Once the truth came out, Gilligan and I thought there must have been some miscommunication. In order to compensate for the oversight, we volunteered to swim another line to the bow, but Bielenda nixed the idea. He didn't want two lines twisting around each other as the boat swung with the tide during the night. That sounded reasonable, so we suggested doing it in the morning. Bielenda wouldn't hear of it. How about if we reset the shackle instead of emplacing a second downline? Not acceptable.

Chowdhury remarked firmly that all the plans the group had made required a starting point closer to the bow. They hadn't studied the stern. Bielenda didn't care. The boat was staying where it was. By this time tempers were heated and no resolution in the charter's behalf was in sight. The brew that simmered over the evening meal was nearly brought to a boil by further insensitivity.

Members of the crew, particularly mate Hank Garvin and second captain Janet Bieser, had kept up a barrage of deprecations while the cave divers were gearing up. They made snide comments about the arrangement of gear with which they were unfamiliar: stage bottles, tank valve protectors, line reels, spare masks, helmets, and helmet-mounted lights with external battery packs that were clipped to waist straps or slung under tank bottoms. They also loudly ridiculed the ability of cave divers to perform very well on shipwrecks.

The customers were insulted by this behavior and by the open display of hostility, which had no place in a service-oriented business. The cave divers didn't help matters, however, by recounting outrageous yarns about their exploits in caves which, true though they were, tended to glamorize their own brand of expertise while denigrating that of wreck-divers. Their attitude was decidedly haughty and high-handed, perhaps even holier-than-thou - although this may have been a mechanism of defense rather than a portrayal of self-perception.

In fact, the two disciplines were equitable only at the most basic level: diving in caves and on wrecks required the inhalation of compressed air. Some of the techniques and equipment modifications were transferable between specialties, but a great deal more was at variance and was dedicated toward meeting the demands of distinct environmental ingredients.

This deplorable state of affairs reached the culmination of absurdity when Chowdhury and I brainstormed an alternative solution to the problem of accessing the opposite end of the wreck while still abiding by Bielenda's personal dictate with respect to boat movement and anchor line placement. Chowdhury would swim to the forward end of the promenade deck, pop a liftbag on a line secured to the wreck, and I would pick him up in the inflatable. Thereafter, I (or Gilligan) would shuttle divers to the liftbag for their descent.

Bielenda wouldn't permit it. Thus he prevented his customers from achieving their intended goals. Gilligan and I felt bad because of our role as unwitting accomplices to Bielenda's nefarious scheme, but there was nothing we could do to alleviate the situation while Bielenda exercised control. We could do little more than offer commiseration. In return, the cave divers became our friends. In later years I went on other trips with Randy Bohrer, Bernie Chowdhury, and John Reekie. And Steve Berman, a cave-diving instructor, invited me to Florida so he could show me the ropes (or guidelines) about diving submerged tunnels and grottoes.

Despite their lack of familiarity with everyday wreck-diving procedures (for instance, they didn't carry emergency decompression reels), the cave divers handled themselves in exemplary fashion, carried along by their sheer wealth of experience and by their evident comfort in deep water and in making lengthy penetrations while deploying guidelines. They also came to accept the fact that the sea is a harsh mistress, and that wrecks are just as unforgiving as underwater caves, but in different ways. They gained a healthy respect for both shipwrecks and wreck-divers.

## THE *MONITOR* AND DRIFT DECOMPRESSION

Then came the Civil War ironclad. I had had my day in court, now I was to have my dive on the wreck. I won't detail the travails of the ten-day trip because I've already done so in *Ironclad Legacy: Battles of the USS Monitor*. But I will recount the salient features that impacted the course of deep wreck-diving.

The major non-bureaucratic obstacles that I had to overcome were getting divers onto the wreck without grappling the fragile hull, and decompressing in the horrendous currents that habitually tore through the area. I used a shot line to get to the bottom, pretty much the same as we did when checking out new sites with Hoffman in the 1970's. Instead of a string connecting a 5-pound lead weight with a bleach bottle, I used 3/8-inch polypropylene line shackled to a 30-pound Danforth anchor that gripped the sand beside the wreck despite the awesome power of the Gulf Stream, and a pair of rubber fishing floats bigger around than a basketball: in essence, a Brobdingnagian shot line. In spite of the buoyancy provided by the floats, or buoys, occasionally they were pulled under the surface and we had to wait for the current to ease up before it was possible to dive. Two floats were better than one not only because the buoyancy was doubled, but because by keeping them a few feet apart with a leader, the first float took the brunt of the force against the downline, allowing the second float (at the end of the leader) to remain on the surface. Of course, a strong enough current could submerge them both.

A suitable decompression methodology required a little more finesse and ingenuity to develop. It would seem that if a diver were strong enough to pull down the line against the current, he should be able to hang on during decompression, especially using a jonline. But the current over the *Monitor* was unpredictable: it could come and go at will depending upon fluctuations in the Gulf Stream. This meant that a mild current could become severe during the course of a long dive. Fighting current is fatiguing, a factor that must be taken into account because severe muscular tension increases the risk of getting bent.

I invented what I called a "breakaway" system. The boat maneuvered alongside the pair of floats and maintained position during deployment of the system. A support diver went into the water holding one end of a 150-foot length of rope. The rope was paid out as the diver descended the mooring line. A fishing float was secured to the other end of the rope. Attached to the fishing float was a separate 50-foot rope to the other end of which a massive weight was secured (a small engine block). The float and weight were kept on the boat until the support diver surfaced and signaled that the rope was tied to the downline at 100 feet. Then the weight and float were thrown overboard, and the diver was picked up on the drift.

Now there were three marker buoys floating on the surface. The second rope rose up from 100 feet, in line with and slightly downcurrent of the downline. When the divers were ready to descend to the wreck, the boat maneuvered upcurrent of the floats. As the boat approached the floats the captain took the engine out of gear and shouted "Go!" The divers immediately rolled into the water, drifted downcurrent, grabbed one of the downlines as they drifted past, then pulled hand-over-hand down to the wreck. The boat drifted clear.

At the end of the dive the divers ascended the downline to decompress. They transferred to the breakaway line when they reached 100 feet, then detached the secondary line from the downline, and went adrift. Once they were drifting with the current, all the strain of hanging onto the line was relieved. It was like holding onto the handrail of an escalator - both handrail and stair treads travel at the same speed. The divers then swam to the weighted line hanging from the float.

When the boat captain perceived that the breakaway buoy was drifting away from the mooring floats, he approached slowly from upwind until the boat lay adjacent to the buoy. A support diver went over the side to ascertain the locations of decompressing divers. The boat eased its hull against the bobbing buoy. Someone on board

# Early 1990's: Mixed-Gas Revolution - 123

grabbed the buoy and held onto it while a line was wrapped around it and secured to a cleat. The engine was shut down. Once the boat, the weighted line, and the divers were united and drifting along together, regulators on long hoses connected to oxygen cylinders on deck were lowered into the water to a depth of 20 feet.

The establishment of the breakaway floating decompression system required the employment of a support diver and strict coordination with the boat captain, but once everyone was briefed on the procedure and understood the operation, execution became routine. I utilized the same system on all future *Monitor* trips.

An inflatable chase boat was part of my original design plan. But two days before the onset of the expedition, the boat captain canceled my charter for unknown reasons and left me hanging - with no dive boat and no chase boat. Roger Huffman was instrumental in making alternative dive boat arrangements, but I wasn't able to obtain another inflatable - and I didn't have the money to buy one.

## THE *MONITOR* IN PERSPECTIVE

Media coverage of the landmark legal victory over NOAA has tended to overshadow consequential events of comparative importance. And reporters invariably focused undeserved attention on the hazards and novelty of the dive. At the expense of appearing to be patronizing, I can't help but point out that a dive to 230 feet in clear, warm water was in no way unduly difficult at this stage in the evolution of deep wreck-diving - at least not among serious wreck-divers.

Novices, aspiring wreck-divers, and the non-diving public might have been duly impressed by the effectuation of such a dive because of the way it was presented, but the true value of the deed was distorted by pandering to popular appeal. Dives to equivalent depths were rather commonplace.

Wreck-divers had been conducting similar dives under far worse conditions for years - a reality to which this book is a testament. Little of this information was in the forefront of the public consciousness, probably because active deep wreck-divers were few in number and because their achievements were largely unknown outside the confines of the wreck-diving community to which they belonged.

This lack of public awareness and the bureaucratic denial of acceptance were the primary obstacles in the path to recognition of wreck-diving as an activity that demanded skill and high performance. The *Monitor* was played up in the press as the culmination of years of underwater endeavor, like the attainment of the Pole or the successful bid for a lofty mountain summit, whereas to those who were intimately involved in wreck-diving the importance of the dive lay not in the exemplification of wreck-diving expertise, but in an historical context - of the ship itself as well as of the groundbreaking issue of permitting public access. Swift current and decompression considerations notwithstanding, on a comparison scale of difficulty a dive to the *Monitor* weighed in like a bantam.

By this I don't mean to denigrate the divers who brought their expertise to the site. I mean merely to observe that in the overall scheme of deep wreck-diving the *Monitor* was just another dive: an enthralling step on the treadmill but in no way a conquest.

## TRIAL RUN FOR THE *OSTFRIESLAND*

Viewed from a different perspective, the *Monitor* trip was an experimental check for the *Ostfriesland* dive and for other high-tech dives down the line. If I was going to rely on Hamilton's computer program for gas mixes and decompression schedules, I wanted to field test it in water shallower than 380 feet. In this respect, my dives on the *Monitor* were relegated to the status of rehearsals for the big dive that was scheduled to take place the following month. On the wreck I concentrated on my photographic objectives: I shot stills for newspapers and magazines, and video for television and a videotape production. At the same time I played with different gas mixes on the bottom and during decompression.

Billy Deans was my fellow guinea pig for these in-water trials. He arrived at expedition headquarters in Hatteras, North Carolina with a pickup truck full of tanks filled with various and sundry mixes and with a fistful of decompression schedules that were generated by DCAP. He had great faith in Hamilton, as did Clayton, who participated in the *Monitor* trip but who chose to dive on air. I was committed to Hamilton's DCAP, but didn't yet have the faith.

Deans and I utilized cave-diving stage-bottle technique by clipping extra tanks to our sides. This enabled us to extend our bottom times at 230 feet to 25 minutes, and allowed plenty of reserve for contingencies and for decompression. Deans repeated his *Araby Maid* scooter performance and zoomed around the wreck with his roving video camera, while I shot stills and slow-panning video from a stationary position.

By 1990, oxygen was in fairly widespread use among deep wreck-divers for in-water decompression, but only as a replacement for air, not to shorten hang time. I brought enough oxygen (ten storage bottles) so that everyone on the trip could breathe it during the final stages of decompression as an added safety factor. Deans and I went further.

On one 25-minute air dive my wrist computer called for 165 minutes of decompression, a schedule that I followed to completion despite breathing oxygen at the twenty- and ten-foot stops. Then I dived the identical profile but followed an accelerated decompression schedule provided by Hamilton. According to DCAP, inspired oxygen reduced the hang time to 95 minutes, thus shaving 70 minutes off a boring decompression. I tried it and didn't get bent. I was sold!

Today this procedure is commonplace on deep air dives, but before the validity of the concept was proven it was a scary proposition to get out of the water with a computer flashing a warning about the deficit.

Deans and I also dived the *Monitor* on trimix-19/30 . By that time in the season I had made so many deep dives on air that I couldn't tell the difference between air and gas: the absence of narcosis felt the same. The DCAP schedule called for both nitrox and oxygen to accelerate decompression, the same as in the schedule devised for the *Ostfriesland* dive. Again the system worked flawlessly and the decompression time was reduced significantly.

By the end of the *Monitor* trip I felt fully confident in Hamilton's computer-generated calculations. This isn't to say that I didn't still have doubts about diving the *Ostfriesland*, only that I no longer felt concern - well, not *too* much concern - about the decompression schedule.

## GEARING UP FOR THE *OSTFRIESLAND*

As the details of preparation were taken care of, the trip coalesced into a quantifiable entity with real rather than imaginary existence. The mountains of gear that we needed to pull off the dive we either bought, borrowed, or rented. None of us owned enough regulators for all the tanks that we had to either carry or stage, and we didn't have enough tanks for all the decompression gases. The tanks we acquired had to be rigged with hose clamps, D-rings, clips, snap hooks, quick links, carabiners, and rubber straps - all commonplace now but new to us at the time.

I had already bought two deep-water regulators to replace my aging ones. Now I purchased two more that I would have to take to depth. My old regulators were relegated to stage bottle status and shallow water use. I also bought extra tank pressure gauges. I haunted the local hardware store for the accessories needed to modify the tanks for side-slung arrangements and to retain all the hoses so they didn't dangle like grappling hooks.

Another major concern I had was buoyancy. I was afraid of a drysuit rupture that would prevent me from getting off the bottom (shades of John Dudas, perhaps). I disdained the use of recreational buoyancy compensators because the fully inflated pockets squeezed against my chest and prevented me from fully expanding my lungs. I also believed that buoyancy compensators were somewhat of a liability when used in combination with a drysuit because the front inflatable pockets covered the drysuit's inflation and exhaust valves, making them difficult to actuate. Now a new type of BC was on the market that was mounted to a backplate between the tanks and the back: they were dubbed "wings" because of the way the inflation pockets flapped. I bought one.

Elsewhere my partners were doing the same. Except for Jon Hulburt. He fully appreciated the risks of the dive and knew that they could be mitigated only by meticulous planning and by a strict regimen of training and mental preparation. But he had situations occurring in his life that demanded much of his attention. With his mind otherwise occupied, with his concentration divided, he couldn't devote the time and energy that was necessitated by an undertaking of such magnitude. Reluctantly, he reached the agonizing decision to drop out of the project.

This was a considerable loss. Hulburt was a great innovator and had always been in the forefront of evolving ideas and technologies. Not only would he have been a valuable asset to the team in the water, but his insights during the planning stages would have been priceless. By way of contribution he did not request, nor would he accept, reimbursement for the one-quarter share he had paid to purchase Hamilton's decompression schedule.

## BOTTOM TIME OPTIONS

Part of the information supplied by Hamilton's DCAP was gas consumption. Everyone breathes at different rates, depending upon lung capacity, physical needs, psychological stress, and training and experience. DCAP calculated gas consumption automatically and compensated for time spent at depth. This was a round figure, of course: an average based upon a moderate breathing rate.

Additionally, Hamilton offered a choice of five bottom times. Allowing five minutes for descent, he provided schedules for exploration on the bottom for three minutes, eight minutes, eleven minutes, fifteen minutes, and twenty minutes. In wreck-diving reckoning, bottom time commences the moment a diver sticks his head under water and ends when he leaves the bottom. When DCAP timetables were converted to actual timetables, bottom times became eight minutes, eleven minutes, sixteen minutes, twenty minutes, and twenty-five minutes.

We didn't believe that we could reach 380 feet in five minutes, especially lugging all that gear. But that was okay. Any extra descent time would be deducted from the bottom time. Thus an eight minute descent on a sixteen minute dive reduced the working time on the bottom from eleven minutes to eight. We would still use the decompression schedule for sixteen minutes, not the one for thirteen minutes.

This was an additional safety margin: less time spent at ultimate depth meant that less inert gas would be absorbed by the blood and tissues.

The shortest bottom time (eight minutes) was a bail-out option in case we had to abort the dive during descent or soon thereafter. The longest bottom time (twenty-five minutes) was meaningless because we couldn't carry enough gas to stay down that long. The second longest bottom time (twenty minutes) was a buffer in case we got delayed in leaving the wreck. Eleven minutes and sixteen minutes were the two bottom times most suitable to our needs for initial exploration, with the sixteen minute duration being our optimum choice - depending, of course, upon the conditions we encountered on the wreck. We selected the sixteen minute schedule for gas consumption purposes, and from this we approximated the quantity of gas to carry, and how much nitrox and oxygen we needed for decompression.

Clayton and Manchee opted to carry two tanks of bottom mix and one tank of travel mix. I didn't feel comfortable with that amount. Because of my injured lung my breathing efficiency was poor. I was "bad on air" as the saying went. I was an "air hog." I decided to carry three tanks of bottom mix and one tank of travel mix: doubles on my back filled with bottom mix, and two tanks clipped to my sides.

## Gas Arrival

Potomac Air Gas notified Clayton that the heliox was ready to dispatch. Where should it be delivered? The problem of transferring the gas from storage cylinders to scuba tanks was solved by Mike Parks, a long-time acquaintance although we had never actually dived together. He was a commercial diver who owned his own commercial diving company in addition to a recreational dive shop, in Baltimore.

He pumped air and sold equipment largely as a side line. Primarily he worked on underwater construction projects, installed cathodic protection on docks, conducted bridge pier inspections, and did interior pipe inspections, welding, and repairs. He held the record for the longest penetration ever made inside a flooded pipeline. He liked diving on shipwrecks, too, and had been involved in more than one salvage operation as an investor as well as a diver. His dual role in the business made his commercial expertise available for recreational purposes. He owned a booster pump and had the esoteric knowledge of how to transfer heliox from storage cylinders to scuba tanks. The gas was delivered to his shop.

## Gas Transfer

Ken Clayton and Pete Manchee had never met. All communication between them was conducted by telephone, and long distance charges were mounting. Now that it was necessary for Manchee to get his tanks to Baltimore, he and Clayton planned a rendezvous at the Virginia-North Carolina border. Manchee drove north and Clayton drove south, and there the twain did meet.

Because I had so many on-going diving commitments my tanks were constantly in use. Clayton let me use one of his sets of doubles - aluminum 100's. He borrowed a couple of singles that I could carry as saddle tanks. He also took it upon himself to supervise the transfilling operation at Parks Diving Services.

Clayton arrived at the shop with all our tanks in his pickup truck: three sets of doubles and one single (for me) to be filled with heliox-12, three singles to be filled with heliox-16, and six singles to be filled with nitrox-36. Parks jury-rigged the hose connections between a storage cylinder, the booster pump, and a scuba tank. The booster pump was not operated by electricity but by high pressure gas that was bootstrapped from the heliox cylinder. In this manner, a small portion of heliox drove the pump that siphoned the rest of the heliox out of the storage cylinder, cycled it through the pump, and compressed it into the scuba tank.

The operation took seven hours to complete. Mike Parks charged us nothing for his time and service.

When the transfilling process was nearly completed Clayton found that there wasn't enough heliox-12 to fill the last set of doubles: his own. By the time the final storage cylinder was sucked dry his tanks held only 2,000 psi. These tanks were made of steel and were rated by the U. S. Department of Transportation to hold 2,640 psi. Cave divers normally doubled the burst disk and overpressurized the tanks to nearly 4,000 psi. Clayton intended to be more conservative and overfill them only to 3,500 psi.

The proper amount of gas had been ordered and delivered, but neither of us had anticipated the loss that was incurred in the transfer process. In fact, we knew nothing at all about booster pump systems. That was not our province.

Clayton left his partially filled doubles at the shop. He called Potomac Air Gas and ordered another storage cylinder. The gas had to be blended and delivered, so it wasn't until later in the week that transfilling the last set of doubles was resumed. By then Clayton had other commitments and didn't have time to go to Parks's shop in Baltimore, an hour and a half drive.

Parks was out on a job. He instructed one of his employees on how to transfer the gas. John Terry, one of our support divers, showed up at the shop to oversee the operation. He then took charge of delivering the tanks to the boat.

## KEEPING IN SHAPE

My concern about my habitually poor air consumption was greater than usual. As the day of the big dive drew near I increased my jogging routine. The heat and humidity of Philadelphia in summer are stifling. I jog just before sunset or in the refreshing coolness of rain. One evening I set out in a thunder storm, luxuriating in the temporary chill of the woods. The trail was a mud slog - challenging but fraught with the risk of turning an ankle. I jogged less than cautiously until I reached the King Kong log.

I call it that because it reminds me of the scene in the 1933 classic, *King Kong*, in which a group of sailors are chased to the edge of a deep ravine by a mad styracosaurus. The men run across a massive tree trunk that spans the ravine, but are turned back before reaching the other side by the roaring, fist-swinging giant ape. In the movie, King Kong shakes the men off the log. They fall hundreds of feet to the boulder-strewn bottom where they are devoured by giant spiders and prehistoric lizards.

The tree trunk on my route spanned a shallow stream which was often dry, but which after a thunder storm gushed with water. The drop was only six feet, but the creek bed was filled with sharp, jagged rocks. I leaped onto the log and balanced myself for the twenty-foot dash across the smooth, barkless top - then paused as I felt my soles slipping on the rain-slickened surface. I realized how easy it would be to break a leg in a fall. That would ruin my chance of participating in a dive that was more than a year in the planning. I backed off the log and took the longer way around.

## LOADING THE BOAT FOR THE BIG ONE

Finally, the departure date arrived.

It was late on a hot Friday afternoon when I pulled into the dock. I had returned from a two-week trip to Maine only five days before. I had barely had enough time at home to mow the lawn, go through mail, pay the bills, make a zillion phone calls, and repack my gear for the big trip. Now it was August 8. The countdown to D-Day (dive day) had officially begun.

The *Miss Lindsey* was a beehive of activity. Long-time acquaintances were renewed, new faces were introduced, scores of tanks and mountains of gear were loaded onto the aluminum deck, bunks were allocated, and personal belongings were stowed where they would be somewhat accessible. Even a boat as large as the *Miss Lindsey* seemed crowded when it was packed to the gunwales with equipment and when its decks and cabin were swarming with people who all seemed to want to be someplace where they were not. The hubbub was perpetual.

Clayton left "my" tanks in his truck for me to unload. My arms were full as I squeezed past passengers and crew. I couldn't shake hands, so I just smiled and introduced myself to strangers whose names I promptly forgot. I've often said that the most commonly heard expression on a dive boat is "excuse me." That's because under the normally cramped conditions people are always bumping into each other - especially in rough seas - or asking them to move in order to get by. I said "excuse me" about a hundred times before we even left the dock.

I felt like a politician building a soapbox single-handedly in the middle of Grand Central Station: all the while he was hammering nails into wood he stopped to pass out campaign literature to every passing voter, and to utter a few words about his platform. Every time I turned around there was someone who wanted to talk with me. Even when I didn't turn around, someone called out my name. I stowed gear and offered greetings and answered questions all at the same time. Because of the constant interruptions I took twice as long to get my gear put away - then couldn't remember exactly where I put it.

Particular spots were allocated for mixed-gas cylinders: doubles, saddle tanks, and stage bottles - all of which had to be doubly secured because the forecast was not promising and rough seas were expected. Everything had to be tied down with rope in addition to the usual bungee cords. The oxygen bottles were laid down on the walkaround adjacent to the central cabin, between the wheel house and the after deck, and were tied with rope to the railing.

Out of consideration for the support and recreational divers, who were participating equally in the adventure and who needed their own personal spaces, it proved impossible to keep all my bags and boxes together. Thus my gear was fragmented about the boat, with stuff squirreled away in odd corners that were difficult to reach. Cases filled with spare equipment that I might not need I stowed behind or underneath the gear of the non-gas divers. Neatness on a dive boat is of paramount importance.

## Departure

In addition to John Terry, our other support divers were Ed Suarez, who had stayed by the ventilator grill on the *Andrea Doria* while Manchee explored the dark compartments beyond, and Glen Klag, owner and operator of American Water Sports in Rockville, Maryland. Klag was lending additional support by supplying us with rental equipment at no cost. Suarez was a long-time friend, Terry I had met twice (in Key West and on an aborted *Washington* trip), and Klag I knew not at all. Nor did I know any of the crew or recreational divers. We were largely a "ship of strangers." Our captain was Ric

## Early 1990's: Mixed-Gas Revolution - 131

Filer.

Amid the confusion of loading gear and exchanging cheerful greetings, a reporter and film crew arrived. Clayton had arranged with a local television network for media coverage of the event. The reporter conducted interviews with the major participants while a camera person videotaped our comments and our specialized equipment. When we left the dock, at around eight o'clock, not only was the boat's departure videotaped, but the reporter and film crew followed the *Miss Lindsey* out the inlet on another boat. While the reporter editorialized about our mission and about Billy Mitchell's historic bombing tests, from the bow of the other boat, the stern of the *Miss Lindsey* was in view over her shoulder.

I appreciated the fascination for deep-sea exploration that was held by the viewing public, particularly for diving exploits in which people pitted themselves against unknown odds and the elements and forces of nature. Danger makes great copy. But by a strange quirk of twisted psychology and boob-tube mentality, that same viewing public applauds failure more than it cheers success - as if the defeat of a genuine adventurer confers a sense of personal triumph to the armchair fantasizer who has never strayed farther from his living room than his own front yard.

I was also mindful of how "adventure" was defined by Lawrence Griswold, a 1930's archaeologist and explorer of tropical jungles: "Adventures, in retrospect, are pieces of extremely bad luck that missed a fatal ending."

The media preys on the public's penchant for morbidity. Success stories are relegated to the back pages of newspapers and magazines and to fillers on newscasts when nothing notorious is happening. Follow-ups rarely appear. But downfall, death, and destruction make banner headlines and often get top billing. Journalists lurk for bad news then bare it exultantly. They knowingly ignore overall achievement and dwell instead on minor miscues that can be embellished out of context into episodes of catastrophe. Forget the truth, forsake veracity - create a story that will sell. Then work it to death like a predator bringing down prey, and don't let go until the prey (or the story) is dead.

If the reader is supposing that this tirade is leading up to something, he'd be right. I've been interviewed enough to wonder how reporters ever managed to graduate grade school. They usually managed to misconstrue the simplest facts. I've read articles about myself and accounts of incidents in which I was a participant, then wondered if I had been involved at all or on another planet.

Given these caveats from prior experience, I am forced to eat the words written above and to retract my negative sentiments - in this particular instance. The network people didn't glamorize the story, they just told it like it was: fairly and simply. And they followed it through for two more years, giving it national coverage.

Clayton confided to me privately that this dive was so important to him that he wanted to accomplish it at any cost, that its successful completion would be the culmination of his ambition. He said that if he got swept off the wreck after touching the *Ostfriesland's* hull, his final thoughts would be happy ones. This sounded too fatalistic for my tastes. There was a lot more of life that I wanted to experience before setting out to discover that final wreck site in the sky. I surmised that he was just waxing philosophical.

On a practical level, what truly bothered me was the mounting pressure to perform. The media made an investment and wanted a return on capital. I just wanted to return. Clayton was pushing for success in order to satisfy deep inner longings. The fraternity of wreck divers watched our proceedings askance: some in disbelief, most with best regards, a few in hopes that we would fail to find the wreck.

Pride and peer pressure have no place in deep wreck-diving. We needed the maturity and self-confidence to call off the dive if the conditions didn't seem right. I don't know if I had those qualities, but I had something far more useful: fear of dying.

## Delay, Doubts, and Second Thoughts

Once the *Miss Lindsey* got under way the commotion subsided. We all began to relax. Anticipation yielded to light-hearted banter. Now there was nothing for us to do but let the crew take charge. I tried to memorize the names that I had forgotten that afternoon. Finally, I crawled into my bunk in the bow compartment and curled up with a book. Later, I couldn't get to sleep at first because the boat kept leaping off the crests. Some time after midnight the seas settled down and I fell into fitful slumber.

Morning found us hooked into the *Ocean Venture*, about halfway to the *Ostfriesland*. It was too rough to go any farther. The forecast called for better weather and moderating seas for the morrow. We decided to get in some recreational dives for the benefit of the non-gas divers. I made the morning dive alone, but opted out of an afternoon descent because I didn't want my body charged with residual inert gas.

Good food and camaraderie were on the menu for the day. We got to know one another better. For the gas divers and three support divers there were endless rounds of discussions on how best to set up the stage bottles and descent lines, who would do what, and in which order. There was no right way or wrong way, no established routine.

There is no operations manual for the exploring the unknown.

By evening the weather improved significantly. There was no reason to believe that we would not get our chance to dive the *Ostfriesland* on Sunday. I would have rested easier had the weather been marginal and the chance of diving unlikely. I twisted and turned in my bunk all night, plagued by nightmares. For the first time in my life I had what psychiatrists call anxiety attacks. I feel sorry for people who have them all the time. The experience was horrible.

I spent most of the night awake and unable to put my racing mind at ease, not because the boat was pounding because it wasn't. I was terrified.

Never before had I felt fear at the prospect of a dive. Several times that night I found myself shaking uncontrollably. Not just my arms and legs, but my whole body: vibrating like a giant electric foot massager. During these many waking moments I kept asking myself if I truly comprehended what I was about to do, how deep I was planning to go. The more I thought about it the more frightened I became.

By morning I was completely strung out. Groggily I dragged myself out of my bunk, praying that something would happen to force the cancellation of the dive: rough seas, strong current, mechanical failure, sickness, injury, anything. I did not want to make this dive. But although the wind was blowing hard the seas were fairly calm. We pulled anchor from the *Ocean Venture* and headed for the *Ostfriesland*.

## Last Minute Preparations

Our intention was to make the dive in the morning so the non-gas divers could make a dive that afternoon. But there were constant delays. It took a long while to find the wreck despite having accurate coordinates. It took even longer to hook the hull. The grapnel kept sliding off or, once hooked, pulling out. Each time the grapnel slipped, more than six hundred feet of anchor line had to be hauled aboard. Hooking a wreck at a depth of 380 was no easy task.

As I dawdled over coffee and toast and tried to clear the cobwebs from my mind, I realized that no one else appeared nervous or demonstrated concern. They were making plans while I was looking for excuses. I felt isolated.

Around mid-morning I realized that I had to start putting my gear together for the dive. This took hours. The wind howled through my hair as I reworked regulator hose configurations to fit the complicated mixed-gas system. In addition to back-mounted doubles I set up four single tanks with regulators and pressure gauges: two filled with heliox to take with me to the bottom, two filled with nitrox to be staged on the line by support divers. The deck around me was such a mass of hoses that I felt like a rabbit in a pit full of vipers. Wrenches and screwdrivers lay strewn over the deck beside borrowed parts and connectors, extra gauges, and a pile of cable ties.

My pony bottle was filled with argon that was not intended for breathing but for drysuit inflation. On the valve I placed a first stage with only a low-pressure inflator hose. Somehow the valve on Manchee's pony bottle had gotten knocked loose, leaving him with less than half a tank of argon. He was wont to slough it off, but instead we used a transfill hose to equalize all three argon bottles then topped them off with air as a way to share the warmth.

I was so intent on my personal tasks that only peripherally was I aware of ongoing events. The grappling process was not my responsibility, nor was setting up traverse lines, stage bottles, the oxygen rig, and so on. Whenever Filer maneuvered the *Miss Lindsey's* stern into the wind, the cloud of diesel fumes threatened to make me sick. I had to go below to escape.

A constant barrage of conversation distracted me from my tasks. Someone needed a tool or a part, I needed to borrow an adaptor, a short hose needed to be exchanged for a longer one - the list of details was endless.

The wreck was eventually hooked, a support diver started running lines, then the grapnel pulled out, and the process had to be started all over again. I wasn't bothered by the delay. I needed all the time I could get.

During this span of busy activity and final preparation, Clayton dropped a bombshell. His doubles weren't full. After putting his regulators on his twin manifold, he turned the isolator valve and heard the high-pitched hiss of gas as it escaped from one tank into the other. The pressure equalized at 2,800 psi. We surmised that Parks's employee had closed the isolator valve prior to topping off the doubles, so that only one tank got filled.

At first this seemed like a serious setback that might compromise the dive. But upon calculation and further consideration the matter was not as bad as it seemed.

In actuality, steel 104's were considered to hold 104 cubic feet of gas when filled to the DOT rated pressure of 2,640 psi. In practice, cave divers commonly overfilled the tanks to much higher pressures in order to extend the range of penetrations. Twin steel 104's filled to 2,800 psi contained 220 cubic feet of gas. This was more gas than I had in my double 100's, although less than my total because I carried a single as a back-up. Manchee carried two tanks of bottom mix, but his consumption rate was much better than ours.

Clayton now possessed 40 cubic feet of gas less than anticipated. Even so, the partial fill was more than double the amount of gas required for a sixteen minute dive - a 100% redundancy factor. And that didn't count the tank full of travel mix, which was far in excess of what was needed for descent and decompression.

In the worst case scenario, the dive would have be shortened from the chosen optimal of sixteen minutes to the less desirable thirteen minutes. (Remember that the durations included descent time.) After lengthy debate, we decided to go with a split objective. There was no way to predict with precision our consumption rates on heliox - thus the rationale for choosing to carry more gas than we actually expected to breathe. The actual judgment call in any case would have to be made on the bottom.

We elected to try for sixteen minutes, with the proviso that if Clayton felt that his gas margin wasn't enough, he would leave early and decompress according to the thirteen minute schedule, while Manchee and I would stay down an additional three minutes on our own.

Manchee had his own problems. He miswielded a knife and cut a deep gash in the fleshy part of his left palm, near the base of the thumb. Blood gushed in a flood that proved difficult to stem. He was determined not to miss the dive, so he wrapped his hand in several layers of duct tape - the universal fix-it material.

The support divers reported that the traverse line was in place and the stage bottles were tied off at the appropriate depths. It was time to suit up.

I had already gone over my gear a dozen times. I went over it again. Everything checked out. I couldn't think of any more excuses for delay, so I climbed into my longjohns and drysuit. I was jittery with stern apprehension, but I no longer had time to think about what I was about to do.

I sat on the bench and slipped into the straps of the tank harness. I stood up, swung my hoses and gauge panel in front of me, fed the waist strap through the buckle, and drew it tight. I donned my weight belt. I had added extra lead to offset the buoyancy of the borrowed aluminum tanks (my own tanks were made of steel). I sat back down.

I arranged my regulators across my chest so that they lay in easy reach. Each mouthpiece hung from my neck by a length of rubber surgical tubing. I breathed off each one as a test. I plugged the low-pressure hose into my drysuit's inflator nozzle and forced in a short blast of argon. I plugged a second low-pressure hose into my buoyancy compensator. I slid my feet into my fins. I clipped the plasticized decompression schedule to the end of my gauge panel, tucked a plastic slate which was my back-up schedule into a pocket. I sealed my mask over my face. I pulled on my mitts. I bent over to pick up my lights, which I clipped to D-rings on my harness.

Except for the addition of written decompression schedules, I hadn't done anything out of the ordinary.

Now I stood up and shuffled toward the opening in the railing. Wearing a diving mask is like wearing horse blinders - one's field of vision is severely limited. I couldn't see what was going on, but the constant chatter between gas-divers and helpers kept me informed of the others' progress.

## THE NEOPRENE MEETS THE WATER

Clayton was ready to go to "where the neoprene meets the water," in his distinctive phraseology. Someone clipped on his single side-slung tank and he jumped into the water.

I held onto the railing while my two saddle tanks were snap-hooked onto the D-rings on my harness and to the bases of my doubles. Someone handed me the video camera, then held onto my shoulders while I let go of the rail in order to snap the carabiner at the end of the lanyard to the metal O-ring on my waist strap. My eyes were more than ten feet above the water. I jumped and hit with a tremendous splash made all the larger by the extra tanks at my side.

Manchee had never dived with tanks hanging from his side, and didn't like the idea. He carried his third tank like a pony bottle: strapped behind his doubles by means of extra long stainless steel hose clamps, creating triples. The argon bottle was then hose-clamped in the notch between the doubles and the single that contained his travel mix. Two deck hands pulled him to his feet and held on to him as he lurched toward the side of the boat and went overboard. He was top heavy in the water, but strong enough to overcome the inherent imbalance of the system.

We wasted no time on assembly. As soon as we hit the water our timers started running, and so did we. We pulled along the traverse line to the anchor line, then angled down the scope through viridescent mist. Clayton led the descent, I chewed his fin tips, Manchee was on my tail. Clayton looked back to see that we were with him. I rolled over and made eye contact with Manchee. He flashed the okay sign.

A calming sense of relief came over me the moment I started down. Gone were the feelings of anxiety and foreboding that had plagued me during the night. Under water I was in my element: confident and assured. The dive was like any other, only deeper: a feeling fed by the non-narcotic attributes of helium. At 200 feet my mind was as clear as it was on the surface. No - clearer. At depth my concentration was total, my focus was supreme. I could have been in the shallow end of a swimming pool.

I must have been overeager. I thought Clayton was going too slow. I kept wanting to pass him. With difficulty I held my zeal in check and maintained my position.

On air dives I could literally taste the change in depth: a function of the increased pressure of nitrogen acting upon the brain. But breathing heliox, my tongue was insensitive to change. The numbers on my gauge were the sole indication of depth.

At 250 feet the only noticeable transformation I observed was the growing loss of light. Turbid green surface water yielded to a darker but clearer shade. This paradox of illumination was occasioned by the thermocline. The turbulence of storm-ridden seas had churned only the surface water to a froth, stirring a planktonic soup where visibility was limited to about twenty feet. Below the warm-water interface few living particulates floated. The temperature dropped 30° in the space of an inch: from 75° to 45°. Poking through the thermocline was like jumping into a glacier fed lake. Sunlight that penetrated the top of the water column was either blocked or absorbed by biota. What little light dodged through the organic obstacle course brightened the emptiness below.

I was imbued with a sense of stark unreality when 275 appeared on my computer's digital display. Without the benefit of nitrogen narcosis the depth seem imaginary - as if I were disconnected from the dive and observing it vicariously on a movie screen. My head wasn't hammering as it had on the *Washington*, my vision wasn't blurred. Yet although my mental processes were not dulled by the depth - felt heightened, in fact, by comparison with past experience - life took on an eerie, phantom existence.

The entire material world was bounded by sparse, surrealistic imagery. Beneath Clayton and above Manchee stretched a clean white rope that vanished in a bleak, pervading, and all-encompassing greenness. Nothing else existed in this mid-ocean miasma. There were no other points of reference. My only anchor on reality was my hand on the line. I felt as if my body was stationary and the rope was moving through my grip. The only other input was the unfamiliar bell-like tinkling of exhaust bubbles rushing past my ears.

I labored under the illusion that the white rope extended no farther than the limit of my perception. That as I moved along its length the rope popped into being before me and ceased to exist behind. I was a disembodied observer.

I was lost in liquid limbo when my gauge registered 300 feet. Below I noticed a dim ghostly outline take shape in the otherwise featureless sea, like a fog coalescing into a faintly recognizable pattern. As I watched, entranced, the shadowy specter took on definite form. It was the hull of a sunken ship.

Media coverage of first ascents often focuses on who was the first to reach the summit. Mountain climbers know that it makes no difference. Members of a climbing team break trail in leapfrog fashion, one after the other. The first one on the summit is the climber who happens to be in front at the moment the mountain runs out of rock. The ultimate attainment is a team effort without individual bias.

When Edmund Hillary and Tenzing Norgay reached the top of

Mount Everest, in 1953, they vowed never to reveal who was wearing the boot that first trod the pinnacle of the world's tallest summit. They climbed the mountain together.

We on the *Ostfriesland* trip made no such prior agreement. Yet because the scope of the anchor line approached the wreck on a catenary, because a mask produces aberrant tunnel vision, and because we were each intensely self-absorbed by the stress and strain of the moment, none was in a position to see the overview of events. All three of us let go of the anchor line and touched down simultaneously.

That is as it should be, for physical contact with the wreck was a meaningless concept. There is more to a discovery dive than touching a section of hull plate.

Six minutes had passed since we left the surface. Six minutes to reach a depth of 320 feet. I switched regulators. Now I was breathing bottom mix - heliox-12 - whose lower partial pressure of oxygen was designed to prevent oxygen toxicity. I made sure that the others switched, and they did the same for me.

What I could see of the hull in the limited ambient light looked like the middle of an immense shoe box - roughly rectangular and extending out of sight in two directions. The anchor line had brought us over the nearer drop-off. The chain lay draped over the farther edge. I wanted to see how well the grapnel was hooked, so I switched on my light, kicked past the edge, and threw myself into freefall.

I dropped like thistledown on a summer breeze. I let out every bit of argon in my drysuit and still I fell no faster than a heavy feather. I hadn't counted on the buoyancy characteristics of helium. I had forgotten some basic physics about atomic density and the compressibility of gases. I felt like such a smack.

Slowly I descended until I reached the grapnel. Two tines were caught in the overlap of a hull plate which was only thinly encrusted. I didn't like the looks of it, but it had held for more than an hour already and resisted my gentle tugs. I pronounced it "probably secure."

Now we each had a job to do. I switched on the video camera and light. The light flashed on for a moment, illuminating fish with an eerie red tint, then went out. The battery was dead. The light worked fine during pre-dive trials. I had borrowed the rig from Smokey Roberts, a commercial videographer. He had told me the battery was fully charged. He must have made a mistake.

Clayton's job was to clip a signal strobe onto a link in the anchor chain. The strobe failed to fire. He tried to get it to work by tapping the unit with his hand and by playing with the switch.

Manchee clipped a wreck reel to the anchor chain. His job was to deploy the line as we moved away from the grapnel. We reasoned that if the grapnel pulled out he would feel the tug on the line and could signal a recall by waving his light. We could then follow the guideline to the grapnel. If we failed to catch up with it, we could ascend as a team without vertical reference till we reached our deepest stop. Then we would pop a liftbag as a marker, and decompress on the emergency line till the boat came and got us.

Clayton was still fussing with the strobe as Manchee and I slowly settled toward the bottom. With only ten minutes remaining there was no time to lose. We had to go for it. Besides, Clayton was in the least danger because he had hold of the anchor line. I let the useless video rig dangle from the D-ring to which it was clipped.

My eyes were adjusting to the darkness, but the side of the wreck we were descending was the shaded side, and it was much darker than above - almost like night. I shone my light toward the wreck, which now passed before my eyes like the wall of a skyscraper. At first I didn't see anything distinctive or distinguishable. Then to my right I saw protruding structure: an undefinable mass of metal. Below I could make out the approaching sand. It seemed near enough to touch. Manchee and I glided down with agonizing slowness.

Came the sound that I most dreaded to hear, the sound that had haunted my dreams: that of the anchor chain rattling across the metal edge of the hull.

The rasping chill that coursed along my spine scraped off an inch of bone. All the back-up plans in the world couldn't reduce the terror I felt at that moment. Now I was grateful for the barely negative buoyancy. I shot upward so fast that the suction pulled Manchee up behind me.

Without the hyperbole, I reached the high side of the wreck and saw that the grapnel was caught on the lip by a single tine. I pressed against the grapnel in order to hold it in place. I was face to face with Clayton. He had finally given up on attaching the signal strobe because he couldn't get it to work. I turned to my right and saw Manchee hovering some fifteen feet away, intent on rewinding the line on the reel.

Technically, none of us was in danger. Clayton had one hand on the chain, I held onto the curve of a tine, and Manchee was attached by means of the reel. Yet at the time I did not feel secure. The grapnel was hooked precariously. Even though we had planned for this contingency, I definitely did not want to conduct a free-floating decompression. But neither was there any way that I would now leave the grapnel to go exploring.

Still, I hesitated about terminating the dive. When I put aside my fears, the feeling of being on a long-lost German battleship was breathtaking (figuratively) and incredibly stimulating, even without the effects of nitrogen narcosis. I didn't want to leave. I wanted to savor every precious minute of bottom time remaining.

Clayton decided otherwise. His situation was different from mine and Manchee's. Because of his short tank fill his mind was pre-set on an early departure. He also knew something that, in my preoccupation, I had failed to notice: one of the stage bottles was missing from the 40-foot stop. He had taken note of it on the way down.

He pointed to his pressure gauge and waved good-bye. I returned the okay sign. Dive masks are like horse blinders. Manchee didn't see this exchange because he was reeling in the wreck line. Clayton maintained eye contact as he moved up the anchor line backwards. The last I saw of him, he was thirty feet up the line and fading into the distance, still looking at me. Then Manchee was by my side and unclipping the wreck reel.

As long as I had a grip on the grapnel I was not overly concerned. By now my eyes were fully adjusted to the low light conditions. Despite a general overtone of murkiness, ambient light visibility approached fifty feet, and may have been greater, but the lack of recognizable features for scale made it difficult to judge. The hull was so vast and flat in front of me that I felt like a football fan watching a game from the fifty yard line. I couldn't see anything immediately downcurrent that might afford purchase for the grapnel.

Then Manchee took the bull by the horns - or in this case, the grapnel by the tines. In a swift, gutsy maneuver he pulled the grapnel free of the wreck and took off to reset it. He didn't even wait for help.

Now we were committed. In my haste to lend assistance I forgot about the video camera dangling on a lanyard from my crotch. The housing snagged below the lip of the wreck, and my forward momentum swung me against the flat of the hull like an upside down pendulum. In retrospect, the situation could be viewed as funny - in a slapstick, Three Stooges way. But I was in no mood for humorous contemplation.

I struggled to release the camera. The deep-water housing and the light on its long arm maintained a firm grip on the edge. Because of the bulky tanks that were clipped to my sides, I pirouetted like hippo mired knee-deep in mud. The video rig kept a grip on the hull with the stubbornness of a rake caught in the branches of a bush. I could not yank the rig loose from downcurrent. I had to kick forward and swim past the corner of the lip, then pull the rig free from the other direction. I felt nearly defeated by an inanimate object.

By the time I finished with my photographic shenanigans, Manchee was more than ten feet away. I had to swim hard to catch up with him. I grabbed the shank of the grapnel, turned around so I was facing him, and kicked upcurrent, hoping to take off some of the strain while we angled along the broad beam of the wreck.

I glanced aside and saw what he was aiming for: a metal cage built of thick rods. It was about the size of a large chair. I have no idea what it was. The cage was thirty feet away and slightly downcurrent, located in the middle of the spread of steel plate. My position was undesirable - the tines pointed toward me - yet I didn't see how I could lend effective aid otherwise. I eased to the side of the grapnel as we neared the cage. Manchee set the hook, it grabbed, and held. The cage seemed to be solidly secured to the hull . . . but for how long?

For the tenth time since the beginning of the dive I checked my pressure gauge and timer. The run time was eleven minutes. I had plenty of gas, but was five minutes long enough to get to the bottom and back? Not in my mind. My chance to touch the sand was lost the moment the grapnel slipped. And although for years afterward I regretted not making the attempt, at that time and depth my audacity was tempered by timidity.

Manchee and I exchanged hand signals and body gestures. Native Americans may have been expert at this kind of silent communication, but not so wreck-divers. Manchee wanted to know what happened to Clayton. He hadn't seen him wave good-bye or ascend the line because he was concentrating totally on reeling in the line. A dive mask permits no peripheral vision. I couldn't understand what Manchee was asking, so I shrugged my shoulders. He interpreted my shrug to mean that I didn't know what happened to Clayton. This miscommunication caused Manchee some awful anxiety.

Now only four minutes bottom time remained. I explored farther to the right of the anchor line, practically hugging the wreck, then drifted downcurrent to the opposite edge. From here I figured I could catch the grapnel if it tore loose from its perch. It would have to drag along the hull in my direction. Manchee and I were in sight of each other, although at times we strayed as far as thirty feet apart.

Further exploration consisted of taking close-up glimpses of the thinly coated hull, looking up to reassure myself that Manchee was nearby, and pirouetting on my fin tips in order to see as much of the wreck as possible.

At sixteen minutes we began our ascent together. I was twenty feet above the hull before I remembered to switch back to travel mix. We paced ourselves according to our timepieces, which told us at what depth we were supposed to be at every minute of the ascent. Actually, we were a minute ahead of schedule. We were supposed to leave the *bottom* at sixteen minutes, not the top of the wreck, and then take one minute to reach 320 feet. This threw us out of kilter but on the conservative side. For nearly three minutes we maintained the prescribed ascent rate of sixty feet per minute.

Manchee was already on tenterhooks about Clayton, now it was my turn to feel suspense. Even though I saw him head for the surface, and knew that he had enough gas to make it on his own, his previous words came back to haunt me because I couldn't see him on the anchor line above. Had he intentionally let go and drifted off into the sunset, letting the first dive to the *Ostfriesland* become his swan song? Such are the vagaries of imagination under pressure - not the pressure of narcosis at depth, but the pressure of uncertainty in the realm of the unknown. Similar alarm was suffered by the support crew when Clayton returned alone.

The overwhelming majority of divers have never gone as deep as 160 feet - yet for us that depth represented only the first of sixteen decompression stops. We dallied there, breathing travel mix, because we were still ahead of schedule according to "run time." When the clock caught up with us we ascended to the next stop ten feet higher. At 110 feet we started breathing nitrox from the bottles that had been staged there. I noticed that only two tanks were clipped off. This gave me hope that Clayton had taken his stage bottle and was farther up the line - beyond the limit of visibility.

The rest is anticlimactic.

Clayton's shortened bottom time required less decompression. He left the two remaining nitrox bottles untouched at forty feet, having enough in the one tank to complete his decompression. He got out of the water a full hour before Manchee and me.

Our decompression went precisely according to schedule - all 155 minutes of it. Support divers came down and removed our side-slungs so we could clip nitrox bottles in their place. The useless video camera was taken away from me. At 40 feet we exchanged the partially used nitrox tanks for fresh ones. At 20 feet we went on pure oxygen. Then we surfaced with no ill effects.

Clayton broke out the "champagne" he had been saving for the occasion. Since none of us tippled alcoholic beverages, the "champagne" was actually grape juice. We needed no other high.

Afterward came long discussions on the conduct and performance of the dive: what went wrong, how to prevent those wrongs from recurring, and what could be improved - both in equipment and technique. We deduced that the lost nitrox bottle had unclipped itself as a result of the anchor line bouncing with the waves - a motion that wreck-divers take quite for granted. The next time, we vowed, we would refine the stage-bottle deployment system.

Clayton's major complaint: once again I went deeper than he did. This was a continuation of the contest between us: a trifling game to me, a serious threat to him.

None of us had set out to prove anything, either to ourselves or to the world. We simply wanted to dive a wreck that no one had ever dived before. In the process of pursuing a personal goal we demonstrated incidentally that the commonly accepted depth barrier was an imaginary construct: a limitation of human perception.

For deep wreck-divers, that barrier was forever broken.

# Early 1990's: Mixed-Gas Revolution - 143

## Concurrent Events

After the thrills of the Dry Tortugas, the penetrations into the *Andrea Doria*, and the groundbreaking dives on the *Monitor* and *Ostfriesland*, one might expect that subsequent activities would be anticlimactic. Not so.

For those who seek challenge and excitement in life, one experience or series of experiences, no matter how provocative, does not culminate with theatrical closure like a motion picture vignette. There are further goals and ventures, each blending into the next storyboard or overlapping others that are contemporaneous. Life is not sequential like a script whose scenes are written for dramatic appeal. Without an author to direct events, real life suffers from staggered continuity. It proceeds in hodgepodge fashion in a seemingly random pattern. The ending often goes unwritten and is usually not foreshadowed - it comes as a surprise to all if and when it ever occurs.

In the small picture, a dive or trip represents a finite quantum that exists within strict boundaries. But singular events accumulate, each contributing a stroke to a much broader canvas that can be viewed on various levels in the process toward completion.

For me, the *Monitor* stood as a testament of personal triumph in the age-old battle against the minions of bureaucracy who sought to pervert political authority for their own evil purposes. On a larger scale, the court decision exemplified the struggle for justice and enlightenment in the progressive course of cultural evolution toward the benediction of human rights superseding the rights of government.

The *Ostfriesland*, a far more challenging dive, captured a smaller portion of the limelight because the issues that it resolved were not appreciated by the masses. The dive may have been the underwater equivalent of the conquest of Mount Everest, but the wreck and what it stood for lacked historic recognition.

The *Monitor* mediated the change in warship hull design from wood to armored steel, and altered the strategy of naval warfare forever (a fact that was instantly acknowledged worldwide and which has been continually maintained), while the *Ostfriesland* brought traditional naval warfare to a premature end (a fact that Billy Mitchell vociferously adduced at the time of the bombing, but which was only accepted grudgingly twenty years later, when Japan attacked Pearl Harbor in its opening bid to take over the world by aggression.)

The *Monitor* was remembered. The *Ostfriesland* lay largely forgotten, or swept under the rug by Navy insolence. By their very existence, the wrecks - more than lumps of inanimate metal - were icons that embodied the essence of turning points in history. Each also represented an important turning point in the field of deep wreck-diving

- turning points that should not be overlooked.

Diving both wrecks within the span of a month brought their images into a contrasting focus that was sharpened by the light of the media attention they received. I can't claim to have perceived this effect in advance. What brought both diving episodes so close in temporal proximity was a combination of many factors, some of which were entirely beyond my control. The best claim I can make is that - for what it is worth - I appreciated the significance of the historic juxtaposition during the sequential unfolding of events. This is a prime example of dramatic appeal in the unwritten screenplay of life.

Other adventures followed. Quite closely, as a matter of fact.

## THE *EMPRESS OF IRELAND*

Because the weather forced us to dive the *Ocean Venture* first, and because the dive to the *Ostfriesland* occurred so late in the day, we didn't reach the dock until after midnight. While helium was still off-gassing from my body, I was already late for my next trip!

I quickly unloaded my gear, said some hasty good-byes, and began the six-hour trek to home and beyond. Actually, I didn't go home. I only passed nearby on my way to Dave Bright's house, which was an hour or so farther. I doused myself with coffee and drove all night alone, fighting desperately to keep from falling asleep at the wheel.

I called Bright from a roadside pay phone about seven o'clock in the morning, and told him I'd be late. I didn't get to his house until well after nine, by which time he and Bart Malone and John Moyer were champing at the bit to get under way. After exchanging curt greetings they helped me transfer my gear to Malone's pickup. Then I collapsed in Bright's car, exhausted, and let him drive, while Malone and Moyer followed for the fifteen hour trip to Rimouski, Quebec. It took two days. The morning after our arrival we began diving for a week on the *Empress of Ireland*, Canada's forgotten tragedy, in the cold St. Lawrence Seaway.

## DEEP DIVE, SHALLOW-WATER TRAGEDY

After that I spent two weeks diving on wrecks in Lake Superior with Mike Parks, Gaye Brown, Betsy Llewellyn, and Ted Bauer. The thousand-mile trip originated in Baltimore. With two vehicles we towed Parks's boat, the *Joss*, and a trailer-mounted compressor. We also took a portable compressor for filling tanks during the week we stayed on Isle Royale, which had no air station. (Or few facilities of any kind, for that matter, except for a restaurant and some cabins near the ranger station.)

Parks and I made a memorable dive on the *Kamloops*, a steel-hulled freighter which disappeared with all hands during a winter storm in 1927, and which wasn't located until fifty years later. The intact wreck lay on its starboard side on a steep rocky slope in green icy water only a few degrees above freezing. The stern rose as high as 180 feet, where the visibility was good and where there was sufficient ambient light to cast the hull in eerie silhouette.

After a tour of the high side we vaulted over the stern deck house and dropped to the lake bed below the fantail. Then we angled down the steep slope past a damaged lifeboat which had not helped any of the freighter's crew to survive. Cargo and broken timbers littered the bottom. In ever darkening gloom we followed the debris field down the slope to the forward cargo compartment, whose contents lay spilled out over the rock. At a depth of 260 feet the Stygian water was the color of pitch. Our lights cleaved the blackness like twin laser beams, and illuminated a pile of candy Lifesavers on whose wrappers the writing was still legible after sixty-three years submerged. The paper was preserved by cold fresh water in which the process of decay was drastically reduced. I wondered about the candy.

We turned around at the tip of the bow, then followed the upper hull back to the downline, which we ascended together in order to conduct our decompression. A couple of weeks later, Mike Parks was dead.

He was working on a pipe inspection job in New York when the accident occurred. He crawled up a water intake in full commercial rig: band mask with umbilical hoses that provided unlimited air and a communication link with his tenders. The water flow pumps were supposed to have been switched off, but somehow one of them was left running. As Parks approached a fork in the conduit, he was suddenly gripped by the suction created by the water being drawn up the side pipe. He was pulled out of his band mask and into the other pipe, where he drowned.

I dwelt upon the tragic irony of his death. He had dived safely and nonchalantly to 260 feet, then was killed during a dive in less than ten feet of water. I suppose this strange dichotomy supports the fatalistic philosophy which proposes that, despite all precaution, when your time is up, it's up. By the same reasoning I suppose I could have died in a rice paddy twenty-three years earlier.

What does it all mean? I haven't got a clue.

For the living, the world goes on.

## THE SECOND MIXED-GAS INITIATIVE

The mixed-gas diving initiative I slapped together prior to the *Ostfriesland* trip met with little more than a yawn, tempered perhaps by a pinch of awe, or worse, by a dash of lunacy. At that time the concept was unproven: a mere flight of fancy from an overactive imagination. Attitudes changed after the successful performance of the dive.

The following winter Steve Gatto arranged a second get-together, one that was attended by more open minds. The gathering was held in the recreation room of his newly purchased home in New Jersey - as before, a central location. This time the yawns were stifled by enthusiasm and an eager desire to learn. Some who brought a view of opposition eventually changed their attitudes.

What was a one-hour drive for me was a three-hour trip for Clayton. Other attendees included Kevin Brennan, Kevin England, Mary Grace Garcia, Jon Hulburt, Rick Jaszyn, Dennis Kessler, Bart Malone, John Moyer, Tom Packer, Jeff Pagano, Gene Peterson, Glen Plokhoy, Pat Rooney, Lou Sarlo, Brad Sheard, John Yurga, and John Chatterton.

I had met Chatterton only once before: at my home when he accompanied Nagle to my impromptu *Andrea Doria* party. As Nagle sank deeper into the dregs of drug abuse and alcoholism, Chatterton and Danny Crowell assumed more of the responsibilities of handling the *Seeker's* charters - often running the boat while Nagle lay stretched out on the wheelhouse bunk in a drunken stupor.

I was fiercely angry at Chatterton. A couple of years before I found an ornate decorative vase in the third class gift shop. I also found a stack of long-playing records in the bar. I packed everything in my goodie bag which Gilligan clipped to his tank harness. On the way up the anchor line the bag somehow unclipped itself. Because the line lay along the hull, we figured it must have landed on the high side of the wreck. This was the last dive of the trip.

Gilligan was duly penitent. With black indelible ink he wrote "I AM A KLUTZ" across his tee shirt, and wore it all the way to the dock. The *Seeker* had a *Doria* trip scheduled before the *Wahoo's* return the following week. I spread tidings of the dropped goodie bag throughout the deep diving community. Most of the *Seeker's* patrons were my friends, and if one of them found the bag I expected to have it returned.

Chatterton found the goodie where it had fallen. I learned about it shortly afterward. Every day I expected him to call or to find a package delivered to my door. I waited in vain. Chatterton threw out the records because they got smashed upon hitting the wreck, but he kept the vase *and* my goodie bag. The goodie bag was personal property that belonged to me, and the vase was mine by virtue of the fact that I had found it. I was livid - not because the vase was valuable or important but because of the principal of the matter.

I intended to confront Chatterton about his breach of wreck-diving etiquette, and mentioned my intention to Gatto. Gatto cautioned me to refrain. He agreed that Chatterton was wrong to have kept the vase, but as far as he knew it was an isolated incident. Chatterton's history contained no other such transgressions. And Gatto had invited Chatterton because he was a highly skilled diver who professed a deep interest in mixed-gas technology.

Everything else I had heard about Chatterton impressed me, especially some of his deep penetrations into the *Andrea Doria*. But that didn't give him the right to break the wreck-diver's code. Despite my bitter feelings I accepted Gatto's plea. I didn't want to subvert the character of the meeting over a personal affair.

Gatto called the meeting to order with a preamble about the purpose of the gathering, then introduced Clayton, who was relatively new to diving and who was therefore unknown (I knew everyone there). Clayton and I described our dive on the *Ostfriesland*, paying particular attention to the gases we breathed, to decompression protocols, to auxiliary tank transportation methods, and to back-up procedures.

Very quickly the exchange of views became a heady free-for-all. Clayton and I were stabbed with pertinent questions which were wonderfully upbeat and hopeful. Instead of having to plead a case on the defensive as I had at the previous meeting, I found myself speaking to willing ears. By the time the meeting was adjourned we had a room full of converts.

## *Wilkes-Barre* Tour Guide

I made three trips to Florida in 1991. Each was memorable in its own way. On the first, in January and February, I went to Key West to dive with Deans and DeMaria. The proposed trip to the Dry Tortugas never materialized because both were too busy to break away from other responsibilities. I spent more time writing on my computer in Deans's house than diving - one of the advantages of having a portable occupation: my time was never wasted. I did have some interesting dives on the *Wilkes-Barre*: two in particular whose relevance stretched into the future.

Diving the wreck for the first time were George Dreyer, who worked for Gene Peterson at Atlantic Divers in New Jersey; and Barb Lander, a housewife from York, Pennsylvania who had been certified by Peterson only two years before. They arrived together in Dreyer's van.

I introduced myself by first name only. I had found that many divers felt intimidated by my reputation, assuming, I suppose, that I might look down upon those with less experience in the water. So I had gotten into the habit of keeping incognito until we got to know each other on an equal basis. Once people realized I was an ordinary person (admittedly, with outstanding deep diving credentials) they felt more at ease in my presence.

(This scheme backfired once when a diver I met on a boat proceeded to brag about how much he knew about decompression line deployment. I listened patiently as he instructed me on procedure, but when he prattled on about the pitfalls of swelling sisal and the rationale for endcaps (which I invented for that reason), and I recognized some of his phrases, I realized that he was quoting from my book, *Advanced Wreck Diving Guide*. The more he talked the more foolish he would feel when he discovered that he was preaching to the author. As delicately as possible I let him know that I had published similar advice. His face reflected horror when his faux pas became evident, but I brushed it off and spoke of related topics until his embarrassment diminished.)

Deans had other customers to watch over, so he asked me to take Dreyer and Lander on a tour of the wreck - not so much to act as a guide and show them the sights as to oversee their safety. Deans never let first-timers explore the wreck on their own. He was far too cautious for that. People often exaggerated their experience and competence, and got themselves into trouble as a consequence. He wanted no accidents on his shop charters.

Deans instructed me not to take the pair deeper than 200 feet - the depth of the main deck. I followed his instructions, but after seeing how comfortable both were at depth, I left them at the railing and dropped over the side to the sandy bottom fifty feet below. We were never out of sight of each other. After we reunited I took them on a grand tour of the stern, both inside and out. Dreyer shot video as I led the way through darkened corridors. At one point we passed through a wash room, where I stopped and posed in pretense in front of a urinal - hamming it up for the camera. A good dive was had by all.

Both of them were exhilarated by the experience and professed a desire to go all the way to the bottom. The satisfaction I received by sharing uncommon adventures was rekindled by their ardor. I was reminded of my dive master days. That night I championed their desire persuasively with Deans. Both had performed well at 200 feet, so I had no qualms against taking them deeper the following day. Deans was reluctant at first, but quickly yielded to my expert opinion.

I did everything by the book. I checked their pressure gauges at the main deck's railing, gave each the okay sign and got it in return, and looked into their eyes for any symptoms of narcosis. They were fine. Slowly we descended the final fifty feet. I landed on the bottom on my knees, then Dreyer and Lander alit facing me. I repeated the safety checks - both had plenty of air and were not overbreathing. They were calm.

We leaned forward so our gauge panels rested on the sea bed. I noted that Dreyer's digital depth gauge registered 250 feet. But the readout on Lander's gauge registered 249 feet. In reality we were all kneeling at the same depth. The difference in displayed depths was merely a function of pressure sensor calibration or sensitivity. The sensor in Lander's gauge may have needed to go only another inch deeper for the fractional increase in pressure to make the readout flip over to the next depth increment.

Numbers rounded to tens possess a fillip of fulfillment. Lander's log book entries had to be based on information provided by her own gauges, not her buddy's. And 249 feet was an unsatisfying depth to record for a deepest dive. As Dreyer looked on aghast, I dug a hole in the sand and shoved Lander's gauge panel into it. I kept digging and shoving until the display registered another foot. *Then* I signaled that it was okay to ascend to the main deck and continue the dive. We laughed about it afterward.

## Introduction to Cave Diving

My next trip to Florida occurred in June. This was when Steve Berman had time in his busy schedule to teach me the rudiments of cave diving - the offer he had made to me the previous year on the *Andrea Doria*. Since he worked as an instructor at Ginnie Springs, that was where he introduced me to the wonders of submerged cave exploration. During the course of half a dozen dives he showed me the ropes - or should I say the lines? Following guidelines and using line reels are the backbone of cave diving. Equipment configurations were slightly different from those employed in wreck diving, as were some of the techniques such as swimming by means of the frog kick in order to reduce the amount of silt that was disturbed, and adhering to the thirds rule: if you turned around when one-third of your air was consumed, it left one-third for the return trip and one-third in reserve.

## The Gorilla Diver Mentality Faces Extinction

During the summer I dived off North Carolina and Virginia, on wrecks which I needed to survey for the Popular Dive Guide Series that I was writing. I also made trips to the *Andrea Doria* and the *Empress of Ireland* - trips that had long since become routine. I did research in Washington, DC. I continued to fight NOAA over the *Monitor* because, despite my victory in court, the agency denied further access. (NOAA bureaucrats were rankled over their loss of eminent domain.) And Clayton and I planned a return trip to the *Ostfriesland*.

The publicity that resulted from the first *Ostfriesland* dive fired the imagination of the wreck-diving community. The possibilities offered by adding helium to a breathing gas were limitless. Not that everyone wanted to dive to 380 feet, but for those with an unusually low tolerance to narcosis, replacing nitrogen with non-narcotic helium enabled them to achieve a potential that was commensurate with their skill and experience. Mixed-gas diving was a Pandora's box, but one that was slow to open and reveal its many treasured contents.

Perhaps the most satisfying development during this early transitional phase was the growing acceptance of deep exploration by the diving community in general, along with the reluctant acknowledgment that deep, decompression divers were not stupid hulking brutes harboring a death wish. (Once after a presentation an attendant asked who took the pictures I showed and who wrote the articles that appeared under my byline - the implication being that deep divers could not possess the artistic sensitivity or the intelligence to do cither.)

In the 1970's, those of us who dived deep and did in-water decompression were branded as crazy gorilla divers. In the 1980's we were associated with the commercial field. Now in the 1990's, due to recognition of the scientific approach to deep, decompression diving, the proliferation of electronic decompression computers, the appearance of sophisticated equipment options, and the adoption of mixed-gas breathing technology, the activity that encompassed these diversified disciplines acquired a label that was not a pejorative.

The descriptive phrases "high-tech diver" and "technical diver" were bandied about for a while. Eventually "technical diver" won out, more likely due to ease in pronunciation than to political correctness, since "requiring specialized knowledge" (the dictionary meaning) refers to all forms of scuba. I suppose that technical diving is akin to technical climbing - a rock climbing subdivision which utilizes specialized equipment (called "aid") that free-climbers deign not to employ.

At last, the activity in which I had been engaged for more than twenty years had been given a proper name and a respectable connotation.

### NOT-SO-INSTANT REPLAY ON THE *OSTFRIESLAND*

As I look back on events, I find it difficult to believe that it took an entire year to organize and implement a second trip to the *Ostfriesland*. Compared to today's pace and established routine, we must appear to have been dragging our fins to get into deep water again. Future generations might not appreciate how unsure we were in those days, how much time was required to create and develop procedures that are now taken for granted, or how difficult it was to organize a trip when so few divers were willing to take the risk or to play the seemingly subsidiary role of support.

By drawing another parallel to Mount Everest, from the time that

## Early 1990's: Mixed-Gas Revolution - 151

mountaineers first attempted to climb the world's highest peak (in the 1920's) until the feat was accomplished (in 1953), thirty years of unsuccessful efforts elapsed. Yet in 1996 alone, eighty-four people reached the summit - many of them non-climbers who were led by experienced guides. This in no way trivializes the achievement in terms of determination and endurance, but it does demonstrate the process of acceleration that occurs between paving the way and following the pavement. Pioneering a route is rigorous and uncertain.

After Mike Parks died, Ray Jarvis came to the rescue by offering to transfer our heliox from the storage cylinders to our scuba tanks. As the owner and operator of the National Diving Center in Washington, DC, he wanted to get into technical diving. He didn't charge us for the transfill work, and he even went as far as to donate all the nitrox we needed for decompression.

Again we left the dock in the evening and traveled all night to the site. I slept much better than I had on the previous trip. Not that I didn't feel strong trepidation, but repeating something I had done already didn't seem nearly as frightening as it had the first time. Our primary objective was to locate and dive the *Frankfurt*, a German cruiser that was sunk in the same bombing test as the *Ostfriesland*. Failing that we would dive the *Ostfriesland*.

Clayton continued to use my reputation as a way to entice people to flesh out the charter, but this time most of our support divers signed on for a different reason entirely: to see how a mixed-gas diving operation was conducted in order to prepare for such a dive themselves. Thus the trip became a hands-on training program for which the charge was the charter fee.

This arrangement appealed to me. If people got turned on to mixed-gas diving and joined the growing peer group, more diving opportunities would become available. I envisioned a time when others would assume the leadership role and I could go on trips as a passenger without responsibilities.

We were only a mile or two from the *Ostfriesland* when the port propeller shaft broke. This time the captain of the *Miss Lindsey* was Mike Hillier, Jr. - the son of the owner. On one engine we limped toward the coordinates that we hoped would be the *Frankfurt*, but couldn't find the wreck.

Should we try to hook the *Ostfriesland* and make the dive with the boat incapacitated, or call it quits and come back another day? It was a difficult decision to make considering all the time, money, and effort that had gone into the trip: a logistical nightmare which I have already described. After due consideration we opted to abort in the interests of safety.

So close, and yet so far.

The return ride was long and tedious and filled with disappointment. After several hours of hard, greasy work and masterful engineering, Hillier Jr. effected temporary repairs, but the loss of time and the mechanical uncertainties precluded reversing our decision to abort the dive.

We scheduled another trip for the following month. That one was blown out by weather - fortunately, before anyone left home and made the long trek to Virginia Beach. If we couldn't reschedule another trip that year, we'd be stuck with helium in our tanks all winter and spring.

Not until October was the *Miss Lindsey* available for another overnight charter. Mike Hillier, Sr. ran the boat because junior was away at college. Hillier Senior thereafter was our captain for all subsequent trips to the Billy Mitchell wrecks off the Virginia Capes, and for other charters as well. He was a never-ending fountain of energy and ebullience that was positively infectious. Not only did he handle the navigation and steering, but he prepared the meals as well. He could flip flapjacks with the best of them and barbecue burgers on the grill with gusto that put most short-order cooks to shame. He was camaraderie all by himself - a pure delight.

Again we searched for the *Frankfurt*. Again we couldn't find it. So we dived the *Ostfriesland*.

I didn't own a still camera housing that could handle extreme depths, so once again I borrowed a video camera from Smokey Roberts. Clayton wanted me to shoot video anyway, rather than stills, because he was pushing for publicity and wanted footage that could be broadcast on television.

Our original trio of gas divers was augmented by one - Ed Suarez, who had been a support diver on the original *Ostfriesland* trip. He was highly skilled in cave diving and wreck-diving, with years of experience in both endeavors. We dived in pairs: Clayton and I went down first, followed twenty minutes later by Manchee and Suarez. The dive was far less harrowing than the original because the grapnel had a firm grip on the wreck, and because I had acquired some confidence from my previous experience.

Nothing untoward occurred except that again the video camera didn't work. I actuated the switch when I reached the wreck, saw the running light blink as it had during pre-dive trials, started shooting, then noticed that the camera had shut down. I fussed with it for several minutes before giving up in frustration. Again the video camera battery had died because Roberts had forgotten to charge it.

Ronny Bell was my support diver. He met me on the anchor line at 110 feet where he relieved me of the malfunctioning video camera and facilitated the exchange of cylinders. The dive went smoothly for Manchee and Suarez as well. Other support divers were Roy Matthews and Harvey Storck.

This brief description gives the dive an appearance of little consequence, but that would be an unfair evaluation. That the dive went without a hitch speaks for itself. What we accomplished was more in the realm of self-education: honing skills, refining techniques, determining which procedures worked better than others.

There were, however, some problems. Manchee didn't like the feel of sling bottles and wanted to wear quads instead: four tanks spread across his back on a plane. I thought it was an interesting configuration but Clayton wouldn't hear of it, claiming that Manchee needed help to stand with three tanks on his back, let alone four. In Manchee's behalf I protested that the weight didn't matter: both configurations weighed the same in the water no matter how they were arranged. And Manchee was strong enough to climb the ladder wearing quads.

Since I had abdicated control of the trips to Clayton, and the boat was chartered in his name, he exercised authority and disallowed the use of quads. "I will not tolerate insubordination," he said firmly.

Furthermore, Manchee and Suarez wanted to extend their bottom time in order to get the most out of the dive. Both were efficient breathers and were willing to decompress longer. Clayton nixed that deviation as well. He gave them a choice: either dive according to his design or don't go on the trip. Manchee and Suarez were mature enough to concede the point.

On a lighter note, I continued to perturb Clayton by getting deeper readings on my depth gauge. He was sensitive to this ever since the *Washington* dive on which I crawled into a washout and got six feet deeper than he did. Because of his short fill on the first *Ostfriesland* trip he had halted his descent on the high side of the wreck while Manchee and I dropped over the edge. This time we were together every moment, but by some quirk of digital depth gauge sensor display he came out a couple of feet shallower. This truly was of little consequence - to everyone except Clayton, who maintained an irrational stance with respect to being outdistanced - or outdepthed, as the case may be. To me it was a joke - one which by continuous jibes I did not let Clayton forget.

## Cave Diving Certification

I returned to Ginnie Spring in December in order to complete my full cave certification with Steve Berman. Ed Dady joined us on the dives. One day they both had other commitments, so they foisted me off on John Reekie, whom I had met on the same *Andrea Doria* trip on which I had met Berman.

Reekie wanted to make a half-mile penetration into Madison Blue. Joining us on the dive was Alex Reyes. Reekie poured on the pressure by making it clear to me that he would be very disappointed if we didn't reach a certain spot at which some rocky fingers overhead guarded a slight restriction that continued down a slope. Because of my bad lung I was a notorious air hog. I did my best to control my breathing rate.

Most of the tunnel curved sinuously through the rock at a depth of 70 feet. Reekie obliged me lead the way along the guideline, ostensibly so I - the beginner - could set the pace. I kept looking back to make sure they were behind me. By the time we reached the vertical opening to the lower level, I was already farther inside a submerged cave than I had ever been before. A couple of decades of wreck-diving experience did not help to calm my overstrung nerves: wreck penetrations were notably short by comparison.

Reekie and Reyes detached their stage bottles and stashed them out of the way. I clung to mine like a child to a favorite stuffed animal. I had long since exceeded my comfort level. In fact, I was scared. I didn't want to be there. But neither did I want to wimp out on them. Reekie gesticulated brusquely, pointing to the hole in the floor that he wanted me to enter: the way to the lower tunnel at 110 feet.

I was overcome with a feeling of dread that he was setting me up, that he was going to cut the permanent guideline as soon as I was out of sight and leave me stranded in the bowels of the cave. I continued to hug my stage bottle, refusing to stash it because I had come to believe that he was going to steal it so I wouldn't have enough air for the return. I lowered myself through the hole but kept the stage bottle with me, and I kept my eyes on the pair who intended to do me in.

I dropped out of the ceiling into the lower tunnel, careful to maintain my buoyancy and not to settle onto the muddy bottom and stir up silt with my fins. Reekie and Reyes came after me. I stayed out of their reach because I now feared that Reekie was going to pull off my mask.

I kept wondering frightfully: what did I do to him to piss him off? It must have been the *Andrea Doria* trip, I fancied. He blamed me for tying into the stern, where Bielenda had maneuvered the grapnel. I had ruined his whole trip. Now he was going to get his revenge by murdering me under circumstances where investigation would ascertain that my death was an accident: a neophyte cave diver who panicked and drowned.

Irrationally, I thought I could placate him by reaching the rocky fingers. We went on, ever on - deeper into the tunnel, deeper into fear. I hoped against hope that one of them would halt the dive by running low on air, but every time I questioned them they returned the okay sign. The passageway angled down. My depth gauge read 120 feet. Finally I could go on no more, I could not attain Reekie's objective. I had consumed one-third of my air so I had to terminate the dive. In cave-diving lingo I "called" the dive, or "turned" the dive.

I gave the thumb-up sign. Both acknowledged by returning it. They turned around. We retreated in reverse order. Now I was last. We frog-kicked along the lower tunnel, ascended through the hole in the ceiling, and regained the upper tunnel. While Reekie and Reyes fumbled with their stage bottles I kept right on going, not getting out of their sight but maintaining a safe distance ahead. With the mild current behind us, we soared through the tunnel with very little effort.

We reached our first decompression stop eight-eight minutes into the dive. Reekie pulled a magazine out of a plastic pouch and settled down to read. Reyes loafed idly. I swam around looking for fossils. After seventy-six minutes of decompression I surfaced, and gloried in seeing the sun that I thought never to see again.

Reekie waxed enthusiastically during the post-dive debriefing. I learned that we had ducked under the rocky fingers and gone a little farther, but I had been so distraught that I had failed to notice that we had passed his prime objective. In his opinion the dive was a great success. But he couldn't understand why I had suffered the drag of my stage bottle all the way to the end. I offered the lame excuse that I didn't want to waste time fumbling with unfamiliar snap hooks. He shrugged.

All my fears were for naught: a manifestation of paranoid delusion. Anxiety from so deep a penetration coupled with inexperience and lack of faith in cave diving techniques distorted my objective view of reality. Irrational fear was a new emotion for me - one that I had never felt before. I didn't admit my dire forebodings to Reekie for more than a year. Then he laughed it off. We've been good friends ever since.

After a few more dives with Berman I earned my official certification. I duly noted the irony of the event. Way back in 1970 I took a basic scuba course so I could extend my exploration of the underground river in Blue Hole, Virginia. Now, twenty-one years later, I had finally gotten certified to do what I had wanted to do originally. (Except that cave diving certifications didn't exist then.) A lot of water had passed under the bridge during the intervening decades. I wonder what caves I would have discovered and explored had I not gotten hooked on shipwrecks.

My life had changed, my goals had undergone revision, my occupation was one that I had not anticipated. Yet I was still the same person, with the same bright outlook on life and the same naive trust in humanity. *Those* are irrational viewpoints that I have been unable to overcome - my greatest weaknesses.

## ULTIMATE WRECK-DIVING GUIDE

Despite all the interest in mixed-gas diving and the hoopla over the *Ostfriesland*, the ocean saw little other mixed-gas diving activity. In two years there had been only a handful of dives to wrecks in extreme depths. Billy Deans was diving the *Wilkes-Barre* on gas, and he and Jim King had dived a cruiser that was deeper. The majority of mixed-gas scuba dives were still being conducted in caves.

Menduno was working hard on *AquaCorps*, motivated by the rush of emerging technological change, but the magazine's circulation was limited to about a thousand. He was getting out the word, but only the few were listening to his gospel. I thought that mixed-gas diving concepts held a broader audience appeal.

So I wrote a pair of articles on nitrox and helium diving, intended for public consumption beyond the readership of *AquaCorps*. Once again I found that not a dive magazine in the business had to guts to publish them. Their editors were afraid to embrace new ideas, timid of publicizing untried techniques, fearful to suffer any change in an industry in which their revenues were based on the status quo. They were fat and comfortable in their editorial chairs, and too wrapped up in resort advertisement to waste column inches on useful information. They thrived instead on ignorance, and neither welcomed advancement nor suffered it to live. Some openly denounced all attempts at progress.

Not one of the nation's certifying agencies would permit the instruction of nitrox - the most innocuous of alternative breathing media - much less condone helium and diving to extreme depths. The Dive Equipment and Manufacturers Association banned any mention of nitrox and all those associated with it from its annual trade show. Those who administrated these profit-making organizations were dead set against any development of the sport lest they lose some control of the market.

This was the same retarded and obstructionist mind set I encountered ten years earlier when I wanted to enlighten the public about decompression methods.

If I couldn't have the information published in the form of articles, I reasoned, I'd publish it in book form. And now that I was in the publishing business with my own small press, no backward editorial attitude could prevent me from doing so - or from writing what I wanted to write. In the spring of 1992 I released the unexpurgated *Ultimate Wreck Diving Guide*. My words were pure heresy to a puritanical diving community bent on halting progress and returning to the dark ages. Nevertheless, the first thousand copies sold within a few months, and the book continued to sell briskly thereafter as more and more people discovered the challenges of technical diving.

## LEAPS AND BOUNDS

By this time, so-called "technical diving" was firmly established as a subspecies of scuba. Practically overnight the progress of deep wreck exploration went from evolution to revolution. As more and more people discovered the existence of technical diving, and became enthralled by the challenges which the activity embraced, the industry expanded geometrically. And technical divers - whose temperament enabled them to grasp new ideas and to originate concepts - took their fate into their own confident hands.

The circulation of *AquaCorps* more than doubled. Menduno added to the momentum by organizing a technical diving conference which achieved great success.

Jon Hulburt and Brad Sheard, both engineers but with different educational backgrounds, wrote their own decompression computer programs. These software programs were a quantum leap beyond my simplistic extrapolations of the Navy Tables. Hulburt based his trimix program on the decompression studies published by A.A. Buhlmann in 1984, and confirmed its validity by diving on the *Choapa* in the Mud Hole on New Years day, 1992.

Soon after the mixed-gas awareness meeting at Gatto's house, Bart Malone and Lou Sarlo - fired by the possibilities of diving deep with a clear head - visited Billy Deans in Key West. Deans showed them how to set up a gas blending station. Malone and Sarlo returned to New Jersey, bought all the necessary tubing and hoses and fittings, compressor and booster pump, oxygen cleaning equipment, oxygen analyzing sensors, and so on, then established a technical dive shop in a corner of Sarlo's warehouse, and constructed a blending facility in a trailer on the property. (Sarlo built roofs over gasoline pumps in front of automobile service stations.) They named the business, appropriately enough, The Gas Station. Clayton, Hulburt, and I were among the first customers.

## *OSTFRIESLAND* THE THIRD TIME AROUND

While 1992 saw an overall expansion in the use of helium mixtures for mid-range depths, Clayton and I managed to increase our annual extreme-depth excursions from one to three. Manchee pulled out his back a few weeks prior to the trip, so he was forced to cancel. Steve Gatto and Tom Packer joined the ranks, and they dived together as a team. We had no trouble finding support personnel. Plus we had a reporter on board to record the event for television.

Once again we couldn't find the *Frankfurt*, so we returned to the *Ostfriesland*.

My dive with Clayton did not go exactly according to plan due to a couple of equipment problems. One of my computers died, and although I was using it only as a time piece, its failure left me without a back-up. Then a fin strap broke. The fin clung tenaciously to my bootee by means of a tight fit and suction, but only because I feathered my stroke. I was afraid that if I kicked too hard I would kick the fin off my foot. This did not encourage me to stray far from the security of the grapnel - which was a shame because for the first time at extreme depth we had ample ambient light and excellent visibility.

On the bottom, Clayton was exasperated because I wouldn't go exploring too far afield. I pointed to my heel but he didn't understand, so he stewed for the next couple of hours until we surfaced and I was able to explain my dilemma. Clayton was a rationalist. As soon as he knew the truth of the situation he stopped blaming me for timidity. On the contrary, once I regained the safety of the boat, I berated myself for not being bolder and for wasting a good opportunity. It was amazing to me how quickly my attitude could change when I didn't have all that water over my head.

Packer experienced a few anxious moments when his support diver didn't show up on time with the bottle of nitrox that was essential for decompression, and left him hanging. The support diver had trouble clearing his ears. After trying valiantly to reach Packer at 130 feet for the exchange, he surfaced and got someone else to take down the bottle. This was a valuable lesson for all of us because it emphasized a weak point in the system.

The alternative was to stage bottles on the down line, but that system had its own disadvantages, brought to prominent attention previously when the hose-clamp assembly broke apart due to the surface chop and we lost a tank and regulator - an expensive lesson that Clayton and I had to pay for by reimbursing the owner of the borrowed equipment. Every dive we fine-tuned the operation.

I shot video footage with a camera that I borrowed from Nike Seamans, who worked as one of our support divers. (She also supported on my *Monitor* trip.) She made sure to charge the battery, and provided a fully charged spare. I had only dive lights for additional illumination since her video lights could not be taken that deep without risk of imploding. I got some usable footage but the images were dark. I presented the tape to the reporter. Creative lighting enhancement made by technicians in the laboratory enabled the footage to be broadcast with the news.

## Mixed-Gas is Aired

Interest in Billy Mitchell and the sunken German warships was waxing, so WAVY-TV decided to give the story full coverage by assigning not only a reporter but a camera crew to accompany us on the next trip. Margaret Douglas conducted live interviews on the ocean, the camera crew recorded the preparation for the dive and the return from the depths, and I shot the underwater footage.

Finally we found what we thought by its position relative to the *Ostfriesland* should be the *Frankfurt*. Clayton and I were the only gas divers on this second trip of the year. But we had a whole host of new support divers, including Barb Lander.

After our meeting in Key West, Lander had engaged me to give an advanced wreck-diving workshop, an introduction to technical diving, and an evening slide presentation that was open to the public - all subsidized by the local dive shop with which she was affiliated. During the weekend lecture tour I stayed at her home with her husband and son. Husband Jeff was a certified diver, but he didn't have time to pursue the activity with enthusiasm, nor did he have the inclination. He was a busy ophthalmologist and a highly skilled surgeon who conducted delicate operations on the human eye. But he did not mind supporting his wife's underwater habit.

Not until this latest excursion did I feel fully confident at extreme depth. The two sling bottles I now took for granted, and the video camera no longer seemed an encumbrance. The sight of the wreck was spectacular, and I am sure that ambient light visibility - which exceeded one hundred feet - helped to bolster my equanimity.

Clayton liked to be in front so he preceded me down the long scope of the line. Every fifty feet or so he glanced over his shoulder to ascertain my proximity. Each time we made eye contact I flashed the okay sign. As we approached the wreck I could clearly see the anchor chain draped over what appeared to be the engine masked under a pile of debris and disarticulated hull plates. I couldn't see the grapnel because it lay hooked on the far side of the mass of metal. If the tines lost their grip, the grapnel would have to drag completely over the steel structure that rose fifteen to twenty feet above the bottom - and then have to drag through a broad field of broken beams before it reached the flat sandy bottom. It was a very secure hook - the kind that instills confidence.

As I passed over the washout along the edge of the wreck I felt a sudden urge to play a trick on my buddy. I dropped off the line some fifty feet before the end and descended straight down into the shallow depression. By shoving my gauge panel flat against the sand I obtained a maximum depth reading of 353 feet. Then I rose slightly and worked my way up and out of the scour and over the collapsed hull to the base of the engine, where the level of debris stood ten feet shallower. I panned the wreckage with the video camera as if I had dropped directly to that spot from the anchor line.

About that time Clayton reached the point where the chain touched the upper steel structure. He saw me below, shooting. He glided down and around me, making sure to dip down deeper than my feet before circumnavigating the taller structure. I could hardly contain the glowing smile behind my mask. I moved back to gain an overview and to get Clayton in the viewing field against the backdrop of the wreck. It was then that the powerful beam illuminated a large curved structure bearing a rectangular hole that looked suspiciously like a gun port. I concluded that it was the crushed remains of a turret.

The dive proceeded with clocklike precision. We traveled a fair distance from the grapnel to look for identifying features. I roamed mostly under the anchor line or not too far from its vertical extension on the bottom. Doubts still lurked in the back of my mind. I figured that if the grapnel broke free, I didn't want to be too far away from its path as it dragged across the wreck.

I turned off the camera when the time came to leave, exchanged signals with Clayton, and began the long ascent. After we settled comfortably into decompression mode, Clayton looked casually at my gauge panel and got prepared to gloat. His face turned into a caricature of a child who had just swallowed a spoonful of cough medicine. My depth gauge read nine feet deeper than his. This time I couldn't help but snicker. Soon I was laughing uproariously, stabbing my finger first at his depth gauge then at mine. Clayton, the serious intellectual, failed to appreciate the humor of the moment.

I shot footage of Clayton and the support divers exchanging tanks. Then, as planned, I handed the still-running camera to Lander in order to preserve the continuity of the footage. She backed away so she could shoot the activity from a distance. She had never shot video before, but she managed to hold the camera steady and get some good ambient light footage - all of it upside down. She didn't know which side of the camera housing was the top!

I emerged from the water a couple of hours later. Still dripping, I faced the topside camera for a post-dive interview with Douglas. I described the dive and the turret. Allowing that positive identification was premature, I surmised that the wreck was probably the *Frankfurt*, and Clayton agreed. In this we were wrong. Not until several years later, when one of the other wrecks in the vicinity proved to be the mammoth hull of the cruiser, did we conclude that we must have dived one of the destroyers, which were smaller - probably the *G-102*.

Again the studio technicians came to the rescue. They inverted Lander's footage so the bubbles did not fall off the bottom of the screen. The footage I shot on the wreck aired nicely. My sole complaint with the final production was that the topside film crew kept the camera rolling while I was horsing around with my equipment after the dive. I took a few breaths of heliox and chirped a tune in the resultant high-pitched Munchkin voice. Juxtaposing sober dialogue and dignified narration, I appeared on national television singing a karaoke rendition of *How Much is That Doggie in the Window?*

## German U-boat Discoveries

In addition to the German warships situated at extraordinary depths, we also conducted a parallel series of historic wreck discoveries: the U-boats that were sunk in the 1921 bombing tests. In 1992 we located the *U-140* at a depth of 266 feet, and the *UB-148* at 274 feet. These were the first U-boats of World War One vintage ever dived in American waters.

The following year we found the *U-117* at 233 feet. We didn't identify it at first because it wasn't where it was supposed to be, and because of the inky blackness on the bottom. I mistook it for a possible scuttled American sub, although Clayton had his suspicions. Not until 1995 did I identify the U-boat to my satisfaction. I did it by comparing features on the wreck with those that were shown on historic photographs that I obtained from the National Archives.

## U.S. Battleship *New Jersey*

There is no need to describe all the subsequent mixed-gas dives unless extraordinary events occurred. The battleship *New Jersey* proved to be such an event. Billy Mitchell continued his aerial bombardment experiments in 1923 by dramatically sinking two obsolete U.S. battleships, the *New Jersey* and the *Virginia*, off the Diamond Shoals of North Carolina.

By autumn 1992, Clayton and I thought that Barb Lander was ready to graduate from support diver to gas diver. She had dived the *Andrea Doria* and some wrecks in the Mud Hole, had acted in a support capacity for two of our mixed-gas ventures, and had demonstrated uncommon initiative and enthusiasm in embracing the challenges of the deep frontier. So we asked her to accompany us to the wreck of the *New Jersey*.

For this trip we chartered the *Margie II*, owned and operated by Artie Kirchner, my long-time friend from EDA days in the early 1970's. He was phasing out his electrical business in New Jersey in order to run a full-time charter business in Hatteras, North Carolina. The boat had been built according to his own specifications, two years before, in Louisiana, and I had helped to drive it around the Florida coast.

Kirchner had his own ideas about how to snag deep-water wrecks without losing expensive grapnels in the process. He proposed to drop a cheap expendable grapnel which was secured to the anchor line by means of a length of thick sisal rope, which he could break by backing the boat. He reasoned that the predominant strong current might prevent the retrieval of the grapnel by the customary method of maneuvering the boat ahead until the grapnel fell free, because the force exerted by the current against the line might create a sweeping arc, in which case the tines could not relax their grip.

The three of us descended through crystalline blue water in which visibility exceeded one hundred feet. When the white sparkling sand hove clearly into view, the wreck was nowhere to be seen. We alit on the bottom at 333 feet. The grapnel appeared to be hooked in the sand - and that by the tip of a single tine. Upon close examination I perceived a tiny shard of metal that must have been the corner of an embedded hull plate - a tenuous grip at best and one that did not instill composure.

At the extreme limit of vision I could barely discern the outline of an inverted intact hull, with a fuzzy notch at one end that must have been the stern upsweep for the propellers. The depth recorder had shown at least a 30-foot spike. I figured that the grapnel must have bounced loose while we were getting dressed for the dive, dragged over the wreck and across the sand, then snagged by the merest chance in this tiny bit of metal that had been blasted off the ship during the bombardment. It was sheer lunacy to contemplate abandoning the line to our decompression gases and swimming along the bottom toward the wreck.

I signaled to Clayton that I was terminating the dive. He acknowledged. I then flashed the sign to Lander. She shrugged her shoulders. That was the wrong answer. I was not asking her a question, I was making a positive statement: the dive was over. I signaled to her again, forcefully. This was neither the time nor the place to hold a discussion. She did not signal consent, but she reluctantly followed us up the line.

We were deeper than 200 feet when I felt a sharp snap and the anchor line went limp. The sisal had broken from the strain - prematurely and only a couple of minutes after we left the bottom. Untethered, the boat was swept along the surface by a combination of wind and wind-induced surface current. We were drawn upward by the pendulum effect, faster than we wanted to ascend. We dumped all our excess buoyancy in order to achieve a semblance of depth control. We were towed by the boat like lures on the end of a fishing line, trolling ever upward from the depths where we needed to decompress.

I wore a drysuit which was inflated with heliox in order to take advantage of the cooling effect of helium in water as warm as 80° F. I squeezed all the gas out of the suit and emptied my BC so I could hang as heavy as possible. We managed to get stabilized after a fashion, holding onto the line like kids at the end of a whip, and ascended at ten-foot increments as prescribed. Support divers pulled themselves down the streaming line to help in switching swing bottles.

All was bearable for a while despite the strain on the arms. Stop by stop we conducted our decompression. We were barely holding our own at 50 feet when a group of support divers swarmed down upon us.

The extra positive buoyancy plus the added dynamic drag coupled with a sudden gust of wind conspired to trip our tenuous trim. In a trice our buoyancy went completely out of control.

The mass of divers skyrocketed to the surface. Warm air and sunbeams splashed against my face. With more than an hour of decompression yet to do, paralysis or painful death lay only minutes away. I was terrified. I screamed at the support divers to get off the line. At the same time I purged my drysuit of all its newly-acquired buoyancy. I inverted myself and kicked down. Clayton and Lander did the same. We barely managed to reach 50 feet. I tried to go deeper, in order to recompress any bubbles that had formed in my blood during the rise, but couldn't get beyond 60 feet - and even that was a struggle.

We spent the remaining decompression in a constant battle with buoyancy. The neoprene material of our drysuits expanded as we ascended to shallower stops. Eventually, someone brought some lead weights and tied them to the anchor line. That helped. I extended my decompression far beyond that required by the short bottom time. I wasn't taking any chances.

Finally we surfaced - with no ill effects. After that fiasco we nixed the idea of a breakaway anchor line. Better to pay for a brand new grapnel than to suffer the possible consequences.

## Mixed Up Gas

The technical diving movement was accelerating so rapidly that silly situations resulted, due largely to uneven and undisciplined expansion. Times of radical change are always crazy, but this particular case of attitude adjustment was worse than most, especially viewed in retrospect.

In the beginning, all technical divers were self-taught. There's nothing wrong with informal education except that subject matter can usually be absorbed more quickly through structured schooling. As soon as there was a market for it, some people in the scuba business began to offer training in mixed-gas diving. An instructor is only as good as his experience. Most of those offering courses and so-called certifications possessed little or no background in either the theory or conduct of technical diving. They threw together a patchwork curriculum with bits of knowledge gleaned from hearsay spiced with misinformed notions, then passed themselves off as "experts."

Whether they did it for money or to stroke their egos is irrelevant. The upshot was that hopeful trainees often received poor guidance from inexperienced, self-styled instructors. The horror stories that ran rampant through the community tolled bitter truths that sounded more like fable.

For example, one instructor whose program required trainees to make two mixed-gas dives before passing the course, descended to the high side of a deep-water wreck with two students in tow. They touched the wreck (but not the sea bed which lay seventy feet deeper), ascended the anchor line to a hypothetical decompression stop, then instantly re-descended for another touch down on the wreck. This action simulated two actual descents in a single dive.

In another case, a student who completed the required course work immediately established himself afterward as a mixed-gas instructor. His only in-water experience with mixed-gas procedure was a single training dive.

Some went so far as to create fictional training agencies that issued certification cards upon graduation. This scenario goes far beyond an act of simple frivolity, for a student who believed he was getting good training could easily get into trouble when he encountered a situation which he was not prepared to handle.

Then there's the sorry plight of the store-bought certification. Yuppies with disposable income were too often convinced by unscrupulous shop owners that they could purchase qualifications without prior experience, as if technical diving skills were products that could be purchased off the shelf. "Certification" should not be confused with "qualification." Experience cannot be bought with money, it is paid for with constant practice. Merit badge collectors didn't realize this. They tried to go from basic scuba to mixed-gas in one or two years, often with dire results - and in the process they swelled the coffers of the shop owners who promoted the evil scheme because of the merchandise they could move.

## Bogus Certifications

Entering the fracas of technical diving absurdity was a new policy from The Gas Station. Malone told me that he would no longer sell helium blends to Clayton and me because we did not hold official mixed-gas diving certifications. The situation was ludicrous on any number of levels. In addition to my comments in the preceding paragraphs, Clayton and I were not just ordinary divers, but were in some measure responsible for the development of technical diving and for the establishment of The Gas Station. Malone and I shared a friendship and diving relationship that spanned more than fifteen years. The Gas Station sold the *Ultimate Wreck Diving Guide* to its customers and touted it as a text book. How could the shop owner acknowledge the expertise of the author, yet claim that he was unqualified to perform the dives that he had been making for years? I could go on, but you get the point.

The fact remains that Clayton and I didn't need certification. We possessed a commodity of far greater value: practical experience.

Furthermore, I didn't *want* certification. I am reminded of the college professor who visited my house and gazed in awe at my vast and varied library. (I have six thousand books, all of which I've read.) He asked, "What kind of degree do *you* have?" I replied, "I don't have any degrees, but I do have the knowledge." As a sad commentary on the American system of academic promotion, he declared, "Knowledge doesn't mean anything. It's the degree that counts."

To be fair, I think Malone was motivated by the issue of liability. To sell breathing gas to uncertified divers might not only set a dangerous precedent, but it might compromise the legal standing of the shop's insurance policy - perhaps with respect to negligent management. Notwithstanding such rationale, I was not about to take a course in order to get certified by someone whose training requirements were based on a book that I had written.

Call me stubborn if you will (I've been called worse), but cards and ribbons and merit badges just didn't mean anything to me. On the contrary, I repudiate the accumulation of such nebulous decorations. Granted that there are those who view technical diving as a way of gaining acceptance, and who collect certifications the way a hunter collects heads for show, but I don't want to be likened with their kind.

As another example of assertive independence, for more than ten years I have rejected entreaties from the Explorer's Club to join their hallowed ranks. At one time the Explorer's Club was an elite organization of geographical explorers, one which people begged to join. The Explorer's Club flag has been flown at the poles, on the world's highest mountains, and in far-flung deserts and jungles.

Today the Explorer's Club is ineffective and effete, a lingering vestige and an empty honorific of past greatness and lost eminence. The majority of those who join the club now do so not because of what they have to contribute, but because of what they hope to gain by advertising their membership. In a way, they seek to create a false image of credibility by associating with royalty.

I permitted the Explorer's Club to fly its flag on my *Monitor* expeditions because some of the participants were members and close friends. The flag has also accompanied some of the trips to the German warships. Clayton applied for membership and found eager and willing sponsors. And while I think he heartily deserved the membership he obtained, I want no part of the club.

I am not a joiner. I am not a belonger. I've always been a maverick who went his own way no matter how difficult the path - I find satisfaction in overcoming unknown obstacles. I joined clubs in the 1970's because it was the only way to get on trips. Once those clubs outlived their usefulness, and I felt shackled by their attitudinal limitations, I left the womb for a position closer to the fore. Unfortunately, one outstanding problem with being in front is that you can't see people shooting at your back.

Be all that as it may, Clayton and I were saved from the disgrace of having to get certified in a field which we had helped to pioneer. For this we had to thank Billy Deans, who was already on his way to becoming the guru of technical dive training. He had joined Dick Rutkowski's organization, the International Association of Nitrox Divers. For years Rutkowski had been certifying divers for nitrox. Now Deans added his mixed-gas expertise and wrote a training course and issued a certification card that held some meaning. Deans grandfathered us into the program and issued our cards ex post facto.

## The Penguin and the Albatross

Technical diving was having other teething troubles, largely from those who wanted the activity prohibited. In a quiet manner, Rutkowski had been fighting off packs of anti-progressive wolves for years by simply going his own way despite growing opposition. Most of the detractors consisted of people who were consumed by envy: they couldn't stand for others to succeed where they themselves had failed.

Leading the witch hunt to stamp out technical diving was *Skin Diver Magazine*. In 1993, editor Bill Gleason published a scathing indictment against nitrox which incensed those who supported its use, and informed those who knew nothing about it. Ironically, by making its subscribers aware of a movement that went beyond so-called "sport diving," the editorial failed miserably in its stated intent and antithetically gave technical diving a boost into fame by advertising its existence.

Suddenly, the concept of technical diving was on everyone's mind. The curious wanted to learn more about it, the venturesome strove to embrace it. In the fallout, even some of the major certifying agencies paid lip service to technical diving by acknowledging that, although the activity violated long established standards, it was a valid force to be reckoned with.

When asked to speak at NAUI's annual convention, I read the following statement:

> The Penguin and the Albatross
>
> Deep diving is an artificial concept created by people without vision.
>
> Like the sound barrier, depth of water is not a barrier of physics, physiology, or human endeavor. It is a barrier of the mind.
>
> Ever since he crawled out of his protective cave, mankind has explored the world in which he lived. He has roamed the vast savannas, braved wild jungles, tramped the floes and

fields of ice, delved into the depths of the sea, soared like a bird in the air, and has begun the investigation of the infinite reaches of outer space. He has done these things because it is his nature to explore. Mankind as a collective entity has an inquiring mind; some men and women, as individuals, have not.

At every stage of human progress there have been those who have tried desperately to halt it. That is the lot of the timid. They are secure with the tried and true, comfortable with the status quo. They abhor change. But change is the way of life. Otherwise lies stagnation.

The situation with respect to diving is the same. We have been here before. People once thought that the crews of the *Nina*, the *Pinta*, and the *Santa Maria* would sail to their deaths in a dark abyss. But the edge of the world did not exist then any more than today exists a bottom to the depths that man can attain.

Diving is no different from any of man's other ventures. He will continue to dive deeper and stay longer, to exert his will against nature, to extend his sphere of influence in the universe. He cannot be stopped, he can only be slowed down - and that only by violating his freedom.

Diving is a matter of personal choice. Those who would pass laws or implement rules to prevent underwater explorers from exercising their basic rights of life would have denied Columbus the opportunity to discover a New World, Magellan from circumnavigating the globe, Amundsen from attaining the South Pole, the Wright Brothers from flying, mankind from reaching out for the stars.

While it is true that we, as individuals, have limitations in what we can accomplish, our personal limitations should not be used as standards for those whose limitations are unequal. One cannot justify grounding the albatross because the penguin cannot fly.

Those who cannot, should not. *But*, those who cannot should not place obstacles in the path of those who can.

I touched a sore spot with a branding iron. Immediately the analogy of the penguin and the albatross became a clarion call to arms. Tourist divers became associated with waddling penguins, and technical divers (who used to be called gorilla divers) flew aloft with the graceful albatross. Academically, deep diving was taking on a whole new look of respectability.

## THE DEPTH OF CORRUPTION

All human endeavors are characterized by the people who indulge in them. Technical diving was no different. Good things were happening - but so were bad things.

Once again, the *Andrea Doria* figured in the middle of a controversy. Once again, Bill Nagle was the instigator. His major claim to fame was organizing the search for the ship's bell, in 1985. The bell that we recovered was located in the stern above the auxiliary steering station. John Moyer, who participated in that expedition, held a nagging doubt that the bow bell - which was missing - was stored somewhere in a forward compartment. The reason for his belief was subsequent research and correspondence with some of the *Andrea Doria's* crew members.

Moyer envisioned a grandiose scheme to vacuum the mud out of the targeted compartment with an air lift. This was quite an undertaking, one that could not be carried out alone. So he put up the money to charter a boat, and sought volunteer help from the wreck-diving community to conduct the actual work. I won't go into all the details here because he and I are writing a book about it, tentatively entitled *Andrea Doria: Floating Art Gallery*.

The feature of that multi-year endeavor that I want to cover within the context of the present volume is not the salvage operation itself, but its effect on the conduct of deep wreck-diving. After Moyer's first expedition concluded unsuccessfully, Nagle called both Moyer and me and told us in no uncertain terms that he did not presume that Moyer's expensive and enormous efforts granted him immunity from competition. Nagle made it quite clear that if we found the bell on the next trip, rigged it with liftbags, then were unable to complete the task of recovery - for example, if foul weather forced us off the site - he would have no qualms about putting the last puff of air into the liftbag and claiming possession of the bell for himself.

This self-indulgent attitude contravened not just the wreck-diver's code of ethics, but *any* sort of ethics. It was also an insult to the basis of my long-time friendship with Nagle. To recapitulate, in the early 1970's Nagle unbolted the compass cover on the *Bass* and exposed the compass. During surface interval, I overheard some opportunists making plans to steal the compass from Nagle, either before he could get into the water again or by letting him unscrew all but the last remaining bolt on the gimbal ring, whereupon they would remove the final obstacle and claim the compass for themselves. I told Nagle what I overheard, then recovered the compass myself and gave it to him. But now that the roles were reversed, he chose to be a victimizer. That Moyer had been his friend for many years, and had volunteered his time to help Nagle with the recovery of the *Andrea*

*Doria's* stern bell, held no special meaning to him. Betrayal was a concept that was not in Nagle's vocabulary.

Sad to say, Nagle had developed an arrogance that is normally associated with the rich and famous, perhaps more so with the noveau riche. For example, instead of parking his Porche within the painted lines in a parking lot, he parked perpendicular to the allocated lanes and took up four spots so that no one could park next to him. This action by itself may seem like harmless eccentricity, but it exemplifies his outlook on life in a world which he perceived existed solely for his amusement.

His solipsistic universe was peopled not with human beings, but with serfs, servants, pawns, and inferiors: mechanical constructs whose purpose was to do his bidding whenever he placed his bid. He thought of himself as a grand chess master who tolerated no wishes from the unwashed masses, and who fostered his inviolable right to push people around like pieces on a game board.

By dint of his neurotic perception of magnificent self-importance, he had become intensely paranoid about protecting his material wealth, about increasing his power and sphere of influence over minions of the lower orders, and about claiming ownership to anything to which he took a fancy. He believed that merely because he wanted something, he had the prerogative and divine right to take it. Thus his concept of possession was limited only by his distorted imagination.

Nagle was not alone in suffering from egocentric perception. I've often heard it said that the rich don't think like the rest of us, and I believe it. The brash pursuit of money and power - which are touted as the American dream of success - too often leads to evil manipulation. Those with affluence or political persuasion have the power to purchase or destroy as they see fit. And they can do it legally, as long as they are willing to sacrifice a portion of their net worth which is beyond the means of the average wage earner to expend in his defense. Those who can't be killed with impunity can at least be demolished financially, then brushed aside like bugs.

Democracy is supposed to prevent unequal treatment of the citizens. Yet the ideal too often fails to succeed when suppressed by those with power who place themselves above the masses. These people are driven maniacally to control the lives of others, and to possess what they, in their delusion, believe they have a right to possess. Humanity, reason, and judgment are fled.

Nagle had gone beyond the bend of reality. He was well on his way toward complete psychosis.

Moyer and I discussed the problem that was posed by Nagle's threat. After long deliberation about the possible consequences of action - and inaction - we agreed that the only way to protect the investment was to stake a salvage claim on the wreck. Commercial salvors and treasure hunters did this regularly in order to prevent claim jumpers from stealing the goods after the finder spent years and millions of dollars to locate a wreck. This practice was unheard of in recreational diving. It had never been done before. Yet no other protective measures were legally available.

Peter Hess agreed to file the claim in court providing that the action would not prevent others from exploring the wreck and from bringing up souvenirs. Moyer had no problem with this. He did not want to alienate the wreck-diving community - his friends - he wanted only to protect his interests.

For the first time in salvage history a claim was filed not on an entire wreck or wreck site and a large surrounding area, but on designated *portions* of a wreck. Nor did the claim seek to prevent others from visiting those areas, only from recovering certain specified items: the bell and selected works of ceramic art thought to still exist.

Moyer's claim created quite a ruckus at first because many of those with a vested interest in the wreck (particularly charter boat captains) jumped to the conclusion of the worst case scenario: that everyone would be excluded from the site. It took some convincing on Hess's part, plus an explanatory letter that he circulated throughout the diving community, that such was not the case. The claim is still in force, but it has not stopped a soul from diving the wreck or recovering artifacts. Hundreds of pieces of china, glassware, and the like have been brought up by divers without any more than a joking murmur from Moyer.

This anecdote demonstrates how the law can be invoked by honest people desiring to protect private interests, without due process being misused or abused in order to violate the rights of others as a way to promote personal large-scale aggrandizement.

We didn't find the bow bell in 1993, but through a major group effort we did manage to recover a pair of half-ton cement partitions, in each of which were embedded grotesque ceramic panels that were the creation of renowned Italian sculptor, Guido Gambone.

## Technical Diving "Standards"

By now mixed-gas was in widespread use as an alternative breathing medium on dives to mid-range depths: that is, for wrecks that had previously been dived on air. (The *Ostfriesland* dive redefined the notion of "deep.") The clearer head that helium provided in the absence of nitrogen narcosis made mid-range dives more memorable - quite literally, as now a diver could make better sense of what he had previously seen through a dull, impaired mind, and could remember more of what he experienced. He could also work more effectively.

## Early 1990's: Mixed-Gas Revolution - 171

Helium opened the gateway to deeper wrecks.

One such deep-water wreck that entered the fold was the *Republic*, an ocean liner that sank in 1909 after colliding with the steamship *Florida*. The depth to the sand was 260 feet. Joel Silverstein, publisher of a slender Long Island magazine called *Sub Aqua Journal*, organized a trip to the site which lay a dozen miles from the *Andrea Doria*. As a crew member, my primary job was to tie in the grapnel. Lander was now my constant companion and tag-along on deep cave and wreck excursions, and helped me set the hook. (Cheryl Novak, my significant other, was proud to be a penguin diver.)

Among others on the trip were Hank Garvin and Lisa Herrera. Garvin sold insurance, Herrera was a chef. Both were experienced deep divers who had touched the *Andrea Doria* more than once. However, neither had ever made a mixed-gas dive, nor did they have any knowledge of the subject. Silverstein made a few desultory comments about the fundamentals of gas switching and nitrox decompression, showed them how to attach sling bottles to D-rings, gave them a print-out of the decompression schedule, and sent them on their way - together and without supervision.

At the risk of sounding impertinent, I was appalled. I didn't doubt Garvin's and Herrera's ability to follow a schedule and to remember to switch gases, but I did fancy that they should have a more structured introduction before descending deeper into the sea than they had ever gone before. I did not object as much to their making the dive as I did to Silverstein's casual invitation to do so, and supplying them with gas. It seemed to me that his attitude of nonchalance was luring them into waters in which they would otherwise fear to tread.

I did not interfere with their plans or in any way try to stop the dive - it was not my place to do so, nor have I ever felt disposed toward telling people what they should or should not do - but I did try to scare them with the potential repercussions of not adhering to the decompression schedule or failing to breathe the appropriate gas at each stage. Both made the dive without a hitch so perhaps my concern was unjustified. Nevertheless, I later went on record by putting my trenchant complaint in writing and by sending it to Silverstein.

Silverstein defended his actions at first, then did a complete turnaround by working energetically toward educating the non-technical diving public on the hazards of being ill-informed. He established a one-person institution which he called the Extended Range Diving Organization (ERDO), through which he sought to implement and seek acceptance for technical diving standards. Moreover, he turned the tables on my criticism by asking me to join ERDO as an officer and technical adviser.

I declined his offer with the following letter:

So there are no misunderstandings, I thought I would explain my position with relation to establishing "standards" by whatever name you choose to call them: guidelines, suggestions, recommendations, or a minimum threshold of experience. I abhor standards of any kind. While the intention of setting standards is usually admirable, once established, standards are too easily abused and too often become "maximums" instead of "minimums." Standards imply limitations, and limitations I cannot accept in any form. Whenever one person or group of people decide to make the rules by which others must abide, a quantum of freedom is lost. In that respect I believe in absolute individualism: let each person decide for himself, based upon his innate ability, what is a sufficient level of training and experience. I know this concept is heretical in a culture that prides itself on rules and regulations, but I must condemn any attempt to force upon people another's ideals.

Education and informed consent I applaud, but abject conformance to an arbitrary assessment of a person's ability to perform goes beyond the pale of instruction. If edification in its purest form is truly your goal, then I wish you success.

Set standards for exploration? Regulate explorers and their activities? If OSHA had existed at the time of Columbus he wouldn't have been permitted to leave the dock.

Technical diving was barely in its infancy, and already the field was fraught with self-proclaimed, self-righteous know-it-alls who wanted to control the course of the activity in their own image. (Silverstein was not one of these. I always felt that his heart was in the right place.) On one hand I objected on purely philosophical grounds to establishing standards and regulations. On the other hand, not only did I think that we didn't have all the answers, I didn't think that we knew all the questions.

The way to improvement was through continued practice, especially practice that stretched one's skills. Do as others do at first, then make refinements or extrapolations as necessary. The process is continual - it will never end because new techniques and better equipment will always come along. Divers entering the technical field should be encouraged to go beyond their teachings, to test fresh ideas, and to try creative ways of dealing with old situations. Social anthropologist Margaret Mead said it all: "Children should be taught *how* to think, not *what* to think."

## DISSENSION IN THE RANKS

Clayton took some flak in this regard. After leading half a dozen trips to the Billy Mitchell wrecks he acquired a small but loyal following. Those who began as support divers were now diving the wrecks on gas: Greg Masi, Doug Sommerhill, Harvey Storck, Ron Tolbert, and others. They were a step or two behind on the technical diving treadmill. Don Koonce became crew chief - relieving Clayton and me of assigning tasks to support divers so we could concentrate more on the conduct of the dive - then joined the ranks as a gas diver. Even the *Miss Lindsey's* captain, Mike Hillier, became a gas diver.

Some of them confided in me about the disharmony that Clayton was causing with his unbending attitude and strict adherence to prototype gear configurations. I acted as mediator and tried to make him understand how he was alienating his supporters. Leadership through example and encouragement is more productive than control through conviction of superiority. A leader should be a catalyst. Clayton had no idea that he was perceived as a martinet until I revealed the fact to him. The logician side of his personality enabled him to see beyond his limited point of view, if not necessarily to change his behavior.

He and I came to grips on another issue. Mike Boring and Doug Buckley made a successful dive on the battleship *New Jersey*, off North Carolina. I heard about it through the grapevine. My friendship with Boring went back to the 1970's. Buckley I knew largely through reputation. We had met once or twice and I was impressed by his expertise. I called Boring and congratulated him, and asked him to extend my congratulations to Buckley.

Clayton was incensed, perhaps apoplectic, over their accomplishment. He said they had no right to make such a dive because they had no previous mixed-gas experience, had not come up through the ranks (meaning *his* ranks), and had not incrementally increased their depth range. I reminded him that Boring had been diving the *Andrea Doria* before Clayton was even certified to dive, that Boring had made many times more deep dives than Clayton had ever made, and that Clayton had dived the *Ostfriesland* without prior mixed-gas experience and without having made intermediate depth excursions. My dialectics fell on deaf ears.

At one time the *Ostfriesland* represented the epitome of deep wreck-diving. It held that position in the limelight for a couple of years. Searching for the remaining German warships was a grand and worthy cause. But by this time in the evolution of technical diving, these wrecks had been pushed onto a sidetrack so the main line could pass. Divers the world over were pulling themselves up by their bootstraps, and they didn't need us to tie their laces.

## The "Hot" Mix Concept

John Chatterton, at first skeptical about mixed-gas diving, was now one of its leading proponents. (To be fair, we were all skeptical at one time.) He organized a three-day search for the *Norness*, a tanker that was torpedoed by a German U-boat in World War Two, south of Long Island, New York. Hang numbers indicated that the wreck lay at a depth of 280 feet.

I showed up on the boat with heliox leftover from a 380-foot dive that had been canceled: one set of doubles and a single sling bottle (each with a capacity of 120 cubic feet). Instead of dumping several hundred dollars worth of gas, then refilling with a mix that was chosen for the shallower depth, I connected a transfill hose between the full set of doubles and an empty set, equalized the pressure, then topped off with air from the boat's compressor. This yielded a trimix blend which, while not the optimum gas for the depth, was adequate. I analyzed for oxygen content, then ran a schedule on a laptop decompression software program - one of several which were now on the market and which were replacing absolute reliance on Hamilton's DCAP.

I used the same procedure on the single tank, equalizing its gas in an empty single, topping off with air, then banding the two singles together so I could dive "splits." Without knowing it, I had just invented the "hot" mix concept. Fortunately, other people on the boat that day were more perceptive than I was, and realized the potential of making mixes that were intentionally "hot" in order to decrease the number of tanks that were needed for overnight dive trips. (John Yurga reminded me of this a couple of years later. He took special notice of what I did as an expedience and helped implement the procedure as a routine.)

By utilizing hot mixes you can make five dives with three sets of doubles. One set contains the precise mix required for the depth, the other two contain mixes which are high in helium and low in oxygen, such that when each "hot" set is equalized with an empty set then topped off with air, the desired final mix is obtained.

## The Utility of Fear

The *Norness* dives did not go well for all. Clayton, Lander, and I planned to dive as a threesome. On the first dive the current was so strong that Clayton exhausted himself before reaching the anchor line and had to abort. On the second dive, Lander had trouble clearing her ears and didn't make it to the bottom. On the third dive, neither one of them made it down, leaving me to explore the wreck alone. No sooner did I switch on my light than it went out. On the previous dive my primary light had failed, and this one was my back-up, supposedly fully charged.

Ambient light on the high side of the wreck was about twenty feet. I had intended to go into the rooms in the after structure. With no light, I made a quick change of plan to check out the debris field below the deck house. A massive fishing net lay draped across the angled deck like the web of a giant, undersea spider. Sticky, clutching strands could easily snare an unwary diver or his "danglies."

At 220 feet I swam slowly over the edge of the hull. I fine-tuned my buoyancy until I hovered perfectly neutral above the net. I burped air out of my BC so that my rate of descent coincided with my forward progress across the deck, which sloped down at about forty-five degrees. There was no perceptible current, yet the net seemed to billow and beckon like a Siren calling the name of her next victim.

I was clear-headed and in control of the situation. But for some reason that I still don't fully understand, I was gripped with a palpable, hypnotic fear - as if I were in the throes of an acute anxiety attack. This feeling of foreboding could not have been induced by narcosis, nor was anything awful occurring that should trip my instinctive defense mechanisms. My best guess is that I was experiencing a manifestation of "potential anticipation." That is, my subconscious mind was preparing to respond to the eventuality of entanglement.

I have often said (and written) that fear is a useful emotion. Many times fear has kept me out of trouble, kept me from being overbold, kept me from entering situations that might be harmful to my well-being. Irrational fear is useless, panic is counterproductive, but rational fear has a definite survival benefit.

My heart pounded like a jack hammer threatening to burst from my chest like the fetal creature in *Alien*. I worked hard to rationalize my fear - neither to yield to it nor to overcome it, but to take advantage of it.

I reached out and touched the thinly encrusted net. My fingertips and outstretched arm kept my body and auxiliary tank a safe distance from the strands. I kept imagining what it would be like if a sudden current shoved me into the mesh - my imagination reran the scene again and again. *That*, at least, was a very real fear. I moved cautiously downward.

I came to the lower edge of the deck, which overhung the bottom because the hull angled back from the perpendicular. Still there was no perceptible current so my fears were in some measure groundless. I dropped straight down to the sand, using the vertical wall of fishing net as a guide. Now my heart pounded harder than before - it was even painful, as if each throb was actually a fierce, external compression. The blood gushed loudly past my ears with every beat, like a slow-motion drum roll.

I checked my gauges and equipment. I had plenty of gas, my regulator was functioning flawlessly, my sling bottle was full of nitrox - I could find no reason for not continuing my exploration. With my heart still pounding painfully, I slipped under the hanging net and moved underneath the hull and deck house. Enough ambient light was reflected off the sand to enable me to see dimly about ten feet ahead. I reached the juncture where the hull touched the sand. The hull was smooth and unbroken. There was no debris field.

I retreated to my touchdown spot, ascended slowly to the net-covered deck, then worked my way diagonally upward and back to the grapnel, where my heart resumed its normal beat and intensity. During the conduct of a flawless dive there was no presentiment of danger. Yet I can't help but admit that, for some unfathomable reason, I was scared half out of my wits. Perhaps the extra caution that was inspired by my fear is what prevented me from getting into trouble.

## OFFER OUT OF THE BLUE

A couple of weeks later I made some dives on the *Monitor* that yielded eminent satisfaction. After years of battling NOAA over the right of public access, NOAA now completely reversed its position by pleading for Peter Hess and me (co-organizers of the expedition) to recover artifacts which a multi-million dollar NOAA-sponsored expedition failed to retrieve. We conceded gracefully. NOAA's only prerequisite was that each item be photographed or videotaped in place before removal.

Because I had an eye for artifacts I became the spotter: I took pictures while others made the recoveries. I didn't bring up any artifacts myself. Even so, I fully savored the irony of the situation. A dinner plate and a batch of mustard bottles were turned over to NOAA for preservation, and ultimately for public display.

Yet all the year's experiences paled by comparison to a phone call I received from a plucky lass in England. Her soft-spoken words led to a new plateau in the annals of wreck-diving history:

"How would you like to dive the *Lusitania*?"

Ken Marschall's artistic rendering depicts how the *Lusitania* appeared in 1993 during the National Geographic film shoot. When The Starfish Enterprise dived the wreck in 1994, the fishing nets held aloft by floats on the port side were no longer upright. The last two letters of the name were covered by the net shown adjacent to the bitts until John Yurga cut it away. (Courtesy of Ken Marschall.)

178 – The *Lusitania* Controversies

Opposite page top: In 1983, when Steve Gatto and I entered the *Andrea Doria's* dining room and recovered this china plate, the first one we spotted, we started a chain reaction that has been growing ever since.
Middle: Bill Nagle torching the shaft of the helm on the *Ioannis P. Goulandris*, in 1985.
Bottom: The author and Bill Nagle in happier times, when we were called Prima Donas. I was Prima, he was Dona. That's the *Andrea Doria's* bell behind Nagle's left shoulder.

This page top: The author and Gary Gilligan holding the plastic slate that *Seeker* patrons cable-tied to the gate that they locked at the entrance to the third class kitchen. We are holding a cup and saucer that we recovered from a nearby storage cabinet. (1989)
Middle: Steve Gatto and Tom Packer - two of wreck-diving's brightest luminaries - and the *Ethel C.'s* hydraulic helm and bronze stand.
Bottom: The rocket I rode on the *U-2513*.

180 – The *Lusitania* Controversies

Above left: Pete Manchee with plates he recovered from the *Andrea Doria*.
Above right: Pete Manchee, the author, Ken Clayton, and the Explorers Club flag over the *Ostfriesland*.
Below: The drift decompression system that I developed for the 1990 *Monitor* expedition. It is more simplistic than the "dec" station we used later on the *Lusitania*.

Photographic Insert – 181

Top: At left is the *Araby Maid's* capstan. Under the thick encrustation lies the brass cover shown at right, recovered in 1990.
Middle: From the *Andrea Doria*, one of two half-ton cement partitions embedded with ceramic panels that were the creation of Guido Gambone.
Bottom: After twenty-one years and a circuitous route through life, I finally obtained my cave diving certification.

## 182 – The *Lusitania* Controversies

Above: Polly and Simon Tapson.
Below: Nick Hope and Christina Campbell.

Photographic Insert - 183

Above: Jamie Powell and Richard Tulley.
Below: Paul Owen and Dave Wilkins.

# 184 – The *Lusitania* Controversies

Top: John Chatterton and the author with the bell from the *Sebastian*, recovered a few weeks after the *Lusitania* trip.

Middle: Barb Lander and John Yurga calculate fractional gas components for nitrox fills.

Bottom: Polly Tapson and Nic Gotto sitting on the engine cover of the *Sundancer II*. That is Gotto's dog in the background.

Photographic Insert – 185

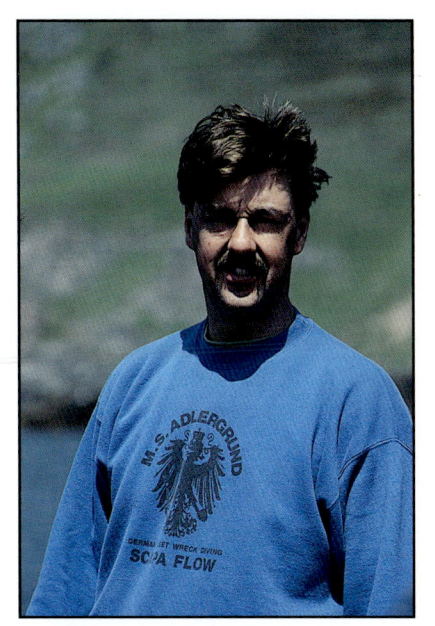

Above: Rich Field and Rob Royle.
Below: Tony Hillgrove and Chris Reynolds (in the red weather suit).
The diver balanced on the transom is waiting for the signal to roll overboard.
Divers had to get their legs and fins out from under the bench before rolling.

## 186 – The *Lusitania* Controversies

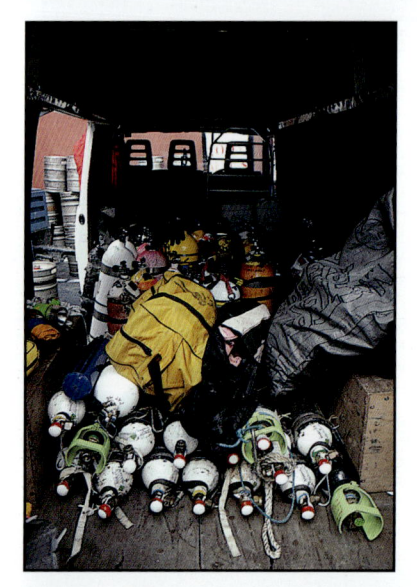

Top: The Peugeot lorry and the Tapson's minivan parked by the courtyard adjacent to Kieran's Folk House Inn.

Middle: Double tanks filled with bottom mix and cylinders of oxygen for decompression are stowed in the Peugeot lorry. Drysuits and kit bags will be packed on top.

Bottom: The gateway to the courtyard. The dive shop entrance is inside and to the right. The compressor is located all the way back, where the white tank is standing.

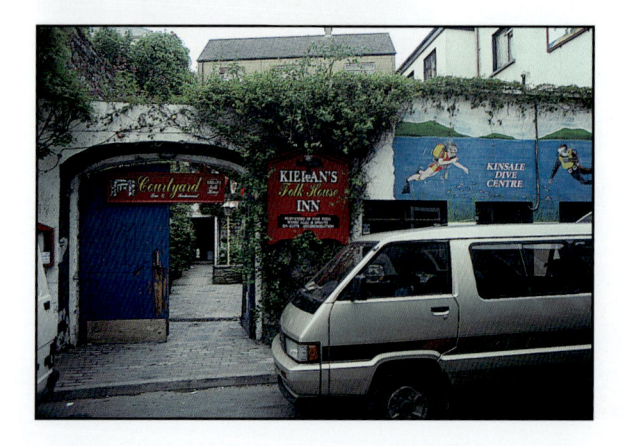

Top: The plywood structure that housed the storage cylinders. Between the open doors and the lorry is the trailer-mounted low-pressure compressor. The aluminum beer kegs are empties from the bar. The gateway to the courtyard is fifty feet to the left of the picture. Tanks being moved between Haskel Land and Compressor Country were carried along the sidewalk.

Middle: Simon Tapson and Nick Hope are transferring gas with the Haskel booster.

Bottom: Initial rigging of the decompression station in the crowded courtyard.

188 – The *Lusitania* Controversies

Top: Kinsale harbor.

Middle: A mud flat at low tide. Be careful where you moor your boat.

Bottom: Mountains of equipment laid out on the concrete launching ramp. The buoys, PVC pipes, and black oxygen bottles were stowed in plastic buckets, then loaded aboard the red rigid inflatable boat. The boat with the black hull and white upper works is the *Sundancer II*. The dark lower portion of the launching ramp is covered with slippery green algae, which caused more than one person to slip on his or her bum.

Photographic Insert – 189

Top: Barb Lander and Paul Owen administer oxygen to Simon Quilligan after he was rescued from the frigid water by the captain of a fishing boat.

Second from top: The rescue boat, carrying the three fisherman who were plucked out of the water and several members of The Starfish Enterprise, glides past an old stone battlement on the way into the harbor. Note that the remains of the house on the hill are completely covered with vegetation.

Third from top: Quilligan was still in the hospital, but the other two survivors stopped by the folk house to have a few drinks with their rescuers. They are posing with Paul Owen and Barb Lander.

Bottom: The overgrown, unkempt cemetery where some of the *Lusitania's* victims are buried. Crouched, John Chatterton photographs one of the mass burial headstones. Looking on are John Yurga and Richard Tulley (holding the baby). Jamie Powell, in the background wearing red, is looking for other headstones.

# 190 – The *Lusitania* Controversies

This page top: High current is indicated by the fact that this recessed porthole is not filled to the top with silt. Note the small fish casting a shadow on the uncracked glass.
Middle: Planks pressed against the inside of this porthole indicate that the interior decks have been squashed together.
Bottom: Most portholes had the glass blown out of them. As a ship sinks, incoming water compresses trapped air which then bursts through the hull's weak points.

Opposite page top: Nick Hope and Christina Campbell shackled the stern shot through the broken glass of this window.
Middle: The remains of a porcelain commode, tile floor, and wooden walls of a rest room.
Bottom: A rectangular window with a ventilation grille and the control rod which opens and closes the grille. The window is upside down. Note the small crab.

Photographic Insert – 191

## 192 – The *Lusitania* Controversies

Photographic Insert – 193

Opposite page top to bottom: Three views, each one closer than the previous, showing the design of the colorful floor tiles of the rest room whose commode is pictured on the previous page. Note the loose cables, dime-sized sea anemones, miniature white starfish, and partially preserved wood. Look for the two-inch flatfish in the middle photo.

This page top and bottom: Two views of the docking telegraph resting on wooden decking in the debris field on the sea bed. In the middle photo, Captain Turner is standing next to the same telegraph on the *Lusitania's* after bridge. (From the author's collection.)

194 – The *Lusitania* Controversies

Above: Diving in less than ideal conditions: considered marginal by some, calling for abort by others.
Below: Different day, better conditions. The five floats are the buoys for the decompression station.

Photographic Insert – 195

Above and below: Climbing the ladder and handing up sling bottles. Swinging around the top of the ladder and over the transom was a tricky maneuver that required a helping hand. That's Polly Tapson and Howard Weston in the RIB, Nick Hope helping divers into the *Sundancer II*.

**196 – The *Lusitania* Controversies**

# Photographic Insert - 197

Opposite page top: Wearing quads, Rob Royle swims toward the downline float.
Middle: The blue rope was tossed to returning divers because the wind blew the boat along faster than they could swim. They held the rope while removing lights, cameras, and sling bottles. The retroreflective tape on the hoods is a safety feature in case of the need for nighttime rescue.
Bottom: The skiff in the background replaced the RIB after it was sabotaged.

This page top: Becalmed by sabotage on a perfect diving day. In the marina, eager divers hang out while Nic Gotto repairs damage to the *Sundancer II* and Howard Weston works on the RIB.
Middle: Nic Gotto works on the engine with help from Rob Royle (looking up). Paul Owen is standing, John Yurga is sitting on the transom.

Left two photos and below: The man who spied on The Starfish Enterprise.

## 198 – The *Lusitania* Controversies

The "dec" station, or decompression station.

Top: Because the corners have not yet been connected, the station lies in a string with the terminal pipes hanging.

Middle: The blue ropes that crisscross from opposite corners prevent the "square" from collapsing into an acute parallelogram. Note the buoys floating on the surface.

Bottom: After the station has been squared off, divers can hang anywhere on the pipes between the corners, or grip the supporting ropes at any desired depth.

Photographic Insert – 199

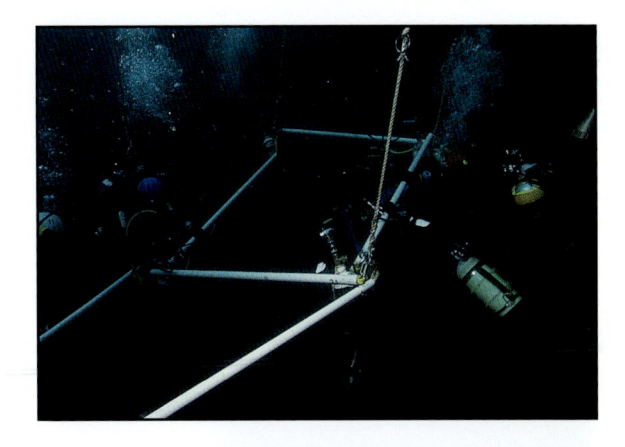

The "dec" station, or decompression station.

Top: In its final configuration, the station looks like the home plate in a baseball diamond. The stainless steel circle ring on the nearest vertical support rope is for clipping off the oxygen cylinder at a shallower depth.

Middle: Each position on the station holds two oxygen cylinders, for one team's use.

Bottom: The triangular extension creates a position for the fifth team.

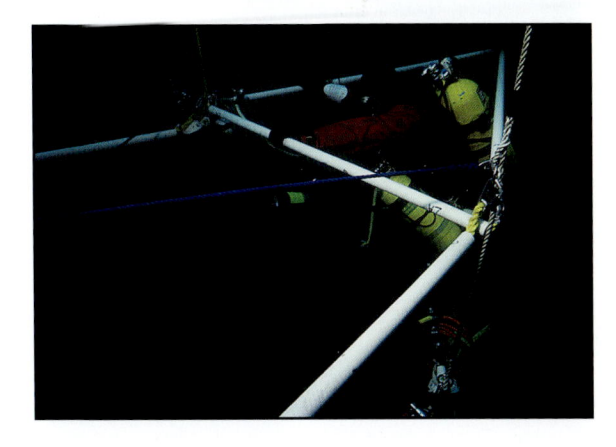

## 200 – The *Lusitania* Controversies

Photographic Insert – 201

Opposite page top: Whoops! A length of pipe from the dec station unclipped itself and fell to the sea bed. The middle and bottom photos show other portholes located at the forward end of the wreck.

This page top: This photo of a net was taken when visibility was only eight feet.
Middle: Less obvious than the orange filter feeders is the small dark crab below them.
Bottom: Nets in the wheel house area were avoidable. Note the red starfish near the guideline's secondary tie-off.

## 202 – The *Lusitania* Controversies

This page top: What looks like a railing might be water pipes for a shower. Note the orange broad-shelled crab.
Middle: Half of a shallow bowl with a decorated rim. Note the orange sea urchin in the right background.
Bottom: A giant shackle on the forecastle deck.

Opposite page top: The letter N from the name on the port bow.
Middle: A swivel-type porthole that rusted out of the superstructure.
Bottom: A loose deadlight.

Photographic Insert – 203

## 204 – The *Lusitania* Controversies

Above: The *Lusitania's* wheel house. (Eric Sauder Collection.) The hydraulic steering mechanism, called a telemotor, is shown behind the wooden door and in the underwater photo below. Note the annunciator in the upper left corner of the topside photo.

Above: The telemotor lies atop the annunciator. Below: A closer view shows the annunciator needle pointing to AHEAD, which is engraved in the lower rim but which is obscured by encrustation. ASTERN, engraved in the upper rim, is clearly legible. These photos establish for the first time that if Captain Turner did order the engines reversed, as he claimed, the order was not acknowledged and was therefore not carried out. The *Lusitania's* high speed contributed to the high death toll because several lifeboats capsized upon launching.

206 – The *Lusitania* Controversies

Above: Group photo taken at the end of the trip. Kneeling from left: Simon Tapson, Jamie Powell, Christina Campbell, Nick Hope, John Yurga, Barb Lander, and the author. Standing from left: John Chatterton, Paul Owen, Richard Tulley, Polly Tapson, and Dave Wilkins.
Below are two artifacts recovered several weeks later: at left is the helm stand from an unidentified freighter off Delaware, at right is the bell from the *Sebastian*.

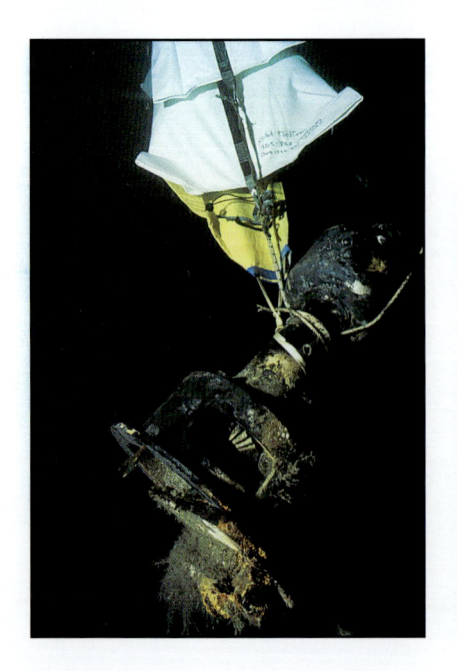

Photographic Insert - 207

Top and middle: These bottles on the *Monitor* were recovered after being photographed and videotaped. The ruler and the PVC pipe were placed there for scale.

Bottom: The results of a firebombing in my driveway in the pre-dawn darkness just hours before I planned to leave for Virginia, where I was slated to testify at the *Lusitania* hearing. The interior of the car was gutted, the glass was shattered by the heat, and the house's plastic storm windows were melted by flames that gushed up the brick wall. My vehicle was parked just a few feet away. Was the bomb intended for me or was the proximity in space and time just a coincidence? Before getting into my vehicle I examined its undercarriage and engine compartment - paying particular attention to the ignition system - and steeled myself for further threats and attacks.

## 208 - The *Lusitania* Controversies

Painting by Ken Marschall.

# Part 8
# Flashback: Historic Interlude

## JOHN LIGHT BUYS THE *LUSITANIA*

Book One left the *Lusitania* in the hands of John Light, in 1962. At that time he vowed to return to the wreck. "We're too close to the end to quit. We're just too close."

Despite Light's sincere pledge, he never did return - nor did anyone else for twenty years.

But Light did not stop dreaming about the *Lusitania*. He continued to explore the ship through the archives. Over the years he amassed a wealth of material which he insisted contained evidence that substantiated a conspiracy to sink the liner. If such a conspiracy existed, it was German, not British. Yet Light persevered in his beliefs. What kind of documentation he uncovered remains unknown to this very day. He never released any of his information, nor did he publish the book that he claimed to be writing. His enduring declaration was that the final, absolute proof of his theories lay within the sunken hull.

One wonders why, if he possessed official documents that confirmed his claims, he needed verification from the wreck.

Be that as it may, in order to ensure his right to pursue an ambition turned obsession, in 1967 Light went so far as to purchase salvage rights to the hull.

Due to unpaid loans, at the time of her loss the *Lusitania* was more than half owned by the British Admiralty. (Cunard still had twelve years out of twenty to pay off the Admiralty subsidy.) Cunard was paid for its portion of the ship's value by the British-based insurance syndicate, Lloyd's of London. Lloyd's was reimbursed by the Liverpool and London War Risks Insurance Association (sometimes referred to as the War Reclamation Board), which was a government-owned instrumentality. Thus ownership of the sunken "hull, engines, tackle, apparel, and appurtenances" devolved upon the British government.

For this giant chunk of rusting metal Light paid the princely sum of £1,000 (less than $2,500 U.S.). That none of this metal was precious was obvious. Said Light to inquisitive and misinformed reporters, "Would I get her for 1,000 pounds if there was gold aboard?"

Light didn't buy the *Lusitania* because the wreck was worth anything monetarily, but because he wanted exclusivity in the pursuit of his single-minded goal: to solve the "mystery" of her sinking.

## The Best Laid Plans . . .

The next step for Light was to obtain financial backing for his exposé. He had already learned that there was little he could accomplish on a shoestring budget. Through contacts in the publishing industry, he managed to convince Holt, Rinehart and Winston to advance the funds necessary to mount a major expedition. For its money, Holt, Rinehart & Winston expected to publish the book that Light agreed to write.

Light had also learned that there was little he could accomplish by diving the wreck while breathing air. At 300 feet, nitrogen narcosis induced mental aberrations that were similar to intoxication: dulled thinking, poor powers of observation, reduced motor skills, and so on.

While air diving trips were simple and inexpensive to mount, they had so far yielded nothing in the way of photographs or useful observations. Meanwhile, helium breathing technology had come into its own. With the money in hand, Light decided that the only way to produce effective results was to dive the wreck while breathing helium.

It would be informative at this point to give a brief overview of helium diving, since Book One left the subject in its primitive state, in 1939, with John D. Craig.

## The Discovery of Helium

In 1868, astronomer Pierre Janssen was conducting a spectroscopic analysis of the sun during a solar eclipse when he noticed a yellow Fraunhofer line that had never been seen before. Later, another astronomer, Norman Lockyer, suggested that the line represented the wavelength of a hitherto unknown element. Lockyer named the element "helium," which is derived from the Greek word for "sun."

Not until 1895, however, was helium discovered on Earth (by William Ramsay) and its existence accepted by the scientific community. The helium atom was found to consist of but two protons and two electrons, and, in its most common isotopic form, the nucleus contains two neutrons.

After hydrogen, helium is the second most abundant element in the Universe. Ten percent of the sun's mass is made up of helium. But on Earth helium is extremely rare. The quantity of helium found in the atmosphere is one part per 200,000. Most helium produced commercially is extracted from natural gas deposits found in the United States. The process of extraction is costly.

Helium is diatomic, which means that two atoms of helium combine to form a molecule in which the electrons are shared. This covalent bond is difficult to break. In its natural state helium is an inert gas: that is, one of the group of noble elements which strongly resist forming bonds with other elements.

Of all the elements in nature, helium has the lowest point of liquefaction. It can be liquefied only under incredible pressure. Liquid helium vaporizes at a temperature less than one degree above absolute zero. (Absolute zero is the point at which all heat energy is lost and molecular motion stops: theoretically, the point at which matter ceases to exist.)

## Heaven-Sent Breathing Gas

The property of helium that is relevant to diving is its disinclination to induce narcosis. Helium has no narcotic effect down to depths of at least 1,000 feet. Thus a diver breathing a mixture of helium and oxygen will suffer no mental impairment from the effect of inert gas on the brain. This attribute was theorized by Elihu Thomson in 1919.

Initial experiments in the breathing of helium were conducted by the U.S. Bureau of Mines, the government agency that controlled the majority of helium available on the world market. The U.S. Navy was brought into the experimental fold in 1924. Laboratory animals breathing a mixture of helium and oxygen showed no detrimental side effects. Later, human subjects were tested in a hyperbaric chamber.

By 1937, divers had reached a simulated depth of 500 feet in a chamber. As related in Book One, John D. Craig heard of these experiments and determined to develop a mixed-gas diving rig to use in the filming and exploration of the *Lusitania*. Progress in this regard was slow due to the paucity of scientific validation and to the primitive state of gas delivery equipment. Then war intervened and the project was canceled.

The first practical application of helium-oxygen breathing mixtures took place in 1939, during the salvage of the submarine USS *Squalus*, which was recovered from a depth of 240 feet. Improvements in understanding and equipment accelerated after the war. Military and civilian testing programs proliferated. Several fatalities occurred as divers attempted to exceed previous depth records.

In 1961, Hannes Keller and *Life* reporter Kenneth MacLeish descended to a new world depth record in Lake Maggiore, Switzerland: 725 feet. The following year, Keller and Peter Small attained a depth of 1,000 feet in the open ocean off the coast of California. Unfortunately, Small died on the way to the surface. (MacLeish is the same reporter who accompanied John Light on his last dive to the *Lusitania*, in 1962, and who was seriously bent. See Book One.)

## Saturation Diving

Divers engaging in deep-diving operations wore the standard hard-hat outfit: a waterproof canvas suit, a breastplate, a copper helmet fitted with glass ports, a pair of lead boots, and a heavy lead belt. Hard-hat divers walk along the bottom, they do not swim above it like scuba divers. Hard-hat divers are fed a continuous supply of gas from the surface, either from a compressor or from high-pressure storage banks. This gas is delivered through an umbilical hose in a never-ending supply.

For shallow depths or for short working times, a hard-hat diver is lowered to the work site on a platform or by a rope. When his stint is done, he is then hauled up to decompress in the water or in a deck decompression chamber.

For deep dives or for long working periods, the decompression becomes so lengthy that it is not practical to decompress after every dive. Divers then "saturate": that is, they stay down until their body and bloodstream are saturated with inert gas, to the point at which no more inert gas will dissolve in the blood or tissues. Instead of decompressing after every dive, the diver instead remains pressurized by "living" inside a bell until the job is done, perhaps for days or weeks. In this manner work shifts can be many hours long. Decompression at the end of the job may require several days inside a chamber.

John Light decided that a saturation diving system would enable him to achieve his objectives with respect to the exploration of the *Lusitania*.

## THE *KINVARRA*

Light's grandiose scheme called for outfitting a salvage vessel to transport the saturation system to the wreck site. With proceeds from his book advance, Light moved to Kinsale, Ireland, purchased the fishing trawler *Kinvarra*, then began converting and outfitting the vessel at the same time that he started designing and building his saturation diving system. For a year and a half Light lived aboard another fishing trawler, the *Doonie Brae*, while as many as a dozen divers and technicians proceeded with the work.

According to Paddy O'Sullivan, a local diver who was well acquainted with Light at the time, one of the first people Light hired was Hannes Keller. Keller brought with him two technicians. "The next two years saw a complete saturation system being built and assembled off the pier in Kinsale under his guidance. A team of men from Verolme dockyard in Cork also worked on the salvage vessel *Kinvarra* to carry out the various modifications necessary to convert the ship for diver support duties on the *Lusitania*. The *Kinvarra* was fitted out with a diving bell and living chamber as well as a lifting gantry to handle the diving bell.

"A helium tank was installed aboard and a helium reclaim system. Vast arrays of gas analyzers, chromatographs, and instrumentation were incorporated in the system. Television cameras were attached to diving equipment and monitors at command centres. U.S. Navy Mark 8 helium back packs were provided for the divers in conjunction with the latest hot water suits."

Many tales have been told about the *Kinvarra's* subsequent activities: most of them sordid, some of them true, few of them recorded, none of them substantiated. This author does not wish to speculate on the accuracy of unwritten accounts. Suffice it to say that the *Kinvarra* did no one any good, that local merchants still complain about the company's unpaid debts, and that the mixed-gas saturation system never got off the ground - or into the water.

In August 1968, a United Press International story commented on Light's enigmatic character, "Although the little seaport town of Kinsale has followed his work with intense interest over the years, his relationship with the folk around here is somewhat touchy. Light minds his own business - and expects everyone else to do the same. He refuses to accept as a fact that the salvage of the *Lusitania*, with all its overtones of drama, is everybody's business. Nor has he been too polite in expressing his viewpoint.

"One notice hung on his trawler sometime ago read: 'Time of Departure - when I'm damn well ready.' Another spelled it out even clearer: 'Time of departure - mind your own damned business.'

"The folk here don't seem to mind this too much. They grin and talk of 'the tough American,' and they wish him luck. When is Light going to begin the actual salvage? 'Any day now,' he tells you. In fact, Light has been saying 'any day now' for many a day but, around the wharves and jetties of the old town, they are beginning to believe something is astir."

Notwithstanding the above, after a couple of years of energetic, almost zealous labor, Light's plans for diving the *Lusitania* on helium dissolved into thin air.

O'Sullivan provided the finishing touch: "Endless teething problems and equipment failures occurred and eventually the venture ended in liquidation without a single saturation dive being made to the *Lusitania*. The *Kinvarra* headed for Amsterdam and the auctioneers hammer."

## Publication Revival

*The Military History of the Lusitania*, by Louis Snyder, was published in 1965. Despite the pretentious title, the 70-odd pages of large-size print merely gloss over the pertinent facts in passing. The text is childishly written, somewhat in the manner of a grade school term paper. Not worth reading.

In 1967, a slender, factual account of the sinking was written by

Donald Barr Chidsey. *The Day They Sank the Lusitania* makes no pretensions of erudition, yet it covers all the main points of the controversial sinking without resorting too much to cliché. The book is readable and entertaining without being thought-provoking. A good minor effort.

Although John Light did not succeed in his plans to dive the wreck in order to answer once and for all the question of the *Lusitania's* supposed mystery, he continued his archival research in London for many years. His dedication exerted an influence on the *Lusitania's* notoriety, and inspired at least one author - with whom Light shared his opinions as well as the results of some of his research - to write the book which Light never got around to penning. The author was Colin Simpson.

Published in 1972, *The Lusitania* was attended by a massive publicity campaign that was aimed toward propogandizing a new and innocent generation that had lost touch with the liner's contrived historical background. *Life* magazine devoted its cover to the book and featured a long excerpt that accentuated Simpson's paranoid delusion of conspiracy. The unfortunate result of this blitz of overwritten exposure was reinforcement in the public consciousness of either inept handling by Admiralty seniors or nefarious deeds performed by willful British warmongers, even in the minds of subscribers who never read the book.

*New York Times* book reviewer Ross Pollack let Simpson off easy when he called the book "flawed," and cited a few instances in which Simpson's unsubstantiated allegations were contradicted by sworn testimony. In my opinion, the book's neurotic view of the sinking and the subsequent controversy was a paean to commercialism and a blatant exploitation of the long-denounced theory of conspiracy. Simpson clutched at the slenderest of straws to make his points, then filled in the logic gaps with wild speculation and twisted extrapolations of coincidental events. The book is a classic example of the hypothesis that ironclad evidence can be defiantly distorted in order to reach totally unfounded conclusions.

Furthermore, the book is confusing because it jumps back and forth between parallel pretexts: if the reader is unable to accept the premise of intentional political intrigue, how about Admiralty bumbling of protection and culpable mishandling of rescue operations?

Adding fuel to Simpson's fire of conspiracy was the 1972 television documentary entitled *Who Sank the Lusitania*, produced by the British Broadcasting Company. It was more of the same.

Thomas Bailey and Paul Ryan were incensed by Simpson's freedom with the facts. In 1975 they published *The Lusitania Disaster*, an admiral refutation of Simpson's convoluted logic, and an honest, workmanlike attempt to put the matter straight. The book was no diatribe, but a reasoned and reasonable affirmation of the truth of the situation. Sadly, their book did not receive the distribution that it so richly deserved. The market was already flooded with Simpson's wretched volume of misinformation and *Lusitania* mythology.

If the reader yearns for a firm acquaintance with the true particulars in the case of the liner's sinking, I highly recommend *The Lusitania Disaster*.

## OTHER PUBLICATIONS

*Seven Days to Disaster*, by Des Hickey and Gus Smith, was published in 1981. The book is a partially fictionalized rendering of the *Lusitania's* final voyage: strictly derivative a la *A Night to Remember*. That is, the primary focus is on the wealthy passengers. Into the mouths of the characters the authors inserted dialogue which was not possible for them to know. Interesting if innocuous.

Perhaps the most insightful chronicle of events leading up to the *Lusitania's* loss is included in Patrick Beesly's *Room 40: British Naval Intelligence 1914-1918*, published in 1982. Beesly served with the Admiralty's Naval Intelligence Division during World War Two. Not only did his service grant him special comprehension into the workings of military intelligence, his position gave him access to documents from World War One. He devoted a 40-page chapter to the *Lusitania* incident, and covered the intelligence aspect from all angles. The book is well worth reading in its entirety, or the *Lusitania* chapter out of context.

*"Lusitania"* (italics included) was published in 1986. It is a large format volume that does not fit on bookshelves - what is referred to as a "coffee table book." The first paragraph, written by editor Mark D. Warren, describes the contents perfectly: "This book is primarily a facsimile reprint of a rare, limited-circulation volume, originally compiled in August 1907 from a series of five articles that had appeared in the magazine *Engineering*\*, and published as a commemorative souvenir to mark the launch and forthcoming maiden voyage of the RMS *Lusitania*. To date, it has remained the most comprehensive published record of the construction and launch of the *Lusitania*. (\**Engineering: An Illustrated Weekly Journal*. Published in London 1 and 8 June 1906, 12 and 19 July and 2 August 1907.)" The book is replete with photographs: exterior views of the hull during various stages of construction; interior shots of the decks, public rooms, and passenger accommodations; and pictures of the machinery before installation; plus deck plans, diagrams, and cutaways. A must for any *Lusitania* aficionado.

*R.M.S. Lusitania: Triumph of the Edwardian Age* is a wonderfully illustrated and vividly described volume by Eric Sauder and Ken

Marschall, published in 1991. The large-format book is a testament of the *Lusitania's* fine appointments: a treatment which is so often overlooked in comparison to the controversial sinking. A collection of rare photographs, many of them interiors, that is definitely worth having.

## OCEANEERING SALVAGES THE *LUSITANIA*

John Light's dream of salvaging the *Lusitania* became a reality in 1982. Unfortunately for Light, he had nothing to do with the project. Perhaps for his peace of mind it was just as well, for the salvors had pinned their hopes for success on the bullion in the specie room and jewelry in the purser's safe.

The salvage operation was funded by a consortium of investors: the British Broadcasting Company, the American Broadcasting Company, the London *Times*, and Oceaneering International. The BBC produced a documentary on the salvage operation; it was called *The Lusitania File*, and aired on television in 1983. ABC produced a segment for *20/20*. The London *Times* reported events for its British readers.

Oceaneering has been touted as "the world's largest underwater contractor," and justifiably so. No job was too big for an outfit that maintained bases all over the world, employed thousands of divers and topside personnel, owned a large fleet of salvage vessels, possessed a mountain of diving equipment, and operated a plethora of remotely operated vehicles. If there was any outfit alive that could succeed where others had failed, Oceaneering was the company that was equal to the task.

For some, a project like salvaging the *Lusitania* would be considered a major operation. For Oceaneering, it was just another job. Oceaneering conducted saturation diving operations all the time. For the company's highly trained commercial divers, the *Lusitania* was not a daunting underwater prospect but rather another day at the office.

## OBSERVATIONS OF THE WRECK

The three-phase operation that reportedly cost £150,000 began in the spring. Phase one was conducted between April 22 and May 1. It consisted of a photographic survey which was carried out by a sophisticated ROV (remotely operated vehicle). The ROV, called *Scorpio*, was operated from the support vessel *Myra Vag*. The mobile underwater platform was equipped with a video camera, high-wattage lights, and a mechanical arm.

Phase two was conducted between July 27 and August 30. During this phase, in addition to continuing the video survey, *Scorpio's* manipulator arm was brought into play. During one of its passes along the bow, *Scorpio's* camera recorded the broken foremast and the crow's nest. Wrote Dan Knowlton, "Amazingly, just above it, still firmly attached by its bracket, was the signal bell, last used on the afternoon of May 7, 1915, by seaman Quinn when he spotted the oncoming torpedo. In a tricky two-hour operation, *Scorpio* succeeded in bringing it to the surface, the first item of the expedition to be recovered." Unfortunately, it wasn't "the engraved ship's bell which was never found."

*Scorpio* also returned a view of the wreck that had not been seen by human eyes in twenty years - and returned that view with clarity unfogged by nitrogen narcosis and unprejudiced by expectancy and preconceived notions. The hull was only lightly encrusted with marine fouling organisms: a veneer so thin that the rivets stood out of the collapsing hull like plump reddish mushrooms. Crabs and conger eels made their homes on the wreck while cod swam leisurely overhead. White, miniature starfish stood out starkly against the rusty, monochromatic background. Dime-sized anemones adhered to the steel plates like large thumbtacks scattered across a rouge, low-pile carpet.

The *Lusitania* was no Caribbean reef effulgent with Technicolor marine life - it was more like a sparsely inhabited desert.

The black and white video images were not as revealing as the visual interpretations made by divers later that summer, during phase three - the salvage phase that was conducted between September 10 and October 25. Six divers went into saturation. They were "stored" at depth inside a chamber that was welded to the deck of the salvage vessel *Archimedes*. The pressurized chamber was their home. A diving bell transferred the divers in pairs to the wreck for eight-hour stints. While one diver worked on the bottom, the other tended his umbilical hoses from inside the bell. Rotating in turns, the three pairs of divers could work literally round the clock.

One diver who was interviewed extensively was Graham Mann. In all, he spent some sixty hours on the wreck and more than twenty days in saturation. He offered some interesting observations and annotations.

"In the physical sense the diving on the *Lusitania* was very easy. It was very comfortable. You went down in your elevator - your bell - you had hot-water suits. . . . Of course, the sitting in the chamber can be pretty monotonous. If you were a very active sort of person you could crack up quite easily as you really are on top of all your fellow friends who are in there.

"The depth to the top of the hull varied on the *Lusitania* from about 235 feet to 245-250 feet. You see the wreck throughout its length is in various stages of collapse. At the moment, if you imagine the wreck to be lying on its side - the starboard side - all the decks would be vertical all the way down the length of the wreck . . . well the only place it is like that is right on the forepeak and right at the after end. In the middle the wreck's got a tendency to sag in slightly.

"The bottoms throughout the rest of the wreck have got a tendency to fall out on the seabed. Consequently your decks are still connected and the bottom's got a tendency to rack over in that direction and everything comes down like a pack of cards. In places the decks, instead of being about nine or ten feet apart, are down to about two feet high.

"We found several very large boxes. Of course underwater it's impossible to tell what they are, but we managed to get one of them out of the for'ard hold and when they opened it on the surface it was full of suspender-belt clips! It was a big box about a metre square, too! These must have been a big demand!

"As the work went on the visibility dropped until, towards the end, it was down to one-and-a-half metres. There's no way, even at the best of times that you can see her as a whole ship. She's huge - and still recognisable as a big ship. She's not that badly damaged. We got quite far inside the wreck on several occasions.

"We found the bridge. We found her steering wheel. We got a foghorn up which still worked. I've done a lot of work on wrecks. This one was interesting in that it was the *Lusitania*, but basically it was no different from any other wreck.

"It was the shoes we found that really made me think of the passengers. So did a pillow. Some glasses. A dress. . . .

"I can tell you one thing. You can't see the torpedo hole, and probably never will. It's on the starboard side - and she's lying on it. There's no way that any of us came off that wreck with any conclusive evidence of massive explosions taking place. . . .

"There were some things up for'ard that didn't perhaps tie in with the ship sinking the way she did. I don't mean the fuses we found. When we opened them on board the ship, they were empty. Those fuses didn't look as though they had ever held explosives to us. We didn't find any explosives on board either, except for some World War II hedgehog depth-charges.

"In the for'ard section there are a few things that don't tie up in the wreckage, but in the war you used to get submarines hiding alongside wrecks in the Western Approaches and consequently they had a lot of depth charges dropped on them. The stern of the *Lusitania* has been blown off, I presume by depth charges - it's too sort of random to be a salvage job. It's certainly been dived before. I pulled a diver's weight-belt out from inside the wreck!

"The wreck at the bow had some odd things about it. There was chain from the chain-locker lying on the seabed. I'm at a bit of a loss to explain that. Usually it stays inside the wreck. The chain-locker itself is a very substantial part of the ship and it's unusual for that to

# Flashback: Historic Interlude – 219

get knocked out on the seabed."

Dan Knowlton, who interviewed Oceaneering executives, wrote that about midships and "about half way down the length of her hull, a crack about two feet wide runs all the way from the sheer line at the top of her hull to her rounded bilge. This is a structural break, the result of settling in the sand.

"Upon reaching the area just below the bridge toward the waterline and below, horribly tortured, jagged and torn plates angrily bend sharply upward from the hull and define a massive hole that extends across the bottom and both sides of the hull for a great distance forward. Crossing over this black cavern, even amidst all this damage, the fine taper of the hull as it narrows towards the bow, can still be distinguished. Moving again toward the top edge of the hull, it is found that the bow lettering has suffered a similar fate as that on the stern."

In the last sentence, Knowlton was referring to a previous statement: "Just below the rail at the extreme stern, holes are found where brass letters once proudly identified her as *Lusitania*. Close examination reveals that electrolysis caused the steel to break down allowing the letters to drop out. No doubt they lie buried in the sand beneath the stern."

At least one of these observations is at variance with the truth. I saw and photographed the letters on the bow in 1994.

Knowlton made other statements that were contradicted by Mann's assertions. Whereas Mann noted that the torpedo hole was hidden because the wreck lay on top of it, Knowlton claimed to have information that "the area damaged by the torpedo and secondary explosion seemed to be centered farther forward than previously believed. . . . the damage seems to be centered a bit forward of the bridge, more in the vicinity of the main forward hold. Something blew out the bottom and a great deal of both sides in this area. Oceaneering's explosives expert, Alf Lynden, said, 'This hole could only have been caused by a massive internal explosion of something stowed in the hold.'"

Dan Knowlton was not involved in the salvage, nor did he visit the site while the operation was in progress. Apparently, he conducted his interviews via telephone with people at Oceaneering's headquarters.

He noted further, "When operating on the foredeck, *Scorpio* came across an 8x8-foot hole that had been cut into the deck. It had apparently been used to access the smaller foremast hold just ahead of the main hold. *Scorpio* was carefully guided down this opening. Upon reaching the bottom, Jim Highlander, the ROV operator topside, exclaimed, 'It's empty. Clean as a whistle! Somebody's been in and cleaned it out!' (Later operations would prove that the area was not barren as first thought although a great deal had been removed.) In addition, a series of triangular marks were scratched deeply into the plating. This was believed to have been caused by a three-toothed salvage grapple."

I do not wish to discount Knowlton's information - just point out that it disagrees with other reports. My personal attempts to access Oceaneering's records have met with failure. The company claimed that all records relating to the *Lusitania* job have been trashed. I wonder what would cause the company to do such a thing.

## THE GOLD BUG

The primary incentive for the salvage operation was the recovery of gold that the investors believed to exist. By a method of negative reasoning, they argued that the gold shipment was left off the cargo manifest purposely so that Germany would not find out about it. This argument is not persuasive with respect to logic - it begins with a conclusion then fits the lack of evidence to support it. By arguing in a circle one can "prove" virtually anything: ESP, UFO's, political honesty, and so on.

The fact is that people continue to believe that there is gold aboard the *Lusitania* because that is what they want to believe. They find comfort in the hope that the wreck is a treasure trove worth millions of dollars or pounds, that it is an investment worth making for the profit that must eventually be realized. They refuse to accept that investing in a sunken ship is much like investing in sinking stocks and bonds.

This is not to say that there are not any wrecks that have yielded a fortune in gold and jewels. But for every treasure salvage operation that has earned a profit for its investors, a thousand have left them at a loss.

## SAVED FOR POSTERITY

As far as the actual work was concerned, the divers wasted a great deal of time in burning and blasting open the specie room. They found no bullion but they did find gold of a sort: buried under a thick layer of mud were 361 brass boxes containing rare and valuable chronometers.

Part of the *Lusitania's* cargo consisted of more than one hundred tons of sheet brass and copper ingots. The divers brought up samples, but with the high cost of daily operations it was not economically feasible to salvage the lot.

Also recovered were portholes, heavy filigreed brass windows, a carved wooden balustrade, dozens of Wedgwood plates bearing the Cunard crest, earthenware, dishes, glassware, chamber pots, legible printed matter, reels of film with visible images, wood, marble pillar bases, a shovel, lead curtain weights, buttons, silver watches with and without covers, several hundred gold watch cases, a lady's pince-nez, knives, forks, teaspoons, vent units, baking dishes, bowls, mugs, bottles, wooden picture frames, wall brackets, brass door frames, a pen holder, patterned glass, a brass mask, floor drain valves, 813 brass fuses, and hundreds of other miscellaneous items of lesser import. Quite a haul.

Perhaps more interesting as well as more valuable were 8,000 souvenir spoons retrieved from the specie room. The spoons had originally been silver plated, but the plating had been dissolved by the corrosive nature of the sea. The spoons were dated 1881, and the handles were embossed with the name "Kitchener" under a bust of the famous British general, statesman, and, at the time of the *Lusitania's* loss, secretary of state. Ironically, as if in tragic premonition, the Kitchener spoons were sent to the bottom only a year before their namesake, Lord Kitchener, was drowned, when the British cruiser *Hampshire*, on which he was traveling on a diplomatic mission to Russia, was torpedoed by a German U-boat, and sunk.

In addition to bringing up thousands of small, hand-picked artifacts, five massive objects were raised - objects that only a big commercial salvor such as Oceaneering had the wherewithal to recover: two 20-ton iron anchors and three of the four 16-ton bronze propellers. Each propeller was separated from its shaft by means of a 30-pound explosive charge. A crane on the *Archimedes* winched the mammoth items onto the deck.

Divers were unable to find the purser's safe due to collapse of the superstructure where the safe was supposed to be located. Wise heads might remark: no loss.

The average depth of the sea bed was given as 315 feet, with 320 being a maximum recorded at high tide in the washouts. Oddly, the divers asserted that the wreck lay on its side and that the decks stood vertical. This assertion is contradicted by a wealth of previous and later observations. In actuality, the liner lies tilted to starboard with a list of less than 45°. Perhaps the superstructure, which was sloughing off, twisted and compressed to such a degree that it made the wreck appear to lie on its side - especially under conditions of limited visibility when only small portions of wreckage could be seen at any one time: like picture puzzle pieces that must be reassembled mentally afterward. Or perhaps the divers, whose expertise lay not in the use of precise descriptive language, did not describe their observations accurately.

Despite Mann's comments and first-hand observation, there were still those who chose to interpret the process of natural deterioration as damage from an explosion, either from an internal source or from Admiralty divers in the 1950's removing incriminating evidence.

In the final analysis, nothing new was learned about the ship and no mysteries were solved - perhaps because no mysteries ever existed except in the minds of the gullible.

## The Auction Block

The value placed upon the items recovered from the *Lusitania* was $3.4 million. And this without gold and jewelry! What an item is truly worth, however, is what the market will bear - not what is printed arbitrarily on a price tag. Treasure salvors are wont to give inflated values to items as a way to convince the purchasing public that they should pay the price being asked, or as a way to attract investors.

All too often, the bids received at auction are much lower than expected (or wanted). Items may take years or decades to sell. An object valued by the salvor at a certain amount may still be in the salvor's possession ten or more years later, because no one can be found who is willing to pay the exorbitant price. A treasure's actual monetary worth lies in how much cash it eventually generates. Nor is this all profit, as it does not take into account the operational expenses or the costs of recovery, preservation, restoration, and marketing.

The unengraved bell was fancied to fetch at least £15,000 at Sotheby's, but it didn't sell. Nor did the ship's steam whistle sell at £3,000.

One value placed upon each bronze propeller was a whopping £50,000. There were no takers.

The public is a fickle buyer. At a private auction held at Oceaneering's headquarters in Aberdeen, Scotland, a number of items were unloaded at prices far below what they were anticipated to bring. The sales figures are a matter of company record and privileged information - not for common consumption. But there is no doubt that for the salvors who invested so much time, effort, money, and expertise, the *Lusitania* salvage project was a constructive loss. The undertaking cost more than the sale of recovered items returned.

Eventually, one of the propellers was donated to the Merseyside County Museum. The second languished in Oceaneering's yard for years, unable to find a home; rumors are rife, but I have not been able to verify exactly what happened to it. Noted *Sports Illustrated* about the third: "Thanks to a company in the British Virgin Islands called Toronto Consolidated Ltd., you can now not only own a piece of history but also chip out of a bunker with it. The firm has melted down a propeller salvaged from the wreck of the British liner *Lusitania* . . . and forged 3,500 sets of golf clubs from it. . . . At $9,000 per set, the clubs have to be one of the most expensive and least appropriate 'memorials' to come along in a while."

Said the managing director of Toronto Consolidated, Alan Koenig,

"We looked into a couple of other things - fireguards for fireplaces with a relief of the *Lusitania* on them, and miniature boats - but golf clubs have a universal appeal. We could have made tacky souvenirs, I guess." How many of these sets of golf clubs were actually sold is a matter of a different company's records.

It seems neither fitting nor proper that such a grand work of endeavor should come to so an ignominious end. As later recorded in Lloyd's Law Reports, "It seems likely that this is a case in which the expenses of the salvors have exceeded the value of the wreck recovered."

In the final analysis, it must be acknowledged that the items recovered from the *Lusitania* found their true evaluation in human terms.

## Status Quo

Thus was the state of affairs with respect to the *Lusitania* when I received the fateful phone call which ended the previous chapter. After the short spell of notoriety that was granted to the wreck by Oceaneering's monumental salvage operation, the famous liner once again slipped into a cocoon of blissful obscurity.

More than a decade passed before the *Lucy* made the headlines again, in yet another bid for immortality.

# Part 9
# Grand Adventure on the *Lusitania*

### THE FAN CLUB

In 1991 I received a letter from Simon Tapson, an executive search consultant in London, requesting information about my books. He had read my *Advanced Wreck Diving Guide* and "applied quite a lot of your wreck diving advice to my own wreck diving in English waters." My other titles were not available in the U.K., so he wanted to know if he could purchase them direct from me. "Of course," I responded.

Months later I got a phone call from his wife, Polly, who was an independent film producer as well as an active wreck-diver. She stroked my ego by claiming to represent the U.K. chapter of the Gary Gentile Fan Club. (The American chapter was unwittingly inaugurated by Cheryl when she issued dive completion certificates to the participants of my 1990 *Monitor* expedition, and signed them "The G.G. Fan Club.") The U.K. chapter, said Polly, consisted of only two members: her and Christina Campbell (a head hunter, but not an anthropophagus). Polly sent me photographs of each of them.

I was delighted by the compliment but did not get a swelled head about it. Polly kept in touch.

### THE STARFISH ENTERPRISE

When the *Ultimate Wreck Diving Guide* was published in the spring of 1992, Polly ordered a copy immediately. Afterward she called a few times just to chat, mostly about wreck-diving and the advent of mixed gas as a way to avert narcosis. One of Polly's group disappeared on a wreck in 190 feet of water. Her body was never recovered. The assumption was that narcosis contributed to her demise.

Then came the *big* call, in August 1993. The first words out of Polly's mouth were "How would you like to dive the *Lusitania*?"

If this were a novel I would have optimized the dramatic effect of Polly's trenchant call by foreshadowing her question with some well-defined characterization about my zealous, life-long ambition to dive the wreck. But this story is nonfiction, and the truth is less theatrical. Any longing to dive the *Lusitania* was lodged covertly in the back of my mind - way back. It was a secret and subconscious longing at best.

As I wrote to Polly later, "I have toyed for years with the idea of diving the wreck, but never considered it seriously due to the logistical impracticalities of arranging an overseas charter. It's not that I've had a burning desire to dive it, but certainly an interest that was more than passing." My research collection consisted of half a dozen books on the *Lucy*, plus a file folder filled with newspaper clippings and magazine articles.

In the foreword to *Andrea Doria: Dive to an Era*, I wrote, "People asked to name some famous shipwrecks invariably rattle off the big three: the *Titanic*, the *Lusitania*, and the *Andrea Doria*. . . . The *Lusitania*, at a depth of 320 feet, has been dived by commercial divers wearing cumbersome and expensive oil rig apparatus, breathing a mixture of helium and oxygen, and tethered to a surface support vessel by means of umbilical hoses. Only a few divers in the world are qualified to operate at this depth."

These less-than-prescient words saw publication in the spring of 1989, at a time when I was already planning to dive the battleship *Washington* at 290 feet; and only a year before the dive on the *Ostfriesland* at 380 feet. In the mean time, after exceeding the depth at which the *Lusitania* lay, I had written the book on mixed-gas diving. Yet such are the foibles and temporal incongruities of events and the human mind that I hadn't even considered the overall possibilities beyond the projects in which I was involved. So many wrecks, so little time . . .

The implacable advance of technical diving was accelerating at an ever quickening pace. Odd as it sounds, although in some respects I was leading the charge, I was so deeply internalized that I lost perspective of the activity's true potential. Or perhaps it isn't possible to be omniscient.

Simon and Polly Tapson were part of an elite group of British divers who had chosen the lofty goal of conducting the first technical diving trip to the *Lusitania* - indeed, the first scuba diving trip of any kind since John Light's zealous endeavors some thirty years earlier.

The name that Polly created for the group was The Starfish Enterprise.

## An Offer I Couldn't Refuse

The *Lusitania* trip had been conceived the year before, when the concept of mixed-gas diving gained publicity and momentum. At first the group contemplated diving to the high side of the wreck on air, and decompressing with an accelerated profile breathing blends of nitrox and oxygen. They nixed that idea when their friend disappeared. Then they thought about using rebreathers, but decided that the field units were too unreliable at the current stage of development.

By the summer of 1993 the approach was fully formulated and the logistics were being researched: cost quotes, helium availability, boat charters, lodging, and so on. At the same time, the members of The Starfish Enterprise dedicated themselves to a year of intensive training. None of them held mixed-gas diving certifications or had ever dived on helium. They felt that suitable instruction was unavailable in England. The best alternative was to teach themselves - not just to dive on helium but also to blend the bottom mixes and decompression gases.

At this point the *Lusitania* trip became a full-scale expedition more like a military operation in its planning and eventual execution. Leading The Starfish Enterprise by the force of her powerful personality was Polly Tapson. She acted and delegated responsibility more like a senior fleet admiral than a young dive master or trip leader, utilizing the same skills in coordinating the project that she used in the complicated production of a film. If the individual members were disparate parts, Polly was the glue that held them together. The Starfish Enterprise coalesced into a cohesive unit in which everyone held equal status and shared the growing work load. Polly handled research, logistics, and organization. She was also the appointed spokesperson.

It was then that Polly called and made an offer I couldn't refuse. The UK GGFC - now expanded to include all the members of The Starfish Enterprise - proposed to pay for my plane ticket if I would join the team for the dive. It was a dream come true.

It didn't take a horse's head in my bed to force my consent.

## Gelling and Culling

Each member of The Starfish Enterprise had a role to play in bringing off the trip. Not so for me. Geography precluded me from acting in any capacity other than advisory - and there was little advice that I could give that the "Brits" (as I called them) needed. They knew what they were about. Perhaps my biggest contribution was sending them my oxygen analyzer when they still had so much equipment to obtain. Or perhaps I offered some inspiration by showing confidence in the project. I felt like an extra thumb with very little to offer.

At this time all the participants had not yet been selected. The core group felt that dedication was a prerequisite. Unless a person was willing to devote sufficient time and energy to the training program - and shoulder his share of the workload - he would not be considered as a finalist for the trip. The Starfish Enterprise wanted no laggards. Nor did they want a stranger taking over.

Out of respect for other trip leaders, it has always been my policy not to interfere with the workings of a trip that was not my own. I maintained that position throughout the planning stages and ultimate execution of the *Lusitania* project. It was *their* expedition. I was an invited guest. But the group held some very real concerns about other divers in England who, because of their reputation in local diving circles, might *expect* to be invited - particularly when a foreigner was invited - and whose ambitions and personalities might predispose them toward assuming control. The core group did not want to have their project subverted by politics and egomania.

Potential conflicts were avoided by adopting a policy of discretion. No public announcements were made, the press was not advised, and all knowledge of the project was held in confidence by the members. The goal of the expedition was to gain personal satisfaction from the accomplishment, not to prove a point or to glory in the limelight.

As the core group of The Starfish Enterprise expanded in order to flesh out the required number of participants needed to handle the workload and make the project economically feasible, Polly let me know that another spot might become available. Could I recommend anyone from the States? If so, this person would become my buddy on the trip.

Naturally, the first person I thought of was Ken Clayton.

## NUMBER ONE CHOICE

Despite our close ties and evident bond of friendship, welded by excursions into deep and troubled waters, Clayton and I had dissimilar personalities. We were also motivated by far different goals.

I was an explorer at heart: diving shipwrecks primarily for the thrill of the adventure, and secondarily for the purposes of discovery and history. In one sense, diving the Billy Mitchell wrecks was a matter of historical research for *Shipwrecks of Virginia*. In another, developing mixed-gas diving techniques and inventing new technologies served as field work for the *Ultimate Wreck Diving Guide*. Those were my rationales: my diving has always been subservient to my writing. But I can't ignore the overriding desire to go where no one has gone before, to see what human eyes have never seen, and to overcome the challenges of my own limitations: a purely personal confrontation with nature.

Depth played no part of the measure of my ambition.

Those who knew Clayton believed that his sole criterion for diving wrecks in the extreme depth range was to go deep for the sake of going deep. In response to this allegation, Clayton always countered by claiming that he was misunderstood, that in actuality he wanted to fathom frontiers that had yet to be explored, that he wanted to extend the boundaries of deep exploration. Sometimes his actions appeared to belie this statement, such as when he remonstrated so defensively whenever my depth gauge registered deeper than his. Or when he emphasized his desire to find a wreck - *any* wreck - that exceeded 500 feet. He openly acknowledged that he was willing to die in the attempt to dive such a wreck. His only concern was that his body might never be found, and that, without his depth gauge for proof, the world would never know how deep he had attained.

Tom Packer spoke for all of us when he responded to this declaration by saying, "I'm not going on any dive that I don't expect to come back from."

In fairness, I think it might be more accurate to say that Clayton had a multifaceted mind, one that wasn't limited by a single driving force, but which synthesized a complex array of goals and inclinations into a single, multipurpose objective.

Further troubling my relationship with Clayton was my greater reputation. Whereas once he relied strictly on my renown to lend authenticity to our endeavors, now he felt that I was overshadowing him. He discussed with me his feelings of disquiet, but I was powerless to help him out of his perception of being slighted, and he refused to do anything to further his cause by writing accounts and giving lectures. He remained publicly silent, grumbling privately instead that others should recognize his accomplishments for what they were

His feathers were ruffled to no end by writers and reporters who interviewed me and published biographical profiles in national newspapers and magazines, because they noted my dives on the Billy Mitchell wrecks without mentioning him as my partner. I was always careful to give him credit, but that information seldom appeared in the final copy. In general, mention of the *Ostfriesland* was merely parenthetical to the main theme of the piece, the thrust focusing on the larger scale of my experiences.

He felt that our measure of recognition should be equal. What he failed to take into account was that my reputation was built on more than a mere handful of mixed-gas dives. I had been making deep, decompression dives for over twenty years, I had been lecturing and writing articles for more than a decade, and tens of thousands of wreck-divers had learned the rudiments of their arcane activity from the *Advanced Wreck Diving Guide*. Either he didn't understand or he refused to accept the true basis of the inequality of our renown.

In my own writings I gave him equal status, both in books and magazine articles (for which he was grateful), but that didn't satisfy him. Nevertheless, we remain good friends to this day because we are so much alike in our ethical and philosophical outlooks.

Clayton declined my offer to dive the *Lusitania*. He pleaded lack of funds along with his intention to continue the search for the remaining Billy Mitchell wrecks - a project which, because of my many other interests and activities, I perceived as a sideline rather than an all-consuming objective. I didn't need to dive every wreck any more than I needed to scale every fourteener in Colorado. How often did climbers overlook lower and more challenging summits simply to achieve an artificial goal? The same concept applied to shallower, more interesting shipwrecks.

## Alternatives

If not Clayton, who?

Manchee was out of the running due to severe financial restraints. I knew from general conversation that most others among our small clique of technical divers were unavailable for a variety of reasons: cash flow shortages, work commitments, limited vacation time, family obligations, and so on. The one person in the group who was not constrained by any of these restrictions was our staunchest supporter and fastest learner, Barb Lander.

Although she was the least experienced with respect to the number of years diving and mixed-gas dives conducted, she was the most enthusiastic. She had become a mainstay on our trips to the Billy Mitchell wrecks, and had come up through the ranks in a remarkably short time. Additionally, she had dived the *Andrea Doria*, the *Norness*, and crewed with me on the *Republic* trip, on which she had helped me set and pull the hook. She was so eager to learn that she dived with me most of the time. She was practically my shadow, which led to no end of innuendoes about the true nature of our relationship.

Lander often mused that the greatest benefits of her marriage were "unlimited checking" and "total free time." (Her words, not mine.) By this she meant that she could spend as much money as she wanted and could travel anywhere, any time. As a child she described herself as a "tomboy"; now she thought of herself as a "jock" (again, her words, not mine). She certainly played the part. She jogged, lifted weights, was highly competitive, and stated desperately and without shame that she wanted to be recognized for her achievements (such as they were).

I discussed her qualifications with Polly, and she agreed that if I had confidence in Lander's ability, she had no objections. It was my call. Lander was enthusiastic about the opportunity to participate. Perhaps ecstatic.

Explicit in my instructions to her was the admonition to avoid all mention of the trip. We did not want to brag about intentions which, due to occurrences beyond our control, might not eventuate. Nor did we wish to tip our hand to the wreck-diving community in general because others - once they realized the practicality of such an endeavor - might beat us to the objective.

## Time Saving Device

As a pre-teen I was shunned by my classmates and neighborhood kids in my age group because of my emotional immaturity and delayed physical development. I had to shift for myself. Before I discovered the satisfying dreamland of science fiction and other escape literature, I spent weekends moping around the house whining to my mother, "I don't have anything to do."

I long since outgrew that predicament. The pendulum swung the other way and I became a confirmed workaholic - this despite nearly fifteen years of "retirement." Between writing, publishing, lecturing, diving, photography, and so on, I never had enough time.

I conducted a time management study and found that I spent an average of two hours a day stuck to my desk because of telephone conversations. This either cut into my writing schedule or left other projects undone. I couldn't live this way. I am a completer: when I start something, I want to finish it - not leave it hanging in limbo. Having a cluttered desk or unfinished projects bothered me enormously.

What saved me from being overcome by the workload and enabled me to wrap up all loose ends was the cordless telephone.

Since people on the other end of the line couldn't see me, they didn't know if I was sitting motionless at my desk or wandering idly about the house while we were talking. I took advantage of this conversational invisibility by washing dishes, dusting, and doing general housework - all the jobs that I hated, that were mindless, and that I considered a waste of time when I could have been doing something more productive. I once cleaned the entire bathroom (toilet, tub, and sink) during a single phone call. Because the conversation captured my interest and attention, when I hung up the phone it appeared as if these chores had been done by magic.

The first caller after lunch and dinner might hear the rush or water and the clink of tableware. When I caught up on household maintenance I worked on other tasks: I clipped my nails, folded clothes, put away groceries, packaged book orders, maintained dive gear - anything that did not require thought. I had buckets of artifacts soaking in fresh water. Within a year I had my entire backlog of artifacts cleaned and put away.

From that point it was but a short step to intentionally setting aside work that could be done during calls. I called these projects "phone jobs." These consisted of sorting slides, filing papers, changing light bulbs, mending, and doing any number of vacuous but necessary functions that did not make noise in the accomplishment.

I could then devote all my other waking moments to thinking and writing. Today, with a cordless headset, I have both hands free.

## NATIONAL GEOGRAPHIC ENTERS THE PICTURE

The Starfish Enterprise was not the only group with designs on the *Lusitania*. National Geographic had been toying with the idea for several years, and by 1993 was finally ready to conduct a videographic survey in order to flesh out the production of a television documentary on the wreck.

Staff members had already made a substantial investment in the project, had done a fair amount of research, and had interviewed aged survivors and locals who vaguely remembered the event. The major item missing from the script was a theatrical hook - some mystery to solve or controversy to resolve - in order to transform a conventional documentary into a vehicle for prime time viewing, where the big money was to be made.

The *Lusitania* was no stranger to television presentation. The topic of her sinking had been covered in numerous broadcasts: as an example of German aggression in network newscasts and documentaries on World War One, as a single segment in shows that covered a broad spectrum of shipwrecks, and as the sole subject matter of half-hour specials. These helped to keep the *Lusitania* alive in the public eye.

To produce a documentary that was different from its predecessors, National Geo decided to spend a couple of million dollars to shoot underwater footage of the wreck. This could conceivably show the *Lucy* in a new light: not klieg lights, but high-intensity video lights mounted on submersibles and remotely operated vehicles to which cameras were secured. Ostensibly, this would permit new discoveries to be made about the wreck - perhaps the reason why the *Lucy* sank so quickly after being struck by a single torpedo.

To elevate the status of the field work in the public eye, National Geo termed the two weeks spent on site an "expedition." However, in the motion picture business this second unit production is more correctly referred to as "shooting on location."

The titular leader of this two-week shoot was Robert Ballard.

## The Risk Factor

Ballard was a geologist turned showman who had forsaken the purity of science for the glibness and glitter of the boob tube. His idea of exploration was viewing objects through the lens of a camera that was deployed in the water on a tether, while he sat in the safety and comfort of a submersible or in the cabin of a surface ship. This telemetered videography he called telepresence. My viewpoints and outlook differ drastically from his.

I want to be there physically.

The non-risk-taker might ask in all fairness: why incur the risks of diving the wreck in person when it is safer and easier to lower a camera down to it, or to pass near the hull in a submersible, as Ballard contemplated doing? Why not wait until the show was televised, when I could explore the *Lucy* from an easy chair in my living room? This temperament toward total safety is not without merit.

I have stated it before but I'll state it again. For an adventurer and a true explorer, diving the *Lusitania* in a submersible is a meaningless experience. One might as well land on the summit of Mount Everest in a helicopter. The challenge is not in *being* there; the challenge is in *getting* there. The mechanical assistance of a vehicle - with no unknown factors and with absolute security - is equivalent to taking an amusement park ride. That way is not for me.

## The Shoot Proceeds

In the *Lusitania's* long history, the National Geo film shoot was only the fifth visitation to the wreck (if all of John Light's day trips are counted as one extended exploration). There was a great deal that could be accomplished with an unlimited budget, sophisticated technology, and round-the-clock bottom time spent watching and videotaping the wreck. Although this type of trip could not qualify as adventure, it had the potential to produce a wealth of high quality images for public consumption, and to learn useful information about the wreck's state of collapse.

Details of the shoot were not immediately forthcoming. I followed media coverage of the event, but nothing previously unknown about the wreck was reported. If the underwater cameras spotted anything new and exciting, or if any unanticipated discoveries were made, National Geographic was holding the scoop for show time.

## Site Reconnaissance

Meanwhile, The Starfish Enterprise was making great strides toward the formation of a full-scale technical diving expedition. Polly and Simon Tapson traveled to Kinsale, Ireland on a four-day "recce," as the British say ("recon" in American slang), in order to finalize arrangements for lodging, the boat charter, gear storage, and so on.

They chose as the base of operations Kieran's Folk House Inn, a quaint, three-story rural hotel that catered to the tourist trade. The folk house was situated in the center of town within walking distance of the docks. (In fact, the town was so small that virtually everything lay within walking distance.) Adjacent to the lobby stretched a large, atmospheric bar, behind which a full-sized restaurant offered exquisite cuisine - for guests as well as for visitors. Upstairs accommodations boasted private bath facilities with showers.

A small courtyard was crowded with tables and foliage. An enclosure in front of the courtyard housed a miniature dive shop that had been established only the year before. An air compressor for filling tanks was located in an alcove at the back of the courtyard.

Denis Kieran was a congenial gentleman who owned and operated the folk house, and who hosted the night club after hours when band music lured locals and tourists to the dance floor and the bar. In charge of the diving concession was Howard Weston. Kieran and Weston also provided a twenty-foot rigid inflatable boat, or RIB, which we could utilize as a chase boat. Weston was the RIB's coxswain (operator). The RIB was equipped with a hand-held radio for communication between boats.

The primary surface support vessel (a pretentious scientific term for "dive boat") was the *Sundancer II*, owned and operated by Captain Nic Gotto. The boat was thirty-five feet long - the same size as Bill Nagle's Main Coaster, which we used when we recovered the *Andrea Doria's* bell. The *Sundancer* boasted twin turbo diesels, a boothlike wheel house, and a head or toilet which sat between a pair of bunks in the forward cabin. Nothing plush - strictly utilitarian.

As navigational equipment the wheelhouse was fitted with radar, GPS, Decca (loran), VHF radio, plotter, and echo sounder. A single burner propane stove could heat water for tea - and was much appreciated after a long cold dive. The engine cover occupied half of the after deck.

Gotto was a gem, both as a person and as a boat handler. He also had *Lusitania* experience. Using the *Sundancer*, he had provided water transportation to National Geographic during the shoot, when the submersible support vessel was anchored over the wreck. This meant that he had the precise coordinates of the *Lusitania's* position. To add a further irony, Gotto was Risdon Beasley's grandson. To recapitulate from Book One, Beasley ran a commercial diving outfit which did salvage work for the Admiralty. After World War Two he was accused of secretly salvaging the *Lusitania*. Beasley denied these charges. Gotto affirmed Beasley's denial.

The Tapsons photographed the boats, the folk house, and the narrow streets of Kinsale, which with few exceptions had not been broadened since the days of horse-drawn conveyance. Polly sent me copies of the pictures along with an update on proceedings and the issues that needed resolving. Chief among the latter were rumors she had unearthed during the course of continued research: there appeared to be some conflict and confusion over the *Lusitania's* ownership.

## OWNERSHIP DISPUTE

Although there was no doubt that John Light purchased the *Lusitania* in the 1960's, rivals now contested his claim and some declared ownership for themselves. One of those contesting ownership was British film maker John Pierce, who in the spring of 1982 supervised the initial video survey by means of robotics. Another was

## 234 - The Lusitania Controversies

Oceaneering International, the commercial diving company that conducted salvage operations on the *Lusitania* a few months later. Another was F. Gregg Bemis, heir to a family-owned multimillion-dollar company and now a millionaire venture capitalist living in Santa Fe, New Mexico. What a tangled skein of intrigue.

"Who is Gregg Bemis?" I asked Polly. In all my reading on the *Lusitania* - in books, newspapers, and magazines - I had never come across the name. "I've never heard of him."

Polly had no idea. She was handling the *Lusitania* project with the same verve and thoroughness with which she handled the production of a film. That meant overcoming obstacles known to exist, as well as seeking out potential obstacles that might crop up later, and settling them now so as not to get caught off guard.

In checking out leads on the issue of ownership, Polly learned that John Light had only recently passed away. This implied that his ownership of the wreck should have passed through inheritance to his widow, Muriel. Yet Pierce, Oceaneering, and Bemis insisted otherwise. Who to believe?

Polly researched all the angles, spoke with government agency officials and civilian marine solicitors, and turned up some interesting if unsubstantiated tidbits. For the 1982 salvage operation, Oceaneering formed a syndicate of which Pierce was a part. Later, according to Oceaneering's ex-chairman Ray Govier, Bemis and George Macomber "weaseled" into the syndicate arrangement by claiming that they had obtained John Light's original ownership rights. (Now, who was Macomber?)

Out of this muddle of tri-partisan claimants to John Light's heritage, Pierce and Oceaneering pooh-poohed the whole idea of *Lusitania* exclusivity. They felt that their rights applied only to the items that were salvaged in 1982. They were not concerned with what The Starfish Enterprise might accomplish. But there arose one who was aggressive and vociferous in his demands: Gregg Bemis. Among baby birds, the one that cheeps the loudest often gets the worm. But in civilized human enterprise, claims must be established in a formal legal fashion.

Polly checked into Bemis's allegations. She soon found that the "Department of Marine in Dublin have no record of anyone owning the wreck. As the Irish authorities over the area they would expect anyone owning a wreck to register with them." Furthermore, "No one in Ireland at the Department of Marine or the Sea Fisheries Office have ever heard of Bemis. He has not registered his so-called ownership of the *Lusitania* in Ireland and until proved in a court of law his claims would not be recognized there." The local Receiver of Wreck reckoned that Bemis "has to prove exclusive title and probably couldn't charge us if we dived it and would have to take an injunction to stop us."

What to do? Should Polly contact Bemis and ask for proof of his claim? I thought it would be a good idea if only to forestall any feelings of ill will. Honesty, I believed, was always the best policy. This proved to be poor advice for which I accept full responsibility and blame.

## False Claim

I will leap ahead of concurrent events, and render fat follow-up correspondence - which transpired over a period of months - into a concise presupposition the way Polly and I understood it. Muriel Light's address was unknown to us, and Polly was unable to track her down. Pierce and Oceaneering professed no interest in our proceedings. But Bemis believed that anyone contemplating a dive on the wreck must obtain his permission to do so. Now what?

Polly and Bemis engaged in correspondence and held telephone conversations. Said Bemis, "John Pierce is a delightful crazy Welshman who came up with a crazy scheme for salvage which of course was out-to-lunch type scheme and he got Oceaneering International involved." He said that Pierce was the instigator and promoter of the salvage operation, but that he "never had a piece of ownership, not even an ounce. He did have a piece of ownership of the salvage that was brought up in '82. He did have a share of that."

Polly kept me apprised of the ongoing issue of ownership. I suggested asking for indisputable evidence: a deed, a bill of sale, any legal document that could establish proof of Bemis's claim. In response to her request, Bemis submitted a copy of the opinion of the Queen's Bench Division (Admiralty Court) on the claim of title to the artifacts that were recovered in 1982.

The plaintiffs in the complaint were given as John Pierce and Barry Lister. The defendants were given as Gregg Bemis, George Macomber, Oceaneering International, Hotforge Ltd., and the Department of Transportation (by which was meant the Crown, or the government). Before the case came to trial, in 1985, the claimants and all the defendants except the Department of Transportation "composed their differences." Wrote Justice Sheen, "They now present a united front. I shall refer to all of them collectively as 'the claimants'. The only issue which the Court now has to decide is an issue between the claimants and the third defendants who can conveniently be referred to as 'the Crown'."

Thus the legal issue that was ultimately placed before the court was whether the Crown had any claim against the salvors on the property that was salvaged - this because in ancient British law, which favored royalty over the common person, abandoned items found on Crown property belonged to the Crown. Crown property included the sea bed under territorial waters, which at the time extended three miles from land (the distance that could be protected by cannon fire at the time the law was enacted).

In deliberating the case, Justice Sheen wrote a 9,000 word opinion which can be summarized simply (without supporting case law): because the wreck was not owned by the Crown, and because the wreck lay in international waters, the Crown could not claim title to the wreck or to anything that was salvaged from the wreck. Therefore, all salvaged items belonged to the salvors.

This was the *only* issue that was decided by the Court. Nowhere did the Court render an opinion on the issue of title to the wreck, nor did the claimants seek such judicial review. The arrangement by which the various claimants and defendants agreed to join forces against the Crown was made among the parties involved. Only the signatories were bound to that out-of-court settlement. (Nor did the Court mention the matter of evidence or information relating to John Light's original purchase of and lien against the wreck.)

Additionally, the Court concurred with an observation noted by the Crown, that the Liverpool and London War Risks Insurance Association could sell title to only that property which it insured: "hull, machinery, appurtenances, fixtures and fittings and the accoutrements, loose equipment, furniture and other goods owned at the time of the loss by the Cunard Steamship Co. Ltd. and used in her operation as a passenger liner. At the time of her loss *Lusitania* contained two further categories of chattel, namely (a) the personal property and effects of her passengers, and (b) a quantity of general cargo. These two categories of chattel will be referred to as 'the contents'."

No one owned the *Lusitania's* contents on the bottom. Anyone could legitimately salvage the contents without the permission of the owner of the hull, in the same manner in which a person may keep items picked up from a public sidewalk next to a privately owned building. Thus anyone could dive on or around the wreck regardless of Bemis's presumption of ownership.

If this court document represented Bemis's sole evidence of title, the validity of his claim was nebulous at best. A cereal box top would have had more snap, crackle, and pop as proof of purchase.

## THE TOLL COLLECTOR PARABLE

During the expansion of the American West a man claimed to own the property that bordered a river ford. He charged people a fee for crossing his land. Most pioneers did not doubt his ownership and paid the toll he demanded. When someone protested the validity of his claim, the man offered as proof the people who had already given him money. In fact the man did not own the land, but by sheer audacity and intimidation he made people believe that he did. He perpetuated his illegal racket because no one had the nerve to challenge his claim or demand actual proof of ownership.

In modern society, such a specious claim must be documented or ascertained in court.

## *Empress of Ireland* Analogy

Consider the case of the *Empress of Ireland*. For years a Quebec diver named Philippe Beaudry kept rivals off the wreck by claiming that he had purchased it from the owners, Canadian Pacific Railway. He shouted his claim so loud and so long that people just assumed he was telling the truth. Even the provincial police accepted his claim without any proof. For years he maintained his presumption of ownership by threats and intimidation.

Beaudry and a couple of friends formed a bogus organization whose pretentious name - The Empress of Ireland Historical Society - was intended to give it an air of respectability. In fact, this "Society" had no official standing or sanction - it existed only in the minds of its creators. Beaudry laid claim to being the Society's president, and promoted his chimera until the name appeared in print. It is an unfortunate foible of human gullibility that any item published in a magazine or newspaper is considered to be true. Thus Beaudry and his friends perpetuated their fiction through media naivety.

Under the guise of the Society, Beaudry recovered artifacts from the wreck when they were lying loose outside - easy pickings. He went so far as to recover two skulls, which he prominently flaunted on his mantelpiece.

(As a side note, the Musee de Mer in Rimouski - a museum dedicated to preserving the history and tragedy of the *Empress of Ireland* - displayed a recovered skull in a lighted plastic case. Elsewhere, such a exhibition might have been vehemently condemned: at the very least considered in poor taste, at worst an effrontery to common respect for the dead. Imagine the outcry that would ensue if a skull were recovered from the *Titanic* or *Arizona*. Different cultures, different values . . .)

Some divers scoffed at Beaudry's claim. These were divers whose skill and penetration techniques enabled them to go where Beaudry feared to tread. They ignored his threats and intimidation, dived the wreck without his blessing - and he didn't do anything about it. The situation escalated for several years while Beaudry fumed in impotence - until he brought the matter to a head that resulted in his downfall.

Beaudry finally pressed his claim of exclusivity with the coast guard and police. If he expected such intervention to scare off those he presumed to be trespassers, he made a serious tactical error. Law enforcement officers acted only within their official capacity, not in Beaudry's behalf. They prevented a group of visiting divers from diving on the wreck - temporarily - until a determination could be rendered by the court.

Beaudry's nefarious plan went awry because the court didn't want to hear his shouts and empty threats - it demanded proof of ownership. Beaudry had none. The case was thrown out, the wreck was found abandoned, and recreational divers were permitted to dive and recover artifacts.

I cited these particulars to Polly. Bemis's unwillingness or inability to provide proof of his claim was directly parallel.

## A Contract with Absurdity

Despite his lack of evidence, Bemis still wanted to collect his toll. He sent Polly a six-page contract which he wanted her and all the expedition members to sign. In addition to acknowledging his claim of ownership, the potential signatories had to agree that "Bemis will own all information, images, data and artifacts obtained during the Expedition."

To this absurd statement was added another that was even more absurd: "As owner of the *Lusitania* and all tangible and intangible interests in the shipwreck site (which Participants acknowledge) Bemis is owner of any tangible and intangible items removed, including images." Against this, "Participants will individually and collectively indemnify and hold Bemis harmless for any claims or liabilities arising out of the Expedition."

In short, Bemis wanted all the rights and none of the responsibilities.

"Bemis will be the owner of all copyrights, as well as all photographs, film, videotape and other sound recordings and images made during the Expedition. . . . And Participants each agree individually and collectively not to publish, cause to be published, or assist in publication of, any book, magazine article, press release, television broadcast, video, movie, any any (*sic*) other publication of any form, regarding the *Lusitania* or the shipwreck without the advance, express, written permission of Bemis. In addition and without limitation, Bemis will also have the sole and exclusive right to use any photographs, images and information of any kind obtained during the Expedition . . . Each Participant will provide Bemis at no charge with one full and complete set of all such photographs, negatives, plates, film, video tapes, sound recordings and other images taken by, or at the direction of, the Participant. . . .

"Bemis will control and manage news coverage about the

Expedition and its results. No Participant will enter into any arrangement with any publication, newsgathering or disseminating agency, radio broadcasting, or television or film company, or anyone else, to release news or information concerning, or to create any audio or audiovisual productions about, the Expedition, and all such rights are, and will remain, in perpetuity, exclusively the property of Bemis as owner of the *Lusitania*."

The dollar amount which Bemis wanted to extract from us was not specified in the contract, a space having been left open to be filled in later. But the amount that he suggested over the phone was in the order of several thousand to tens of thousands of dollars.

I told Polly that only a fool would sign such a contract. She readily agreed. Our consensus of opinion was to completely ignore Bemis and his unsubstantiated claims. We were not going to kowtow to Bemis or be intimidated by his threats.

## A Case of the Blabs

Despite my voiced imperative to keep the existence of the *Lusitania* project in confidence, Lander blabbed - and she wasn't even tortured! This betrayal of confidence occurred on a dive boat. Once the balloon was punctured and the mixed-gas had escaped, there was no way to mend the hole and re-inflate the balloon. Soon the projected technical dive became an open secret among wreck-divers.

"Loose lips sink ships," as the World War Two poster proclaimed. They can also torpedo an expedition.

What Polly feared, came true, and almost immediately divers were professing their interest in getting on the trip. No one can be blamed for wanting to get involved. After all, wreck-divers are an adventurous breed, and the opportunity to dive a most historic wreck wasn't one to be passed up lightly. However, the upshot wasn't nearly as bad as Polly had anticipated. She wasn't besieged by phone calls, although she did have to suffer through one onerous interview.

Dave Bright contacted her in order to arrange a meeting. He worked for a pharmaceutical company that did business in Europe. He modified his schedule on an overseas flight in order to stop over in London and meet with the Tapsons. Prior to his arrival, Polly asked if I knew him and what I knew about him.

Bright was an experienced deep wreck-diver with great penetration skills and numerous dives on the *Doria* to his credit. I had invited him to dive on the *Monitor*, which he did. But he was not a technical diver. On the *Monitor* trip I introduced him to the efficiency of accelerated decompression. He pretended to grasp the concept, and breathed oxygen during decompression, but was afraid to surface before his air decompression profile permitted. He lacked faith in the system.

The story he told the Tapsons was entirely different. He claimed that he had made more than ninety dives on the *Doria*, had been certified to dive on mixed gas for over four years, had made some seventy-odd trimix dives, and, in relation to boat handling skills, "he was involved with Howard Klein and Billy Deans moving the dive boat from Long Island to Key West."

His first claim was an exaggeration of at least one hundred percent. His second was an obvious fabrication since mixed-gas certifications didn't exist four years before. The third was pure imagination. The fourth was a blatant lie - I called Klein and asked. When Polly questioned Bright about the name of the wreck on which he had made his check-out mixed-gas dive, he couldn't remember, but thought it might have been a German submarine. There was more, but you get the picture.

It was the concoction of experience that worried Polly with respect to soliciting other divers to join The Starfish Enterprise. She wanted divers who were competent in the water, with established expertise, not those who would try to bluff their way onto the expedition with lies and misrepresentation. That would lead only to trouble.

Polly initially figured on ten divers to share expenses and handle the workload once we arrived on site. When it developed that twelve would make a more equitable distribution, she asked if I could recommend two more. Now Lander's unauthorized disclosure became helpful.

## Filling the Quota

I didn't have far to look. Through Lander I learned that John Chatterton's desire to get on the trip was fervid, zealous, perhaps even hysterical. His customary dive buddy, John Yurga, felt a similar amount of excitement at the prospect but presented a modicum of reserve. Both had outstanding qualifications. They were prime choices to fill the last two spots, not only because of their backgrounds and occupations (Chatterton was a commercial diver, Yurga a dive shop manager) but because of their deep-water experience and enthusiasm. They wanted more out of the trip than to "tag" a famous shipwreck in order to add it to their resumes.

I didn't know Yurga well and didn't have his phone number, so I called Chatterton and caught him completely off guard - he never expected to be asked, especially by me. He verified his interest and that of Yurga, but said that before he could commit himself he had to check with his boss about getting the time off. He would also discuss the proposition with Yurga. I told him to get back to me soon. He did - about ten minutes later. "I'm on."

# Grand Adventure on the *Lusitania* – 241

"How did you get in touch with your boss so fast?" (It was ten o'clock at night.)

He didn't. He decided that if his boss didn't like it, he'd get another job when the trip was over. This demonstrated enthusiasm that was greater than my own. Chatterton couldn't speak for Yurga, but thought that he would make the same decision. He said that he would call Yurga and put forth my proposition.

Ironically, Chatterton and Yurga worked for the same boss: Floyd van Name. This was the same van Name I had known since the '70's, when I chartered his boat, the *Kiwi,* then its larger replacement, the *Jackpot.* Now he owned a recreational dive shop and a commercial diving company, both located on Staten Island. He consented to letting his two prime employees take the time they needed for the trip.

## Friendly Faux Pas

In light of my earlier conflicts with Chatterton, one might wonder why I invited him and his partner rather than friends of long standing whose fraternity was more meaningful to me, and with whom I would have preferred to share the experience. As I've already stated, I knew how my friends were situated: none of them were available.

Furthermore, the enthusiasm expressed by the two Johns took me back to my dive master days, and the satisfaction I gained by making opportunities available to those who would appreciate the experience.

Because of the now "open secret," two of my best friends and dive buddies heard about the trip through the grapevine. Steve Gatto and Tom Packer were hurt because I hadn't asked them to participate. I was horrified. Were they available? They said no, but that wasn't the point. The point was that they hadn't been asked.

I must have wilted in their presence like a dying flower. I had made a terrible blunder: the error of assumption. I realized at once that they were right. I should have asked them first and let them decline. My act of omission was a serious breach of friendship that showed lack of sensitivity. I had no reasonable defense to offer.

I was so busy with my writing, my business, and a heavy diving schedule; I was so swept up in other aspects of the project; I was so intent on maintaining Polly's protocol of secrecy; that it never occurred to me how others might perceive my actions. I begged forgiveness.

Because Gatto and Packer were friends and good people, they forgave me.

## Turning Points

Death is inevitable. Sometime, somewhere, eventually, everyone dies. Death can no more be prevented than the setting of the sun. Yet some deaths occur quite avoidably premature.

An untimely end may be the result of accident, untreatable disease, or intentional action. Not all self-inflicted expirations are considered suicide, either by legal definition or in the commonly accepted meaning of the word - but they may be suicide just the same.

On November 15, 1993, Bill Nagle died. He was forty-two years old. When the final moment came he was alone in his rented apartment. His passing was observed only by some of his prize possessions: artifacts that he had recovered throughout the years. As the end approached, I wonder if he felt lonely, or morbid, or helpless, or sad. Did he feel cheated by life? Or was his mind so obliterated by alcohol and drugs that he had no awareness of death portending? No one will ever know.

The death certificate may have given the cause as incipient renal failure or a burst pharyngeal clot, but the plain fact of the matter was that he drank himself to death. He had no one to blame but himself.

It is impossible to determine when Nagle's slow dissolution began. He had been dying for years - first emotionally, then mentally, then physically. Many people tried to save him after I gave him up as a lost cause. None was able to penetrate the strong mental block he had erected in order to blind himself to reality. Eventually, he wore them out the way he wore me out.

Nagle did not have a sense of introspection. He repressed everything that was too painful to bear. He could not recognize his weaknesses. His strength lay in denial.

As he accelerated along the path to self destruction, blinded to reality, he made feeble and insincere attempts to straighten out his life. On more than one occasion he checked himself into a drug and alcohol rehabilitation center - but it was all a sham. He was looking for penance, not rehabilitation. He put in his time like a criminal serving a sentence, then began drinking again as soon as he was released.

He habitually stopped at a liquor store on the way home from the detox unit, drank himself into oblivion, and arrived at his apartment barely conscious. On one of these binges he was stopped by the police because his truck was weaving from one side of the road to the other. He was arrested for drunk driving. Kevin England bailed him out of jail and gave him a stern talking-to. It did no good. Nagle wanted England to take him to his truck so he could drive to another liquor store for booze. England refused, and took him home instead.

Nagle chose not to change the course that he had plotted for himself. His moments of lucidity were rare. Most of the time he was fallen-down drunk, incapable of getting out of his bed or bunk, certainly not competent to run his boat or business. John Chatterton and Danny Crowell operated the boat while Nagle lay prostrate on the wheel house bunk. The business fell apart.

By 1993, Nagle was a pale and wasted image of his former self. His skin was yellow with jaundice, and it hung in loose folds about his emaciated body. People who had known him for years didn't recognize him. The doctors gave him six months to live. Anyone who looked at him knew that the doctors spoke the truth, or possibly exaggerated. His kidneys, pickled in alcohol, were failing. He entered the hospital several times for temporary relief, but he was only prolonging the agony. Despite stern medical advice, he continued to drink at least one quart of alcohol per day, including the days on which he was released from the hospital.

I like to believe that I had a genuine and authoritative insight into Nagle's character, into the workings of his mind. I had known him intimately for two decades, I had observed his deeds and actions, I had listened to him for hours on end as he spouted his decadent views on life.

He believed that the world existed for his pleasure, that objects existed for his possession, that people existed for him to control. His measure of success was a bank book - a kind of score card in which the bigger the number, the better the person. He believed that his inherited wealth made him superior to the poorer masses, that it granted him power and authority over them, that he was more worthy than anyone else and therefore privileged to look down upon them as servants to do his bidding.

He never understood that human life is transitory, that personal possessions are chattels that are not truly owned but merely borrowed for the duration of our visit to this world - that they are redistributed upon demise. People are respected not for what they owned but for what they created or produced. Thus the only commodity that has any value or meaning in human terms is that which a person adds to the great gestalt of humanity. Vanity dies, legacy abides. Nagle's legacy was one of his own choosing - one whose endurance ended abruptly upon his death.

I attended his funeral filled with staunch contempt, perhaps even a touch of hauteur. There was no pensiveness in my mood. Yet as soon as I saw his casket being pushed along the aisle of the church, I burst into tears. I cried loudly and continuously throughout the service. Even today I barely understand why.

Nagle was once my friend. We had often laughed together, and had gone through many great adventures together. But somewhere along the path of life he took a wrong turn. He embraced the dark and arrogant side of human nature. The Nagle I once knew and loved died long ago. The person who took over his body was a loathsome stranger to me. I suppose I cried for the person Nagle used to be, not for the one he had chosen to become.

Outside, in the light of day, reality reappeared. Family and friends (mostly dive buddies) chatted in small groups. Also attending the services was George Hoffman. He and I had not spoken to each other for more than a decade - since we had been pall bearers holding opposite sides of John Dudas's coffin. Once again I took the initiative.

I approached Hoffman in the cool autumnal air amid a flurry of fallen leaves. I spoke without inflection. "Hello, George."

At that moment the church grounds and parking lot were charged with intense and painful silence. Steve Gatto told me afterward that he heard every word of the resulting conversation although he stood more than thirty feet away. Every diver there knew of the long standing feud; every eye turned toward the inevitable confrontation; everyone expected wild fireworks to fly. Including me.

Hoffman astounded us all. He looked up at me without a hint of disequilibrium. He appeared calm and in control of all his faculties. "It's been a long time. I think we should bury the hatchet. I've heard good things about you, about what you've accomplished. I'd like to believe that I had something to do with the making of your success."

This was quite a statement to make completely unrehearsed. I was dumbfounded. And, if I may be permitted a touch of vanity, Hoffman's evaluation struck me as crudely arrogant. I perceived Hoffman's contribution to the success of my diving career in the same light in which I attributed my financial success to the bus driver who dropped me off at the brokerage firm where I purchased my stocks and bonds. Each took me to a destination to which I paid him to take me, but I made the dives and the investments. If this broad assessment sounds conceited on my part, so be it.

My convictions notwithstanding, the long years of dispute had come to an end.

Within the span of minutes both a friend and a hatchet were buried. What a day.

## THE BRITS IN TRAINING

On the other side of the Atlantic, the Brits were going full speed with their dive training program. They were mixing gas, running decompression schedules, and diving in extremely deep quarries on a weekly basis. The program that began in earnest in the autumn continued throughout the winter and spring. These Brits were dedicated. They even made dives when the ground was covered with snow and the temperature of the air stood far below freezing. Nor was the water much warmer.

The quarry they dived had a remarkable depth of 300 feet, and they managed to find a spot that went as deep as 330 feet. They conducted long decompressions in the frigid water - not simulated decompressions, but obligatory decompressions based upon bottom time.

They practiced decompressing on various nitrox blends, then breathed oxygen at the 20- and 10-foot stops. This gave them the opportunity to fine-tune gear configurations.

There would be no surprises for them when they dived the *Lusitania*. They were confident in their equipment and in their ability to handle whatever occurred on the bottom.

## POLLY PAYS A VISIT

Polly Tapson kept me apprised of the British group's progress and sent me minutes of the meetings. I in turned passed this information on to Lander and the Johns. Just before Christmas we each received a surprise package from Polly: a complete, up-to-date report on the expedition's status. The word "Confidential" was stamped boldly in red across the cover, followed by "for your eyes only."

The report was a complete operational design. It gave detailed information on the charter boats, the folk house, gas delivery, booster and compressor rentals, transportation logistics, cost breakdowns, decompression methodology, chamber facility, back-ups, and so on. It also covered those issues not yet resolved, items that needed further research, and the situation with respect to Bemis's unsubstantiated claim. (We had long since decided to ignore his persistent entreaties.) The report was a masterful piece of work: more like a corporation's annual report than a dive plan. I was impressed. It demonstrated admirably the dedication and professionalism with which the Brits were treating the project.

Polly prepared a liability release for the members of The Starfish Enterprise to sign. This legal document was a standard instrument among dive shops and dive boats. It stated merely that all participants were engaging voluntarily in an activity which entailed a certain degree of risk, and that none could be held responsible for accident or death. Polly asked if she could put my name down as co-leader. I agreed, but there was a tacit understanding between us that this was for liability purposes only. She was still the leader, I an invited guest.

Mike Menduno asked me to lecture and give an advanced wreck-diving workshop at *AquaCorps'* second symposium, to be held in New Orleans in January 1994. Polly Tapson planned to attend. It was the perfect opportunity for her to meet the American contingent - the Yanks - since Lander, Chatterton, and Yurga also planned to be there.

I was as eager to meet Polly as she was anxious to meet me. I found her waiting in the hotel lobby - I recognized her from her photograph. She was shy, reticent, perhaps somewhat awkward. I tried to overcome my embarrassment at meeting my number one British fan. We talked late into the night. Polly spoke in a firm, mellifluous voice, and with a lilting London accent that I found captivating - or was the accent mine, since she was pronouncing the King's original English?

The next day I introduced her to the rest of the Yanks as they arrived. We all had a grand time together. There was endless conversation and exchange of ideas. When the weekend was over Polly's shyness was gone. Also gone were any misconceptions she might have held about our implied superiority. She may even have been disappointed at our penchant to horse around, turn every comment into a joke, and take nothing too seriously - in stark contrast to the British stiff upper lip.

Despite our shortcomings, the Yanks and their British leader were firmly molded in their resolve to overcome all obstacles in bringing the expedition to successful completion. No one was going to stand in our way.

## Keeping Fit

When my activity level drops, so does my appetite. Every winter I lose weight, muscle tone, and upper body strength: the result of not carrying dive gear and of exercising nothing but my fingers on the keyboard. To compensate, I jog more frequently and do push-ups and sit-ups in the house.

The winter of '94 was one of the worst in Philadelphia's history. The city was hit by a succession of snow and ice storms, with no warm respites in between. Twice the entire city was shut down for two days, during which time the streets were buried in snow, cars couldn't move, public transportation was at a standstill, all stores, businesses, and government offices were closed, and no mail was delivered. (The new motto of the post office was "Through rain, through hail, through sleet or snow - we don't go.")

Only the main arteries were plowed. Suburban side streets remained snowed under while the temperature never reached the double digits. Inflation's numbers rose higher than Fahrenheit's. When the city wasn't completely paralyzed, traffic moved slowly and onerously because the streets were paved with solid sheets of ice. This put a damper on my outdoor aerobics. The sidewalks and roadways were so slick and slippery that it was practically impossible to walk, let alone jog. I came up with some unique solutions to the problem of physical fitness.

My garage and basement were full of books that were packed in shipping cartons - the way they were delivered from the printer. Lander had demonstrated on her barbell bench how to lift weights in sets or repetitions. I took to curling a box of books as if it were an iron weight. I also joined two boxes with a broom handle so I could lift them like a dumbbell. I called it "pumping paper."

Instead of jogging shoes I donned my winter mountaineering boots complete with twelve-point crampons. Now I could run on ice with perfect traction - to the amazement of people I passed on the street. Probably they had never seen such a strange device, and couldn't understand how I managed to be so sure-footed.

The first time I jogged in crampons I ran my routine winter route: from my house to the park, through the woods on the frozen bike trail, then back home - a circular course of about seven miles. What I did not take into account was the weight of the contraptions on my feet: each boot and crampon totaled five pounds. By the time I reached the end of the bike trail, which was four miles from home, I could hardly lift my legs. The last three miles I was in agony.

As soon as I slowed to a walk I was bitten by the bitter cold, which was about twenty-five degrees below freezing. To make matters worse, a fierce nighttime head wind cut through my sweat suit as if it were the Emperor's new clothes. After walking only a hundred yards I was forced to jog in order to keep warm against the wind chill. All too quickly my legs gave out, reducing my gait to a painful stroll. Then came the final agony: searing pain followed by numbness in an important part of my anatomy.

Fearing that the next phase was frostbite, I tucked one hand down the front of my pants to provide extra warmth, and alternated between jogging and walking fast. By the time I reached home I had lost all feeling in my pelvic extremity. Then came the painful ignominy of leaning over the bathroom sink and pouring lukewarm water over the frostbitten member. Feeling eventually returned. After drinking a cup of hot coffee - a good diuretic - I ascertained that the kidney connection still worked. Not until I visited Cheryl several weeks later did I determine that the plumbing was fully functional.

## Deadly Intent

After this I had a more sobering jogging encounter. A fresh snowfall covered the ice to a depth of several inches, enabling me to run without the crampons. It was late at night and the streets were deserted. I was on the last leg of the route, jogging along the right lane because plows had pushed tall mounds of snow onto the sidewalks.

A city jogger needs to be aware of his surroundings. All my senses were always on the alert. I saw headlight beams pierce past me from the rear, heard the steady crunching of granular snow beneath rubber tires. Then I noticed a change in the regular drone. Instantly, subliminally, I knew the car was crossing the low ridge of snow created by the edge of the plow blade . . .

I glanced over my shoulder in time to see the car right behind me. I swiveled my hips, then was struck on the thigh by the metal fender.

## 248 - The *Lusitania* Controversies

The glancing blow knocked me aside - with incredible good fortune, into the narrow space between two parked cars. I went down on all fours - hard.

Although this was my closest call it was not an isolated incident. I don't have enough fingers and toes to count the number of times that vehicles have swerved to hit me - intentionally. You might think the drivers were juvenile delinquents who were out to cause murder and mayhem, but most of the time you'd be wrong. Men wearing suits, teenage girls, a car full of senior women, mothers with children, families - these were the kinds of people I've dodged in order to save my life. And that's to say nothing of the dogs that have attacked me.

Had I not been fully attentive, had I not taken instant action, the sharp metal edge of the headlamp assembly would have caught me in the pelvis and done considerable damage. Plus I might have been run over by the car. Or I could have ricocheted off a parked vehicle and then crushed by the car's rear tires. I was lucky only to have been sideswiped.

The fender hit me in the worst possible place: on the still sensitive bullet wound. I was practically paralyzed by pain. Nonetheless, I picked myself up and stumbled fanatically after the car. At the next intersection it sped through the red light without ever slowing down. I didn't get close enough to read the license number - or to kick in the window and pull the bastard out by his hair.

My adrenaline rush over, I limped home in pain and seething frustration.

The world is full of jerks.

## COMPETITION

The Starfish Enterprise was not the only group that intended to dive the *Lusitania* in 1994. An Irish group had gotten the idea and they were close upon our fin tips. This group called themselves ITD, for Irish Technical Diving. Since they did not have either deep diving experience or technical dive training, they hired Robert Palmer to teach them what they needed to know.

Palmer was a respected British diver who had done extensive deep penetrations into Caribbean blue holes. He had written several books about his exploits - books that were both erudite and exciting to read: a rare combination. As a long-time scuba instructor his qualifications were impeccable. With the advent of technical diving he established himself as a mixed-gas instructor.

All eight members of ITD entered Palmer's rigorous training program. Four of them flunked out - Palmer refused to issue them certification cards - and the other four he passed as "marginal," along with the advice not to attempt to dive a wreck as challenging as the

# Grand Adventure on the *Lusitania* - 249

*Lusitania* until they had gotten some deep-water wreck-diving experience under their belts. That they did not heed his advice proved disastrous.

ITD contacted Bemis with their proposition to dive the wreck. When Bemis declared his presumption of ownership and demanded that the members sign away all their rights by contract, they acquiesced to his extortion without protest.

Following our lead every step of the way, ITD also chartered the *Sundancer*. Nic Gotto did not mind accepting a charter from another group - after all, business was business, and he owed no one allegiance - but through some mechanism unknown to us, Bemis learned that Gotto was also planning to take The Starfish Enterprise to the wreck.

Bemis wrote to Gotto, "There is a 'rogue' group from England that is talking about an illegal dive in early June. This we want to prevent." Bemis went further and tried to coerce Gotto into canceling our charter.

Gotto was incensed. He didn't like Bemis to begin with. (We didn't know it at the time, but Bemis had wangled his way onto the National Geographic ship during the film shoot, and Gotto had run him out to the site and back. Gotto said that Bemis annoyed him.) Now Gotto was annoyed even more. He was not one to take strong-arm tactics lightly; no one pushed him around. He told Bemis bluntly that he would in no way consider backing out of his obligation to us. He said that this made Bemis angry and even more persistent in his efforts to put pressure against him, but to no avail.

## LAST VOYAGE OF THE LUSITANIA

The television show that we were all waiting to see was aired in April, in time for us to learn something from it about the wreck before the trip. In that regard we were sorely disappointed. Only four minutes of footage showed the wreck under water.

Except for a few scenes, such as those showing the bow and the name board, the wreck footage was shown in isolated segments of short duration without locational correlation, leaving the viewer clueless as to where on the immense hull the scenes were shot. This editing method worked well stylistically for the market for which it was intended - it was only wreck-divers who found it so unsatisfying.

The way the story was presented it was pure pabulum for the public, and blatantly contrived. The show was wonderfully directed, to be sure, with some telling and memorable moments the way the awful tragedy was portrayed. But the dramatic effect was off-balanced by interviews with people who barely remembered the event - were children, in fact, at the time the sinking occurred - and who had little if anything to offer in the way of enlightenment other than to evoke an

## 250 - The *Lusitania* Controversies

emotional connection with the past.

Some viewers thought the show dwelled overlong on file-photo close-ups of the bodies and faces of drowned women and children, garishly frozen in rigor mortis. Yet the brutal imagery of distorted, cherubic faces left a haunting impression that I found difficult to dispel. Violent death personifies the true horror of war much more vividly than the mere destruction of property. The *Lusitania* was a machine surrounded by inanimate steel. The tragedy of her sinking was not the loss of the ship but the loss of the passengers and crew. Perhaps in that context the pictures of those who perished presented a fitting and appropriate view of the what war is truly about.

On-location scenes of the surface support vessel, the crew, the submersibles being lowered into the water, and so on, added nothing to the story - they merely slowed the pacing and dragged out the length to fill a one-hour slot. Ballard's great "discovery" - the hook for the show - was a lump of coal on the sea bed. This lump of coal was presented as proof of what was portrayed as Ballard's novel solution to the mystery of the second explosion - the reason why the ship went down so fast: a coal dust explosion that blew out the lower hull plates.

This tired explanation was first noted by Captain Schwieger in the *U-20's* war diary, then by manifold marine and naval architects of the time. Throughout the years, only conspiracymongers discounted the coal dust theory as the true explanation for the *Lusitania's* rapid sinking. To present such a theory as new was, at best, ignoring the facts of history; at worst, taking credit for the concepts and performance of others. Not unexpectedly, the show broke no new ground in historical perspective because there was no new ground to brake.

The telecast was ambiguous in that there was more than twice as much air time of Ballard than there was of the *Lusitania* (Ballard was on for nine and a half minutes, the wreck for only four.) Since wreck footage played such a minor part of the show - second fiddle, so to speak - one might wonder what the show was truly about. The final product would have been more honest, more cohesive, and less costly to produce without submersible operations and a concocted coal dust explanation that did not give proper credit . The four minutes of wreck footage was more of a distraction than an informative adjunct.

The shoot did, however, reaffirm what Oceaneering's divers noted in 1982: that the bow was intact and not "nearly severed from the main hull" as John Light claimed. Nor was the ammunition storage compartment, located forward of the bridge, blown out - thus refuting Light's thesis that the secondary explosion was caused by unregistered explosives that were stowed there. Furthermore, no guns were found or photographed anywhere on the wreck or in the debris field - again reconfirming what Oceaneering's divers noted. This supports the fact that none of the passengers or crew or any of the thousands of spectators, including journalists, reported seeing naval gun barrels protruding from the hull at any time during the *Lucy's* career or final days.

The show was a mixed bag of fact, fiction, and speculation that didn't work on all levels. Had the historical narrative been extracted and focused upon, it might have succeeded as a worthwhile documentary: a remake and higher quality production of earlier versions. Instead it was diluted by commonplace and by a blatant attempt at historical misdirection, and made offensive by Ballard's disinclination to use the first person pronoun in its plural form.

## A Clearer Picture

I learned more about the wreck from two of National Geo's participants than from the television show: Ken Marschall and Eric Sauder. Marschall was a noted marine artist whose primary occupation was painting backdrops for Hollywood movies. Sauder was a banker whose ardent avocation was ocean liner history. Both held huge collections of photographs of the *Titanic*, *Lusitania*, and other famous liners. Together they produced a lavishly illustrated book, *R.M.S. Lusitania: Triumph of the Edwardian Age*.

I had never met Marschall or Sauder, both of whom lived in California, but we had corresponded and had spoken on the telephone - exchanging information about shipwrecks. Marschall was on board in order to obtain visual references for the paintings that he was contracted to do for the video jacket, and for the inevitable book that would be produced as an adjunct to the film project. (National Geographic did not lack in sound business management - in every commercial enterprise they eked profits from a multitude of media presentations.)

Sauder was invited as an historical consultant. Their expertise enabled them to interpret images of the wreck and to identify objects in the debris field so that submersible operators, photographers, and members of the film crew - who knew little or nothing about the *Lusitania* in particular or about ships in general - might understand what they were looking at and where the submersibles were located with respect to the immense length of the hull. The other participants would have been lost without their input.

Marschall and I had previously worked together on a painting that he did of the *Andrea Doria*. Using photos from my book and my personal observations, he painted an elegant, lifelike color visualization of the way the *Andrea Doria* appeared on the bottom - not as a ship in harbor but as a wreck that had undergone collapse and deterioration. He painted with absolute precision, not in approximation with stylistic interpretation.

Each rough sketch of the *Doria* posed dozens of questions about current structural integrity, the condition of the hull and upper decks, the location of trawler nets, even whether a particular door was opened or closed. He was a stickler for exactitude. Since he relied solely on me for details of the wreck's actual appearance, any inaccuracies in the final product were attributable only to me.

Both Sauder and Marschall imparted to me many invaluable insights about the *Lusitania's* state of collapse and the location of certain items which held photographic potential. Furthermore, Marschall asked me to make specific observations which the National Geographic team had ignored. In addition, he asked me to take a number of souvenir postcards down to the wreck, so they could be authenticated as having been down there.

These postcards were illustrated with his painting of the *Lusitania* backing away from her New York dock immediately prior to her final passage. He had taken some of these postcards down with him in the submersible, but had given them all away. He sent me two hundred postcards - one hundred that he wanted returned, and one hundred for me to keep.

## Overseas Logistics

The Brits didn't own enough gear to equip the Yanks, nor was there any way to rent or borrow doubles. We each had to supply our own. This meant that the Yanks had to send ahead or carry on board all the gear that was needed for the trip. Tanks and valves we shipped in advance, courtesy of British Airways.

The youngest member of The Starfish Enterprise was Jamie Powell. He was a commercial shipping agent for British Airways, while his mother held a more influential position. Powell's mother arranged to have our tanks flown as freight at a special rate: nothing. We didn't argue.

We - the Yanks - each broke down our doubles, packed them in individual cardboard boxes, boxed two singles to be used as sling bottles, and packed the valves and manifolds separately. Lander and Yurga delivered all the boxes to the British Airways terminal at the JFK Airport on Long Island. Powell picked up everything at the other end.

This process was reversed at the conclusion of the trip.

## Get the Picture?

My primary objective on the *Lusitania* was photography. To do this I needed an underwater camera housing that could withstand the depth; my old plastic housing wouldn't do. After researching the options I decided upon a Tussey aluminum housing that was rated to

350 feet. The only thing I didn't like about the system was that it was designed strictly for Nikon - and I'm a confirmed Canon person.

So I had to purchase a fully automatic Nikon 8008 to go inside. The total cost for the camera and housing was $4,000 - more than the cost of the trip, which was expected to be about $2,500, exclusive of air fare for the Yanks (about $500). As much as the investment set me back, I've been happy with the system - once I got the bugs worked out.

I intended to familiarize myself with the housing and learn the camera's idiosyncrasies in the spring, but with one thing and another, I didn't have the opportunity until the middle of May. At that time I was in North Carolina, spending some time with Cheryl before the big trip, while completing the paste-up for *Wreck Diving Adventures*. By previous arrangement, Mike Moore trailered his boat to Southport, from where we planned to use the good weather days to dive wrecks in the vicinity.

On the day of arrival the winds picked up. Big breakers crashed upon the beach, and the offshore wave heights ranged from four to six feet - far too rough for Moore's little boat. Every day the winds blew stronger, the waves grew higher. The trees were on their knees. To make a long story short, we got blown out for eleven days straight - a record in my career. By the end of that time the wave heights ranged from ten to twelve feet. I completed my book but never got in the water.

## Losing Through Intimidation

While I was in North Carolina I received a call from Gregg Bemis. He had phoned my home and left a message on my answering machine.

Drew Maser - no longer just my accountant and shipper, but manager, booking agent, and general factotum - retrieved my messages so he could handle the book orders and miscellaneous business. When he heard the message referring to the *Lusitania* trip, he called to tell me about it.

The gist of Bemis's message was that the Brits didn't have his permission to dive the wreck. He recommended that I cancel off the trip. This made me wonder: how did he know me? He was not a diver, and as far as I knew he had no connections in the wreck-diving community. I also wondered too how he knew that I had anything to do with The Starfish Enterprise? And how did he know where I lived? What kind of espionage network did he have working for him?

The mystery deepened when Peter Hess called Maser to find out where I was, then called me in North Carolina. Bemis's attorney, Rick Robol, had called Hess and strongly suggested that I divorce myself from the British group. There was a veiled threat of legal action should I not comply with his wishes. How did Robol know that Hess and I were friends? And how did he know how to get in touch with Hess?

Granted that I didn't lead a secret life, and that any good detective or investigative agency could ascertain the facts, but why bother? And at what cost? Curiouser and curiouser . . .

I thought about the implications of this two-pronged attempt at intimidation. What was Bemis afraid of? What did he think we might accomplish on the wreck? National Geographic, with unlimited funding and sophisticated machinery, accomplished nothing. They certainly hadn't found any gold. Their findings were specious, not in specie. So how could anything that we might see or do on scuba have any adverse affect on the wreck, or diminish Bemis's imaginary prerogative?

And why hadn't we heard any more from John Pierce? Or Oceaneering? Or Muriel Light? Or George Macomber?

Months before, I suggested to Polly that she ask Bemis to produce documentation of ownership - to put up or shut up. He did neither. Instead of providing proof of his claim - a simple expedient if true - he sent threatening letters and made harassing phone calls. That in itself was telling.

Bemis's desideratum was illegitimate in light of his absence of proof - nor was he acting sensibly under the circumstances. Desire does not equate to entitlement. We all want things, but that doesn't meant that we have right to take them.

Bemis's lack of frankness and underhanded methods rubbed me the wrong way. They only firmed my resolve to dig in my heels. All the members of The Starfish Enterprise - Brits and Yanks alike - felt the same. Whether his flagrant attempt at intimidation was merely a tactical error or in keeping with his character, it served only to erect an impenetrable barrier rather than to facilitate understanding. You get more bees with honey . . .

Meaningful communication between two people requires that both of them spend time listening. All too often, one person talks *at* another instead talking *with* him. Diplomacy works only when diametrically opposed parties possess a willingness to consider alternative points of view, and can demonstrate an ability to accept differences of opinion, from which compromise may be reached. Only a fool tries to reason with a lion that is leaping at his throat.

Based on Bemis's past behavior, my assessment of his potential receptivity to meet on middle ground was unconditionally negative. He was a toll collector who wanted his toll. From his language - and from that of his attorney - my interpretation of Bemis's purpose in calling was to frighten me off the trip, possibly in the hope that my withdrawal would scare the rest into doing likewise. If he believed that I was the linchpin in the trip's implementation, he was sorely mistaken. I was but one player among a group of strong-willed individualists. My lack of support would account for nothing.

To return his phone call would be either pointless or counterproductive: an exercise in futility. So I exercised a conscious policy of what is known in counseling as "confrontational avoidance."

Little did I know or suspect the clandestine machinations that Bemis had already put in motion.

## The Yanks Are Coming

Lander and I flew to London a week ahead of the Johns. This was so we could dive with the Brits off the south coast of England over the three-day weekend. Polly Tapson met us at the airport. The smile on her face could have lighted an auditorium; the enthusiasm she generated could have powered a large building.

The Tapsons owned a three-story row house in London with several extra rooms. This was the headquarters of The Starfish Enterprise. The large basement "rec" room did multiple service: Polly employed one side as a home office, while the other side was used as a gas mixing station complete with storage cylinders full of helium and oxygen; a small adjacent kitchen provided cooking facilities. Living room and bedrooms were upstairs. My lament and complaint: there was no shower, only a bathtub. I hated baths.

In addition to her home office, Polly shared a downtown office with her business partner, so she was gone part of the time. Simon worked all day in a posh executive suite. I saw him only in the evenings. He reminded me of a British commando who had gotten out of uniform after the raid on St. Nazaire.

## The Crew of The Starfish Enterprise

Jamie Powell delivered our tanks and introduced himself. His pronunciation, while distinctly British, can be described without offense against the native English language as noticeably accented. For example, he pronounced his name "Jymie." If that was how he called himself, who was I to dispute him. Jymie he was, and that was how I called him ever after - to everyone's delight.

During the week there was hardly any time to explore London as a tourist because there was so much work to do: reassemble tanks, re-rig equipment, discuss gas options, run decompression schedules, study logistics, and so on. I honestly cannot remember all the things that we did - only that I was busy from morning till night either working on my gear or learning minute but important details of the forthcoming expedition with which the Brits were already familiar.

I also had to learn the quirks of the language: a truck is called a lorry, a toilet is a loo, the subway is the tube, a bus is a tram, and "brilliant" meant "great" or "very good." A quid is to a pound what a buck is to a dollar. Some ordinary words had totally different meanings, such as "bum." And whatever you do, Polly warned me, be careful of what you call those small packs that are worn on the lower back.

That week and weekend I met the rest of gang. Technically, Christina Campbell wasn't British because she had been born and raised in Scotland. She spoke with a wonderful lilt that was every bit as Scottish as Polly's lilt was British. I delighted in listening to both of them speak. I didn't care what they said, just so long as I could listen to the crisp tonal qualities of their language. American enunciation is crude and slurred by comparison.

Nick Hope was a design engineer who, with Simon, had become proficient in the fine art of gas blending and boosting. Richard Tulley was a lecturer and computer researcher who was brilliant in the American meaning of the word; he designed and built the floating decompression station, which for convenience was called the "dec" station (pronounced "deck"). Dave Wilkins was a civil engineer whose specialty with respect to the expedition was decompression software. Paul Owen was a police inspector who had just wound up a serial rape and murder case that he had been investigating for several years. His accent was so harsh and guttural that only by listening carefully could I understand one word in three. It made for interesting conversation; I said "What?" a lot.

## CHECK-OUT DIVE

I was happy to get away from the frenetic pace and swarming crowds of London. The south coast of England was more to my liking. Rolling hills that were green with crops stretched as far as the eye could see: a motif of bucolic splendor. The countryside was a patchwork quilt in which macadam took the place of stitch lines. Off the main thoroughfares the roads grew narrower until they became country lanes - the quaint British phrase for slender passageways between fields, and barely wide enough for a single automobile. When two vehicles met, one had to back up to a shoulder - perhaps hundreds of yards - and let the other one by. (This system could never have worked in New York; it required common courtesy.) These country lanes were sunken several feet below the surface of adjacent fields, which were grown high with grain. I felt like a bug in a rut in an unmown lawn.

We had a package deal for the weekend: lodging at a modest motel that also included meals; the motel's owner ran a dive boat called the *Panther*. Not too early on Saturday morning we chugged out of the harbor under a warm sunny sky. The coastline consisted of jagged rocks whose upthrust facade curved back to green plateaus that stood hundreds of feet above the water. Grazing sheep climbed down to the ocean's edge where they gamboled among the boulders. A spectacular panorama.

Also aboard for the weekend were Rich Field and Rob Royle, long-time fellow divers of the Brits, and Tony Hillgrove. Royle astounded us by wearing quadruple tanks on his back. The inner tanks consisted of a set of doubles that were filled with bottom mix. The outer tanks, containing decompression gas, were secured to the doubles by means of Velcro straps - one tank to the outside of each inner tank. The valves of the outer tanks pointed down. He admitted that the cave rig was hard to handle on a boat.

Hillgrove was lean and wore a close-cropped beard that gave him the grizzled look of a nineteenth-century cowboy. He was crewing for the weekend, and we were glad to have him aboard. He went down first on every dive to check the condition of the hook. His friendly personality endeared him to all of us.

We passed the rusted remains of giant freighter that lay aground on protruding pinnacles. The wreck had been reduced to a broken-down hulk by the action of the waves. It was a visible reminder that, despite modern aids to navigation, ships could still go afoul of Britain's hostile and treacherous coast.

Our destination that day was called the Ten Mile Wreck, as yet unidentified. The ocean off the south coast of England is tidal, so the best time to dive was during slack. We arrived on site a bit late, as the tide was beginning to turn. The captain did not hook the boat into the wreck. Instead, he dropped a grapnel that was connected to a surface float by a thick nylon line. We "kitted up" and dropped over the side, swam downcurrent to the float, and descended. The boat hovered nearby, waiting to pick us up after the dive.

It was an easy pull down the gentle scope. Lander was right behind me. On the bottom I saw that the grapnel wasn't hooked into the wreck, but was snagged on the wire of a trawler net. The wire was pulled taut in a V. I could barely make out wreckage some twenty feet away across a white sandy bottom. The depth was 235 feet. About that time my regulator began acting up.

Experience told me never to leave the security of the grapnel under the circumstances. The wire might break under the strain, or the net could pull out of the wreck. I decided to reset the hook. Try as I might, I couldn't pull it forward off the wire. The rise of the hull was within sight but not within reach. I began to gasp from the struggle. And to make matters worse, my regulator freeflowed after every inhalation. I was ten minutes into the dive.

I signaled to Lander that I was having trouble with my regulator. She acknowledged, then swam off along the wire. I terminated the dive and began my ascent. I got only thirty feet off the bottom when my regulator failed completely. As air gushed out of the mouthpiece, I switched to my back-up, loosened my waist strap, hiked the tanks over my head, reached over my shoulder, and closed the valve to which the offending regulator was attached. My waist strap pulled out of the buckle.

I hadn't lost a significant amount of gas. I wasn't in immediate danger. I breathed easy from the functioning redundant regulator; plus I had a pony bottle. I pulled myself up at the proper rate of ascent to my first decompression stop.

Because I didn't have enough lead on my borrowed weight belt, I couldn't get stabilized. I forced all the gas out of my BC and drysuit but I was still positively buoyant. Compounding my flotation problem was the fact that my tanks were loose: they kept floating off my shoulders, and the valves kept banging me in the head. I wrapped my legs firmly around the ascent line in order to have my hands free to search for the waist straps and to reeve one end through the buckle, but as soon as I took my hands off the rope I slid up toward the surface. This forced me to grasp the rope again and pull myself back down.

One of the other teams on the line (I think it was Powell and Tulley but it might have been Field and Royle) observed my antics and gestures for several minutes before I was able to make them understand my problem. They held onto to me and helped rethread the strap. I was still too light, especially at the 10-foot stop, so I still had to fight to cling to the line so I wouldn't shoot to the surface. All in all, I did not make a good first impression on my fan club.

To make matters worse, during post-dive debriefing I realized that I had made an erroneous assumption on the bottom. My fear - that the descent line might pull the net wire free - was groundless. The only strain on the line was the strain imposed by the tide against the buoy and the rope, not against the boat! I had treated the situation in terms of anchoring procedures at home. Even though my response had been based upon years of ingrained habit, I felt foolish.

## BENT OUT OF SHAPE

My embarrassment was soon overshadowed by a problem then developing: Polly felt tingling and numbness in her legs, as if they were going to sleep. She and Simon had dived the same profile, had breathed the same bottom mix and decompression gases, and had ascended together. Yet Simon showed no signs of the bends, proving once again that decompression injuries were impossible to predict.

In the 1990's, wreck-divers recognized symptoms of the bends much better than when I had last taken a hit some twenty years earlier. They also acted more decisively in response. Polly tried feebly to deny her condition, hoping perhaps that it would go away - as it often did. But not this time. We didn't let the patient prescribe her own medicine. We took charge.

Polly's infirmity worsened despite treatment with oxygen. By the time we reached the dock she was having trouble standing. We all commiserated with her about her unfortunate accident - for accident it surely was. She hadn't done anything wrong. We also impressed upon her the need to take a chamber ride. The most insistent was Hillgrove. He fussed over Polly like a mother hen attending her chick.

Simon and Hillgrove drove Polly straight to the recompression chamber. The rest of us unloaded their gear and retired to the motel to await word of her condition. Polly underwent two and a half hours of decompression therapy, but emerged from the chamber on still shaky legs. The doctors kept her in hospital and scheduled another treatment for later that night.

In the mean time Simon and Hillgrove returned to the motel. Simon's concern for his wife was evident on his face. He planned to go to the hospital again after dinner, timing his arrival for when she was due to exit the chamber. I asked if I could go along. Simon may have been her husband, but I was her idol and she was my number one British fan. I thought some words of encouragement from me might lift her spirits.

The hospital was forty-five minutes away by car. I couldn't share the driving because my Pennsylvania license wasn't valid in England. Besides, I'd never get used to driving on the wrong side of the road.

Polly emerged smiling but unsteady on her feet. The tingles were gone yet some of the numbness remained. The doctors wanted her to stay overnight so she could be treated again the next day. It was sound medical advice. They also advised her not to dive again for several weeks, perhaps a month. This was a serious setback for her.

Polly had invested vast amounts of time and energy in The Starfish Enterprise. She was primarily responsible for bringing the *Lusitania* project together. It was her hard work, dedication, and enthusiasm that fueled the project. It was *her* expedition. For her not to make the dive would be a devastating disappointment. I pulled her aside from Simon so we could talk privately in a corner of the chamber complex. We discussed the implications of the medical opinion.

Only she could decide whether or not to heed the doctor's advice. It was a personal resolution. I told her this. But I also told her that I would support her position if she chose to dive the *Lusitania*. Even if she didn't make full-length dives, or if she held off diving until the latter part of the trip, I thought it was critically important for her to realize the dream that she had invited others to share.

The literature on the bends contained no empirical evidence of increased susceptibility after a hit. Doctors recommended abstinence

## 260 – The *Lusitania* Controversies

because they gave that kind of advice after any kind of injury. No one knew how long after a cure was effected it was safe to go back under water.

Polly told the doctors that within a week she intended to dive to 300 feet, so if they believed that further recompression therapy might make her less susceptible to the bends, they'd better schedule more treatments. I take no credit for her resolve. Polly was a strong-willed woman who needed no one's approval for her actions.

The second day we dived the British destroyer *Penylan*, at 230 feet. The wreck was considerably broken up. It was a good dive. Nothing unusual occurred.

Polly rejoined us at the motel that night. She appeared to be healthy and in good spirits and eager to talk about her chamber episodes. The multiple treatments had resulted in complete recovery and left no residual effects. She was somewhat cowed by the experience and what it portended with respect to the *Lusitania*, but she had plenty of time to make a decision in that regard. She went out on the boat the next day in the ignominious role of a bubble watcher.

This time we dived the *Medina*, a huge ocean liner that lay substantially intact at a depth of 206 feet. There were lots of openings and rust holes in the hull, so I wasted no time in shooting inside. Visibility was about twenty feet. Standing bulkheads sectioned off the interior into good-sized compartments. I picked my way from room to room, careful not to disturb the ultrafine white silt that lay many inches thick. During my wandering I took note of alternative exits, so I was not much concerned about finding my way back.

I was surprised to stumble upon a china dinner plate in plain view. I scooped it right up, but because I wasn't carrying a goodie bag I had no place to put it. I swam out of the wreck, reoriented myself, and headed toward the anchor line. On the way I came across Richard Tulley, who had a mesh bag, and convinced him to take charge of the plate for me.

Everyone oohed and ahed over my find. On the border of the plate was a crest that consisted of a rising sun and the letters "P & O": the logo of the Peninsular and Oriental Line. It was the first one of its kind that the Brits had ever seen.

"But it was lying right out in the open," I protested. Simon said that as far as he knew, no one had ever gone inside before.

## And Now a Word from Our Sponsors

The Johns arrived in the middle of the week. They made themselves at home in the Tapson house, which now looked like a tenement. Luggage, clothing, and dive gear occupied every nook and cranny. The basement rec room was filled with double tanks, singles, storage cylinders, and two booster pumps.

Nick Hope's responsibility was caring for the booster pumps. By means of intense self-study, Hope became an expert in breaking down, rebuilding, and repairing Haskel boosters. The Brits purchased outright a second-hand booster pump, for which Hope built a wooden frame. Haskel very kindly loaned us a brand new booster pump that was delivered already mounted in a metal transport frame.

This brings up the issue of sponsorship. Polly contacted several manufacturers in hopes that they might help support our venture with the loan of equipment, or by selling or renting equipment to us at a reduced price. Several manufacturers came to our aid besides Haskel.

Peugeot loaned us a heavy-duty van that was large enough to carry most of the heavy hardware. This was quite generous on their part. Paul Owen had a special operator's license, making him the designated driver.

Graseby loaned us a full-face mask and communications unit. We intended the in-water support diver to wear it for topside communication. Richard Tulley experimented with the unit during our Memorial Day weekend dives. It worked fine, but we elected not to use it on the *Lusitania*.

Diving Unlimited International offered to sell us drysuits and thermal underwear at far below cost. I already owned a new DUI suit, but the deal was so good that I purchased another as a back-up. Dick Long, president of the company, took a personal interest in our endeavor by facilitating delivery from the company's U.K. division.

## LAST MINUTE DETAILS

What we did those last few days before leaving for Ireland I have no idea. So much minutia needed so much attention that all I can remember is a blur of activity. A lot of what we did was purely mental: exchanging information and planning for contingencies. There was hardly a moment to rest. I can best describe our frenetic proceedings as "controlled confusion."

Each of us was harried by personal concerns as well as by group responsibilities. Polly assigned jobs to everyone, and woe to the laggard who failed to complete his tasks in a timely fashion. Somehow she managed to keep the progress of events in her head. She always knew what was left unaccomplished - and who had unaccomplished it!

I scurried about like an ant whose nest had been disturbed by a demon. The pace was so hectic that I often found myself doing several things at once, jumping from one assignment to another in a round robin fashion. I was exhausted before the trip even began.

One day I had a few hours to myself. I chose to take the tube downtown and explore second-hand bookstores - one of my great loves. I didn't have time to see the famous Crystal Skull - that strange archaeological relic unearthed in Central America and placed on display in the Museum of Man. Nor was I able to visit Sherlock Holmes's historic dwelling at 221-B Baker Street. For me, these were two of London's greatest attractions.

## FROM ROGUES TO PIRATES

Then came the crowning outrage. The night before we were scheduled to leave London, Bemis sent Polly a fax that arrived while we were enjoying a celebration dinner - it was my birthday. Bemis's timing was uncanny, as if he knew our precise timetable - and perhaps somehow he did, as later events suggested. This message came from the office of his attorney, Richard Robol.

Once again he warned us that we did not have *his* permission to dive the *Lusitania*, and that to do so he "believed" might be "possibly an act of international piracy." I've been called a lot of names in my life; never a pirate. Did this mean that I had to wear an eye patch?

To this absurd notion Robol added, "you are also on notice that, in addition to civil consequences, in the event a court of competent jurisdiction finds that a violation has occurred, under the law of some jurisdictions, the court may also impose not only fines and imprisonment, but also forfeiture of any instrumentalities, equipment, vessels and the like used in conducting criminal acitivities (*sic*)."

This was a typical lawyer scare tactic, and, as we were later to learn, a tactic that Robol used consistently whenever his legal grounds were uncertain. Robol's accusations were carefully couched in catch phrases such as "believe" and "in the event" - they were not backed by fact or legitimate evidence. Robol's rhetoric brought howls of laughter. Instead of frightening and intimidating us, he had provided us with a humorous nightcap with which to laugh ourselves to sleep.

Coincidentally, late that night a prowler broke into the house through the front door and stole Simon's wallet from a shelf in the entranceway. Gone were his credit cards, driver's license, other important documents, and a small amount of cash. The wailing of the burglar alarm and Polly's scatological curses - spouted as she pounded down the stairs - scared off the miscreant before he got any farther inside. Nothing like this had ever happened before in the neighborhood. Ironically, it was the first time ever that Simon left his wallet on that shelf, and afterward he couldn't think of any good reason why he left it there.

## On the Way at Last

Then came the day of packing the mountain of equipment into the vehicles. We formed a caravan with the loaned Peugeot van, the Tapsons' minivan, and two cars. Richard Tulley was accompanied by his wife and months-old baby.

From London we drove to Swansea, in Wales, then boarded the ferry for Cork. Due to a misunderstanding, we Yanks expected the ten-hour crossing to be made during the day. We had not reserved staterooms. But the crossing was made overnight. We slept on benches or on the deck under bright lights while the Brits slept cozily in bunks.

Polly warned us wisecracking Americans about the serious volatility of the Irish political situation. She sounded quite sincere when she said that offhanded snide remarks that were overheard by the wrong people could result in a bombing or a shooting. Ireland was a hotbed of aggressive guerrilla warfare. We accepted her admonition with great solemnity.

From Cork it was only a forty-five minute drive to Kinsale. We checked into Kieran's Folk House and immediately got to work. We were planning to dive the very next day, so we set up all the equipment for a dry run in the courtyard: personal gear, oxygen cylinders, sling bottles, the dec station complete with weights, floats, and lines, and so on. The only job that we didn't have to do was to decant gas, because we arrived with our primary tanks already filled with trimix.

Denis Kieran had had a temporary plywood addition tacked onto one end of the Folk House in the parking lot facing the street. Two "quads" of premixed trimix had been delivered and forklifted into the storage unit and were locked behind plywood doors. This was to prevent vandals from wrecking the expedition by opening the valves on the storage cylinders and letting out the gas. (A quad is a pallet holding sixteen cylinders arranged in four rows of four.)

The low-pressure compressor had also been delivered and was parked in front of the addition. This compressor was the size of a small car, was driven by a gasoline engine, and was built into a trailer for ease in towing. It was used to drive the booster pumps. The boosters were set up, the hoses were connected to the compressor, and the system was checked out so that everything would be ready to use the next afternoon after the dive.

All that day and into the night we were engaged in continuous activity. To the uninitiated we must have appeared like a colony of ants dashing aimlessly about - but we each had specific jobs to do and we all knew what we were doing. The only respite after a long and tiring day was dinner - a leisurely indulgence that became a daily anticipation.

I brought out the one hundred postcards which Ken Marschall

gave me for my personal use, and distributed eight cards to each member of The Starfish Enterprise. When I explained that Marschall wanted me to take *his* cards taken down to the wreck, and that I was going to do the same with my own, everyone decided to follow suit. We did this by wrapping the postcards in plastic and sticking the packet inside our drysuits. I couldn't take all of mine and Marschall's down at once because the packet was too thick. I took down two packets on each dive until they had all been on the *Lusitania*.

That night as I lay in bed I had time for introspection. I looked forward to the morrow with eagerness and a little trepidation. No longer did I shake with the abject terror that preceded my first descent to the *Ostfriesland* four years earlier. Nevertheless, not lightly did I face a 300-foot dive no matter what the circumstances or my level of experience. Cockiness leads inevitably to misadventure.

The street below was filled with the noise of Saturday night revelers. Because my second-story window faced the narrow thoroughfare, I could hear every word of the song and dialogue of the younger set as they left the downstairs bar unabashedly inebriated. Despite the volume and occasional outbursts, I didn't stay awake for very long.

## THE MORNING ROUTINE

The 6:30 wake-up call came too early - especially for an author used to a somewhat Bohemian lifestyle. It was the duty of the dive marshals (dive masters in the States) to wake up the team members via telephone. For the length of the project, Christina Campbell was liaison with the folk house staff for arranging meals. Since the restaurant didn't open till 8 a.m., she was responsible for setting out the simple fare of buns, cold cereal, and coffee - unless she could convince Kieran to have his cook come in early in order to prepare a more substantial hot meal consisting of eggs, sausage, and French toast. She also had to ensure that the box lunches were packed. I'm not normally a breakfast person, but I forced myself to stoke up on food because of the prodigious amount of energy I expected to expend before lunch time.

Dive marshals for the first day were Christina Campbell and Nick Hope. They were responsible for organizational details as well as for maintaining the timetable. Slowpokes were goaded verbally.

The first morning was the easiest because our gear was already packed. On succeeding mornings we analyzed our gases after breakfast, then loaded the tanks and kit bags into the Peugeot for Owen to drive to the dock. It was a pleasant half-mile walk through town and along the quay side to the marina. Kinsale was a clean, quiet, rural community as full of happy faces as it is of people. No one bustled. On week days, students stood patiently on street corners waiting for the bus, pedestrians strolled along the sidewalks on their way to work, clerks and shop owners greeted tourists cheerfully.

What surprised me was the thick Irish brogue. Most people talked in comprehensible English, but some affected an accent that was so heavily broken that I couldn't understand a word they were saying. It was as if they were conversing in a foreign tongue. What shocked me most about the language, however, was the lilting, songlike pattern which many people pattered. With my eyes closed I could imagine myself surrounded by a host of happy leprechauns. I didn't know that the Irish actually spoke that way - I thought the melodious inflection given to cartoon characters was an invention of Disney animation.

Depending upon the tide, the inner harbor might look like a mud flat or an idyllic basin for sailboats. When the tide was out, boats canted over in the soft black ooze; when the tide was in, the boats floated idly at their moorings. The difference in height was ten feet.

Nic Gotto and Howard Weston greeted us at the quay side. They tied the boats to the wharf, and we began the onerous process of unloading several tons of gear from the vehicles and loading it onto the boats.

Each team was given a number for the day - in accordance with the order in which they were to enter the water - and each stowed his gear accordingly on the *Sundancer.* Tanks and sling bottles were stowed with shoehorn precision. Position number one was situated against the transom, position number two was split on either side abaft the engine cover, positions three and four were squeezed in the space between the sides of the engine cover and the gunwales, position five ranged along the port gunwale between the leading edge of the engine cover and the rear wall of the cabin, and along the rear wall of the cabin. Dive marshals doubled as support divers - they did not have double tanks and sling bottles, just singles.

Regulators were emplaced before the boat got under way, and all systems were checked for leaks and malfunctions. Each team was responsible for rigging his oxygen bottles on the dec station, which was then packed in buckets and stowed aboard the inflatable. Richard Tulley oversaw the operation to make sure that the dec station was ready for instant deployment.

The process of loading the *Sundancer* was one of near anarchy. Because the gear was crammed so tight in the limited space available, great effort was required to move about the after deck - stepping over tanks, kit bags, loose paraphernalia, and fellow divers. If someone needed an item such as a tool or a piece of equipment, it was easier to ask the person standing next to it to pass it over than to clamber for it. The problem that this expedient created was twofold: it was difficult to get someone's attention due to the requestee's preoccupation, and it interrupted the requestee's routine and train of thought.

I often found myself shouting a person's name at the top of my lungs from less than arm's length away, and had to call out multiple times before I got that person's attention. When people yelled my name and eventually broke through my barrier of absorption, I responded dazedly; then, when I returned to the task at hand, I had to reorient my thoughts. If I was in the middle of a sequence of executions, I had to start over again in order be sure of not missing an important step.

For example, I might be checking my tank pressures when Tulley wanted to load the dec station. I had to stop what I was doing, find the regulator for the oxygen bottle, fight my way through people and equipment on the boat in order to regain the dock, emplace the regulator, check that the tank was full, rig the lines from the float to the cylinder to the weight, then wait for Lander to do the same before I could stow the weights in the bottom of the bucket, put the two cylinders in the bucket, then loop the line carefully so it would deploy without snagging. If Lander was otherwise occupied at the moment, I might go back to checking my tank pressures and emplacing my primary regulators, then get stopped by someone who wanted a screwdriver that was on the shelf behind me, then discover a minor leak that required a wrench to tighten a hose connector. Meanwhile Lander, out of phase with me and Tulley, might need help with loading our dec station onto the inflatable. I would stick the wrench in my pocket, clamber off the boat, and get stopped on the way to the inflatable by someone who wanted to borrow an oxygen analyzer.

This was the same "controlled confusion" that I mentioned earlier. My voice was but one among the loud general hubbub, and my personal occupation was somewhat subservient to the needs of the group. In the example above, think how difficult it was for Tulley to attend to his own equipment while directing nine distracted divers to assemble their dec stations so he could supervise the loading.

## First Day on the Wreck

After an hour of intense work and total concentration we were ready to get under way. The two boats pushed off from the wharf. Campbell rode in the RIB with Weston. The rest of us were crammed into the wheel house of the *Sundancer*, in imitation of college pranksters trying to fit the entire student body inside a phone booth. A few began the trip by remaining on the deck, but when the boat left the protected waters of the harbor and bounced head-on into the waves, the salty spray that washed over the cabin roof chased them back inside. On future excursions, some people donned their drysuits and sat on the engine cover.

Weston and Campbell took a beating in the RIB. Weston was

wearing rain gear, Campbell her drysuit. The boat pounded over two- and three-foot waves with teeth-loosening violence, drenching the occupants. The ride was exhausting, Campbell was soon worn out. After her feedback that night at dinner, we decided that Weston would have to suffer alone. The support divers had to maintain their energy levels in order to help as they were supposed to.

The ride to the wreck should have taken no longer than an hour and a half - the distance from the dock to the site was sixteen miles. But several times Gotto had to idle the *Sundancer* in order to let the inflatable catch up. On flat seas the RIB could easily outflank the *Sundancer*, but it couldn't handle waves as well as the larger and heavier vessel.

There were no whitecaps, just long heaving swells and lots of wave action. Visibility deteriorated the farther we got to sea. We soon lost sight of the Old Head of Kinsale, and twice lost sight of the RIB in the fog, forcing Gotto to turn the boat around in order to search for the inflatable.

Once near the site, Gotto switched on the depth recorder and passed back and forth over the wreck. Within minutes he dropped the shot line.

Tides in the area flowed to several knots - impossible to swim against and likely impossible to decompress in while holding onto a static line. Thus the wherefore of the drift decompression system. Dives could be conducted only during slack: the time when the tide changed direction. Slack tide occurred about four times per day - or more precisely, twice during daylight hours and twice at night. The duration of slack was forty-five minutes. After three-quarters of an hour the tide accelerated: slowly at first then with ever quickening speed. For this reason the window of opportunity was strictly limited.

Polly and Simon Tapson were designated as team one. They sat on the transom fully kitted except for sling bottles. When Gotto gave the word to get ready, Nick Hope and others picked up the sling bottles and helped to clip them onto the Tapsons' harness D-rings. When the Tapsons gave the ready signal, Gotto maneuvered the boat close to the marker float, and shouted, "Go!"

The Tapsons rolled backward into the water. They swam to the marker float, grabbed the descent line, and disappeared beneath the choppy surface of the green speckled sea. Team number two - Dave Wilkins and Paul Owen - moved immediately from the rear of the engine cover to the transom; they were already wearing their doubles. Team three - Richard Tulley and Jamie Powell - scooted up to the rear of the engine cover and began to don their tanks.

The Tapsons landed on the sea bed some 300 feet beneath the surface. The bottom was dark and visibility was limited. They found themselves in the wreck's debris field but the hull was nowhere in sight. They searched the vicinity for several minutes. Debris lay strewn all over: broken beams, twisted sheets of metal, unidentifiable junk. They knew they were close to the hull but somehow they seemed to be skirting the edge of it. Finally, failing to locate the main wreckage - to which they were supposed to secure the shot line - they ascended. They did not release the "pellet," as the Brits called the marker buoy that was a rigid, non-crushable float, because that was the signal that they had tied the shot to the wreck.

Campbell deployed the Tapsons' dec station. She also ascertained that they had not tied in the shot line. While the RIB drifted with the dec station's marker float, to which the Tapson's had transferred, Gotto maneuvered the *Sundancer* back to the wreck and tossed another shot line overboard. Wilkens and Owen rolled into the water. This time the shot weight landed directly on the hull. They secured the line to the wreck, and had ample time leftover for exploration before beginning their ascent.

Team three - Richard Tulley and Jamie Powell -were kitted and eager to hit the water. They teetered on the transom's edge, waiting for the signal from the captain to roll in. But now there were two teams decompressing in different locations. The Tapsons were adrift, Wilkins and Owen were holding onto the shot line and waiting for their oxygen supply. This logistical situation had to be managed with due deliberation. Furthermore, so much time had elapsed since the Tapson's initial entry that slack tide was coming to an end.

Dive marshals Hope and Campbell put a halt to further diving, a difficult and unpopular call to make. It was a bitter disappointment for the six of us who stood so close to the *Lusitania* - temperamentally as well as physically - and whose opportunity to achieve a culminating ambition was suddenly taken away. Yet we all concurred in the decision because it was proper under the circumstances. Safety was paramount and superseded any chance of increased risk.

A couple of hours passed as we waited anxiously for the four lucky divers to surface, and to tell us how it felt to have dived the *Lusitania*. They were ecstatic.

Unfortunately, their efforts went for naught, for the shot line chafed through.

## Dinner Discussion

The evening meal was debriefing time. After placing our orders we discussed the events of the day. The primary focus was on what went wrong (if anything), what might have gone wrong, and how to improve the efficiency of the diving regimen. Campbell addressed her concerns about dive marshals riding in the RIB - intended to make more space available on the *Sundancer* - and how fatiguing it was. No one objected to her suggestion that we scrap the idea and let both dive marshals ride in the *Sundancer*.

The time loss that prevented three teams from diving was unavoidable. Tie-ins seldom go like clockwork, and the delay between the entries of teams number one and two was caused by the Tapsons alighting in the debris field instead of on the hull - certainly no one's fault. We had to wait for them to complete their bottom time before letting the second team enter the water. Once the downline was secured, we expected the dives to proceed more efficiently.

## Day Two - Ante Meridiem

The morning routine was changed primarily by the designation of different dive marshals. According to the rotation protocol, the previous day's dive marshals became team number five, team number one became surface support, then team number two became team number one, three became two, and so on. (The convention was overruled in the beginning in order to give the Yanks more time to familiarize themselves with the decompression system and the in-water method of assembly. That's why Hope and Campbell tied in on the second diving day.)

The only other modification to the agenda was the dive schedule. Tides are caused by the force of gravity exerted by the moon (and to a lesser extent, the sun). Because the moon revolves around the Earth in the same direction in which the Earth rotates on its axis, the moon constantly falls behind: it doesn't reach the spot over which it hung twenty-four hours earlier. This is why the moon rises fifty minutes later each day. As a consequence of the tides rising and falling in conjunction with the phases of the moon, the diurnal cycle of slack between tides progresses predictably. The lag is identical and ongoing: fifty minutes per day. We didn't get to sleep any longer, but we weren't always so rushed in the morning.

The day was again warm and sunny but a bit more blustery. How windy we didn't realize until we passed the Old Head of Kinsale and ran head on into mountainous seas that washed over the bow and against the wheelhouse windows. It was worse for Weston in the inflatable. We stuck it out for a couple of miles, hoping that as we got farther offshore the seas would moderate, but they only got worse. Gotto was willing to go for it, but he suggested that we turn back and call off diving for the day. Reluctantly, we concurred.

On the way back to Kinsale an urgent message came over the radio. A fisherman was calling for help. At first we couldn't make out what he was saying. Irish brogue and static conspired to obfuscate his diction. Eventually we deciphered that a fishing boat had sunk and that men were in the water. But where?

As it turned out, we were passing close by the fishing boat at that very moment. The fisherman left his radio in the cabin and ran out on the deck and waved to us. Gotto turned the *Sundancer* immediately in his direction. The boat lay half a mile away. Then the fisherman stopped signaling in order to pull a man over the gunwale. When he got back on the radio he said that he had three men on board and that they were suffering from hypothermia.

What better group to come to their rescue than a boat load of divers who had bags full of thermal garments and insulated underwear? How fortuitous that we had turned back at a time and on a course that coincided with their needs.

The plight of the fishermen was desperate. One huddled in the miniature cabin, another shook violently outside the door, the third lay on the deck unconscious. The *Sundancer* bristled with rescuers who were prepared for immediate action, complete with a boarding party and plenty of drysuit underwear. When Gotto brought the *Sundancer* alongside the fishing boat, several of us held the boats together while others swarmed aboard to lend assistance.

No time was wasted in stripping down the fishermen to their underwear and stuffing them into one-piece Thinsulate suits. Dry socks and woolen caps completed their new ensemble. The man in the cabin rallied quickly, for he had been the first to be pulled out of the frigid water. The one sitting outside the door soon came around, although he continued to shiver for quite some time. The one lying on the deck was near death: his skin was blue with cyanosis and his cardiac ailment had been aggravated by the trauma of cold water immersion. Worse, his heart medication had gone down with the boat. Our emergency oxygen kit was handed over. A steady supply of oxygen was delivered through a mask that covered his mouth and nose. He soon roused and told us what medication he needed.

Gotto radioed ahead to have an ambulance and medication waiting at the dock. As soon as the fishermen were stabilized, both boats got under way. A crowd gathered in the marina: curious onlookers who had been alerted by the arrival of the police, an ambulance, and a medical team. The medicos took charge of the patient most in need, and carried him off the boat in a stretcher. The other two men were able to walk. Simon and Polly drove them to the hospital in their minivan, where they were evaluated and treated for hypothermia.

We had been in Ireland for two days, and I had yet to achieve the purpose of my being there. Yet at that moment I felt that the trip was already worthwhile - even if I never got to dive the *Lusitania* - because we had helped in the saving of life. In a moment of spiritual introspection, I imagined that perhaps in some grand, cosmic scheme, the true purpose of The Starfish Enterprise was to be on the scene when some fellow human beings were in need.

Now we got the full tale of events from the lone fisherman who had initiated the rescue. The three fishermen were setting out their net when a wave hit their small boat broadside and swamped it. As the boat went down, the net floated off the deck and ensnared the men in the water. Entangled, they were dragged down, and it was only by struggling with every ounce of strength they possessed that they managed to keep their heads above the surface.

The lone fisherman fortuitously noticed their plight. He raced to where the men bobbed and were struggling to save their lives. Single-handedly, he managed to pull the first fisherman into the boat and release him from the net. He also pulled in the second fisherman by himself. But the third - the one with the heart ailment - was rotund and overweight and proved too much for him to heave aboard alone, especially as his legs were more entangled in the net than the others. Fortunately, by that time the first man rescued had recovered enough to help, and together they hauled in the third victim.

We divers certainly lent some much needed assistance, provided warm clothing for the sufferers, and saved from death at least one and possibly two of the fisherman - those who might have passed beyond the recovery stage without warmth and medical attention. But the real hero of the day was the lone fisherman, whose extraordinary solo efforts set the course for events that followed.

## Day Two - Post Meridiem

I think of naps as sleep for children and lazy adults. But so busy were my days since my arrival in London that already I was worn out. Given the afternoon off I fell promptly into a deep slumber. Simon woke me up in the late afternoon with an offer to go to Cobh to visit the cemetery where the *Lusitania's* victims were buried, and to see the memorial in the town square. Several of us went.

That evening, back at the folk house, two of the rescued fishermen came to thank us for our efforts and to drink a toast or two or three on our behalf. Their friend was doing well in hospital, but was being held overnight for observation. The fishermen were a jovial lot whose language might as well have been Greek for all I could understand.

Word travels fast in small communities. It was no secret that we had come to Kinsale to dive the *Lusitania*. Nor did the locals have a problem with that. They seemed to be thrilled by the fact that outsiders had an interest in their heritage. After the rescue and the appearance of the story in the newspapers, we became celebrities and were treated with fond regard. During our perambulations in and around town, the locals seemed to know who we were, and on numerous occasions we were recognized - singly or in groups - as the divers who came to see the *Lusitania*. I felt as if I had become part of the clan.

My afternoon nap did not prevent me from sleeping that night.

## Day Three – The *Lucy* at Last

The sea was pale green like a crude piece of jade. Plankton floated in the liquid realm much as intrusions marring a crystal lattice, causing the light to play tricks upon the eye. Distances were difficult to estimate because of the lack of perspective: the yellow polypropylene rope dropped away beneath me like a solitary strand from a giant spider's web, fading into the haze.

As I plunged through the stratum of floating micro-organisms the color of the ocean darkened. The water below was eclipsed by the living bulk adrift in the broad Irish sea. Down I went into the biotic miasma, going deeper, growing darker. I felt no sense of weight, or of cold, or of depth - only of the gathering gloom. Not until my eardrums grew taut did I have an indication of descent. I worked my jaw to relieve the unequal pressure.

I peered below anxiously. My controlled breathing pattern was timed to the cadence of my mitted hands pulling my body and five compressed gas cylinders down the limp shot line. Inhale, pull, pull; exhale, pull, pull; inhale, pull, pull . . .

The blackness became absolute at about 200 feet. No ambient light reached these depths. Despite the sun shining bright in the cool cloudless sky, this was essentially a night dive. I could *feel* the "poly" rope but I could no longer see it. Masochistically, I withheld switching on my light in order to savor the unique experience of sensory deprivation. Then came a short burst of brilliance from far below, followed rhythmically by evenly spaced flashes. One of the teams before me had secured a signal strobe to the chain at the base of the shot line.

I was alone. For reasons of her own, Lander had opted not to dive.

I let go of the rope as I dropped past the strobe clipped to the terminal chain. *Then* I turned on my light and took a good hard look at my gauge panel. It had taken only four minutes to reach the wreck. With no tide or current, descending the shot line was as easy as sliding down a firehouse pole. Hope and Campbell had carried the chain to a rectangular window whose glass was conveniently shattered, looped the chain through the brass frame, then secured the shackle and released the pellet.

I slid down the angled deck and settled onto the rock-strewn sea bed. Because I was breathing a mixture of helium and air, I did not feel the narcotic effect that was induced by too much nitrogen under pressure. I was as clear-headed as if I were in 150 feet of water. Yet my heartbeat quickened with trepidation for, although I could not sense the depth as I generally could when breathing air, I knew intellectually that I was far beyond the pale of ordinary scuba.

I illuminated the area about me and had my first glimpse of the wreck of the *Lusitania*. Almost immediately I spotted a porthole lying loose in the debris field, then another, then one that was firmly attached to a steel plate. Because the wreck lay at a sharp starboard angle, the superstructure had weakened and sloughed off into a confused heap, much as a wooden house that had cascaded down the side of a hill in a California mud slide. Most of the rivets and thin metal plating had long since rusted away, leaving behind a jumbled pile of beams and broken bulkheads.

In the near distance the hull rose upward. I could clearly see two light beams thirty feet away, carving white swaths through the blackness like laser swords in a space epic. Since my projected bottom time was limited to twenty minutes, every moment was precious, and I did not want to waste even a second. I moved closer to Chatterton and Yurga, ostensibly my buddies on this dive but in actuality I was a tag-along and went my own way.

I had gone only a few feet from the shot when my attention was attracted to a curious brass cylinder the size of a partially smoked stogie. It was threaded at one end. I knew its import immediately, relegated the information to the back of my mind, then continued on my way to make the best of my remaining time. The hull loomed above me beyond the slanted deck where once walked passengers through Edwardian era splendor . . .

. . . but I saw no majesty here. Instead I was faced with a collapsing, shrunken ruin whose original beauty survived only in memory or imagination. The port rail that should have towered more than eighty feet above the sea bed reached no higher than thirty feet. The interior decks that should have stood neatly atop each other like floors in an office building were squished together and buckled up and down like a squeezed accordion. My impression was that a huge hydraulic foot had stood upon the wreck and tried to crush it flat - and very nearly succeeded.

Keeping an eye on the flashing strobe and the lights of the two Johns, I followed the debris field along the starboard edge. I didn't have my camera with me. This was an orientation dive. Disarticulated hull plates gave no clue of order or logic in construction. The steel sheets lay scattered about like discarded chips of wood from an axman's handiwork. The farther I strayed from the shot the more apprehensive I became. It was paramount that I return to the surface up the "poly" rope in order to reach the decompression station and the oxygen cylinders which were hanging at twenty feet. When I got out of range of the signal strobe I was ready to turn back.

For a moment I was caught between increasing alarm and wild exhilaration. I was a long way from the surface both in distance and in time. Every minute down here added eight minutes to my decompression schedule, and time passed quickly at 300 feet. My gas supply was dwindling fast. I checked my gauges to make sure I could remain a few moments longer.

Finally, the anxiety became too much for me. Although I wanted to continue my exploration, I turned and swam back in the direction of the shot. In my personal absorption I had lost sight of my fellow divers.

My heart was pounding hard despite the facts that my equipment was functioning perfectly and that I had sufficient gas to complete my allotted time on the bottom. I felt so far from help, so far from the surface and sunlight and the open air. I was in a constant state of apprehension. At this depth, where things can go wrong in a hurry, I needed to be on edge. Yet, although I recognized the necessity of threat-induced alertness, I did not enjoy the feeling. Even backtracking over familiar pieces of wreckage - symbolic bread crumbs in the trail to the shot - my inner doubts were not dispelled. This kind of diving was not to be taken lightly. Nor did I.

When I reached the point at which I should have seen the comforting flash of light from the signal strobe, and did not see it, I was gripped with the same measure of fear that must have gripped those who saw the wake of an oncoming torpedo that fateful afternoon of May 7, 1915 . . .

By now my gas supply was nearing the lower margin of safety. I was eighteen minutes into the dive, with only two minutes to spare if I were to maintain my planned schedule. An extra five minutes of bottom time would incur forty more minutes of decompression, and would seriously compromise my deco gas supply. I rose several feet above the seabed and worked my way aft. In total darkness I played my light on the wreckage and looked for the yellow poly.

Then I saw the welcome flash of the strobe. Due to the angle of my return, a large section of hull stood between me and the shot, placing the ascent line temporarily out of my field of view. Now, instead of being behind schedule, I had a minute or so to spare, so I hovered in the vicinity and soaked in as much of the experience as I could.

If only I could relocate that curious bronze "stogie" I had spotted in the debris field, and which I had recognized immediately upon picking it up and noticing the threaded end. It was a primer, or detonator: the explosive device that screws into the base of a cannon shell.

It was time to ascend. Slowly I rose up the yellow shot line. At 250 feet the water turned from stygian black to somber dark green. At 180 feet I could read my gauges by ambient light. At 150 feet I paused for my first decompression stop. I was still breathing trimix-14/43 from the tanks on my back. At 130 feet I switched to my first sling bottle, which contained nitrox-32. Many minutes later, at 60 feet, I switched

## Grand Adventure on the *Lusitania* – 275

to my other sling bottle, which contained nitrox-50. Increasing the partial pressure of oxygen in my breathing mix accelerated the decompression, which is inordinately long when breathing helium mixtures because helium is absorbed by the tissues faster than nitrogen. Decompression on bottom mix would take twice as long.

I read the names on the slate at 60 feet. Only three names remained unchecked: mine, Owen's, and Wilkins's. I used the attached pencil to place a check mark next to my name, then moved up to my next stop at 50 feet by leaving the shot line and moving across the breakaway line (which for some reason I could never fathom, the Brits called the lazy shot.) This breakaway system was identical to the system that I used on the *Monitor* to make the drift decompression possible.

The last team was responsible for ascertaining that no one else was still down, then unclipping the carabiner that secured the decompression station to the shot line. By that time the tide was already picking up and putting a fair strain on my arms as I hung on. The tension ended abruptly. The entire team of divers then drifted effortlessly with the tide, like a group of travelers on an airport slidewalk.

After forty minutes of staged decompression, ascending slowly at ten-foot increments, I reached the dec station and found my oxygen cylinder. I pulled loose the extra-long hose, placed the regulator in my mouth, and for the next hour I floated next to the PVC pipe which separated my station from the two adjacent stations. Here above the thermocline the temperature was 52° (it was 47° on the bottom). Because I was using argon for drysuit inflation, I was comfortably warm.

The Tapsons were dive marshals for the day. Simon finned past and flashed a questioning signal with his hand, the thumb and forefinger touching. I returned the okay sign. He was breathing from a single tank of air. Clipped to his harness was an emergency oxygen bottle complete with regulator, to be used by anyone who ran out of oxygen or who had a regulator malfunction. Polly directed retrieval operations on the surface.

I broke into the air after more than two hours of total elapsed time. The chase boat sped to my side and the coxswain asked if I was okay. I told Weston I was. In the distance I could see the *Sundancer* and someone climbing up the ladder.

After recovering the diver who had surfaced previous to me, the *Sundancer* turned in my direction. Gotto pivoted the boat expertly during the final approach. He idled the engine and flipped the ladder over the transom.

I was still on the lower rungs, unclipping my sling bottles, when Polly leaned over the side. "How was your dive?"

I gave her my typical enthusiastic response, "*Un*-believable," for truly it was.

## Location, Location, Location

I won't describe every dive I made in such detail, only how they differed from my first and what follow-up observations were made by me and my fellow divers.

Initially we didn't know where the downline was tied to the wreck. The consensus was that it was positioned somewhere amidship because no one reached an end (bow or stern). My job was to debrief everyone with respect to their observations, piece together the reports, and determine where on the wreck the shot was located. This was not as easy as it sounds.

I have already compared a diver on the hull of the *Andrea Doria* to a flea on an elephant's back. In the manner in which a flea must determine its whereabouts by identifying distinctive parts of an elephant's anatomy, a diver needs to recognize a shipwreck's characteristics: hull features, superstructure, artifacts, and so on.

A diver on the *Lusitania* is more like a blind flea on an overgrown, mutated mastodon. The *Lucy* is larger than the *Doria*, the deterioration is more pronounced, and the blackness is nearly absolute. Only in the narrow cone of artificial light can a diver see wreckage and debris. Then he has to utilize all his wreck-diving experience to make sense out of seeming disarray.

Each dive is like a snapshot. Each snapshot is equivalent to a puzzle piece. When a collection of snapshots is assembled as a mosaic, a larger picture is formed. This synthesized image is what a wreck-diver creates in his mind. As a group, we collated our mental snapshots and overlaid them on the ship's plans and what we knew about the wreck's state of collapse. Much debate ensued. Even so, several days passed before I was sure of our location on the hull. In retrospect - when I had fuller knowledge - I felt as if I should have figured it out sooner.

The two whose powers of observation I found the most remarkable were Dave Wilkins and Jamie Powell. Each seemed to have a photographic memory of his dive, and could describe the wreckage in the minutest detail. I made sketches of the hull on which I superimposed objects which were noted and identified, by me as well as by others.

I find it difficult to believe that I was confused for a while by a capstan and a set of double bitts that I noticed about halfway down the deck. Instead of jumping to the obvious conclusion that we were diving close to the bow or stern, where a capstan and bitts ought to be (as depicted on pictures and plans), I wondered what those items were doing amidship - another example of how a preconceived notion can cloud otherwise sound judgment. Slowly the pieces of the puzzle came together, and I was able to visualize the wreck as a whole.

## A Picture of Despair

The keel was contiguous but the hull was broken by two major cracks: one amidship and one just forward of the fantail. Our shot was secured between the cracks, and closer to the sternmost crack. Forward of the sternmost crack the curvature of the hull was evident as it formed the counter stern. The crack was actually a gap some twenty feet across going all the way to the sea bed.

Abaft the sternmost crack lay a cylindrical steel casing some three to four feet in diameter, looking very much like a mast but in the wrong place to be the after mast. Besides, the cylinder was located only about ten feet below the port bulwark - a mast was placed on a ship's centerline. This cylinder angled downward toward the stern for a length of twenty or thirty feet.

The fantail appeared to have been largely demolished, likely by Oceaneering when the saturation divers cut into the specie room or blew off the propellers.

Slightly below and forward of the shot line clung the remains of lavatories: the broken stubs of porcelain sinks and commodes protruding from a floor of patterned tiles. The boundaries of the lavatories were delineated by being raised slightly above the deck, and by the existence of boards which were the bottom planks of the walls.

Farther below the shot and some sixty feet forward yawned a rectangular hatchway offset to port. About halfway to the sea bed, on the centerline, and slightly forward of the hatchway gaped a circular opening which was the well over the saloon and second class dining room. This opening beckoned to Dave Wilkins. He swam into the dark interior for about ten feet and emerged from the rectangular hatchway. During this transit he noticed steam pipes which may have been condenser lines in the engine room.

More lavatories were located forward of these deck openings.

The sea bed directly below the shot was covered with debris: crunched steel plates containing portholes (round and rectangular), many of which were unattached; square brass frames, each with four eighteen-inch panes; framing, I-beams, and partially intact sections of superstructure.

At this point the deck's angle of incline was 55° to 60° off vertical.

Any doubt about the location of the shot was removed when word came back that the stern docking telegraph had been discovered lying loose on a section of decking on the sea bed. Not in my wildest dreams did I imagine that so prime an artifact would remain after Oceaneering's thorough salvage of the wreck. Saturation divers had unlimited bottom time in which to collect anything that lay loose - and to disconnect anything that was attached. Yet they left this rare and exciting item behind.

Not only was there one docking telegraph, there were two - plus the railing that encircled the docking bridge.

## Compression of the Hull

More dramatic than specific artifacts left behind was the condition of the hull. The *Lusitania's* original beam was 88 feet; her height from keel to skylight was over 100 feet. With the main hull lying at an angle of approximately 45°, the longest vertical extension became the diagonal. The minimum bottom depth was 305 feet (10 feet deeper at high tide, and deeper still at exceptional spring tides). Thus the wreck might originally have risen to a depth of some 200 feet. Even allowing for some crushing of the starboard turn of the bilge - due to the massive weight of the hull that it now supported, and which it was not designed to support - the least depth should have been in the neighborhood of 220 feet.

John Light reported in the 1960's that the least depth over the wreck was 240 feet. The highest point that we found in 1994 was 270 feet. Nor was there any reason to believe that the wreck will not continue to collapse. That is what shipwrecks do - from the time of sinking until they no longer exist. Ballard's description of deteriorating shipwrecks being "preserved in various stages of destruction and decay" is utter nonsense: an absurd notion that can result only from scant experience with shipwrecks. Deterioration is a process, a progression, the downhill slide of ever-increasing entropy; it is not a state of inertness. A wreck does not attain stasis until it lies flat and dissolves into nothingness.

The shortening of the wreck's vertical dimension was the most obvious and distinctive feature of the progression of collapse. The decks have been squeezed together by the breakdown of their supports: that is, the hull and vertical beams. Because of the wreck's list, sheer forces caused the supports to collapse laterally at the same time that they were weakened by structural failure. Picture a free-standing metal-frame storage shelf that is bent so far sideways that the shelves practically touch.

This condition was most noticeable when I peered through portholes in the hull - portholes from which the glass was missing. Instead of seeing the vast empty space of an interior compartment, I saw thick wooden deck beams pressed tight against the inside rim of the porthole. The decks lay atop each other like playing cards in a stack - as if a tall house of cards had collapsed, as indeed it had.

It appeared to me as if the wreck originally landed neither on its hull nor on its side, but at some angle in between - probably favoring more than 45° from vertical. Due to the interior collapse described in the previous two paragraphs (the loss of transverse support) the port hull has caved in to the point at which in many places it now rests almost horizontal, giving the appearance of a nearly level deck.

## Daily Drudgery

I am not exaggerating when I remark that practically the only rest I got during the trip was on the dec station, while drifting effortlessly in neutral buoyancy. After climbing aboard the *Sundancer*, the first priority was to stow gear, doff drysuits and longjohns, don street clothes (while maintaining propriety in the mixed-sex setting), and gulp a cup of hot tea that Gotto prepared on the hot plate. Add exhilaration to the rush of activity as everyone tried to talk at once about what was seen and done during the dive, and there is little wonder that the ride back to the dock was not relaxing. I was too stimulated by the experience to take it in stride.

Ashore, the real chores began. Every bit of equipment had to be unloaded from the boat and reloaded into the vehicles, to be taken back to the folk house where it all had to be unloaded. The tanks were deposited outside the boosting station. Simon Tapson and Nick Hope cranked up the compressor, connected the booster pumps, and started filling tanks: doubles with trimix, singles with oxygen. Everything else, including personal gear, was taken into the courtyard for rinsing and reassembly.

The in-water oxygen bottles were stripped of adapters and hose-clamp assemblies, then marked and put aside. Full oxygen bottles were rigged for the next day's use. Each diver was responsible for stripping and rigging his own tanks under the supervision of Dave Wilkins. He also had the responsibility of cutting new decompression schedules when mixes did not analyze to predicted percentages of oxygen.

The original thirty-two bottles of bottom gas were pre-mixed at the factory. It was a simple matter to cascade and boost trimix from the storage cylinders to the scuba tanks. After the pre-mix was gone, Polly ordered a quad of helium and more tanks of oxygen, both deco bottles and storage cylinders. Then, Simon and Hope had the additional task of blending and analyzing bottom mix.

Every day Richard Tulley untangled the mess that the deco system had become after retrieval from the water. He straightened out the ropes and reorganized the PVC spacers, then stowed each of the five stations in its plastic bucket. He was ably assisted in this endeavor by Jamie Powell.

The Johns were in charge of the air compressor station and nitrox blending. This job took so long that often they were still filling tanks when the dinner call came at eight o'clock. They would stop to eat and add their input for the debriefing session, then excuse themselves early and go back to pumping air, sometimes till ten or eleven o'clock at night. Whatever wasn't finished by bed time they completed in the morning.

Their workload increased about halfway through the trip when we ran out of pre-mix. Then, in addition to making nitrox, they had to top off tanks of bottom mix too. This was because after Simon and Hope put oxygen and helium into the doubles, the Johns had to add air up to a precise final pressure in order to achieve the desired blend.

The distance between the booster pump station by the street ("Haskel Land") and the air compressor alcove in back of the courtyard ("Compressor Country") was a couple of hundred feet. Doubles and singles were transported between locations by means of manual labor: a back-breaking job at best and one that fell especially upon my shoulders. This was because my chief responsibility - debriefing divers with respect to their observations of the wreck - was non-essential as far as preparing for the next day's dive, and therefore was necessarily subservient to the more important tasks at hand.

This is not to imply that I was alone in doing grunt work. When a person's primary duty was done, he or she didn't have the rest of the day off to relax while others were working. Instead, each individual asked what needed to be done, then pitched in to help wherever help was needed. The Starfish Enterprise included no laggards.

## Camera Quandary

I took the camera on my second dive. It wouldn't work.

It functioned when I tested it the night before. But at 300 feet, no matter how often I tried, I couldn't get the shutter to trip.

The newfangled camera was fully automatic but its mechanisms were complicated and the functions difficult to understand, what with all the modes and settings and the integrated autofocusing lens. I read the instruction manual thoroughly - twice. I used the camera topside without any glitches. I placed it in the housing and checked the strobe synchronization. The system worked fine until I pressed the shutter release on the bottom. Then nothing.

I was fussing with the control knobs when Lander appeared. She had decided to dive this day, but took her time making her descent - dawdling and hanging back when she should have been charging for the wreck. I waited for her at the bottom of the shot line, toying with my camera. When she finally showed up she didn't look at me, she didn't acknowledge me; she just sped past and disappeared in the gloom. I was nonplused. She didn't reappear until it was time to ascend. In the meantime, I scouted the wreck in frustration and noted subjects to photograph on subsequent dives - assuming that I got the camera to work.

Oddly enough, the camera began functioning when I tried it again on the dec station. I shot a roll of film of divers decompressing.

That night I reread the instruction manual - all eighty-eight pages. I tested the camera both in and out of the housing. It worked perfectly. I made sure all the connections were tight, the batteries were fresh, the housing levers aligned. As a precaution, I disabled the automatic mechanisms and followed the instructions in the Tussey housing manual for setting the focus and strobe synchronization.

The next day I tested the system on the boat immediately prior to entering the water. No problems. This time it worked on the bottom. I cruised around the wreck and shot the subjects I had noted the day before: the window to which the shot line was secured, portholes in the hull, broken toilets in the remains of a bathroom (or water closet), and colorful mosaic tiles.

## CLOSE CALL

The *Sundancer* barely tied up to the wharf before smoke commenced curling out from under the engine cover. We hastily moved tanks and cleared the cover so that Gotto could unfasten the latches. Thick smoke enveloped the deck when he threw back the covers, accompanied by the smell of burning oil and insulation. A fire extinguisher quickly doused the flames.

When the air cleared, Gotto inspected the engine compartment in order to assess the damage. The electrical wiring on both engines was burned to a crisp, as was one of the starter motors. The apparent cause of the fire was a malfunction in the float switch, which failed to turn on the automatic bilge pumps. Rising water created electrical shorts when it flowed over the contacts of the starter motors.

This was a major catastrophe. In the span of a few seconds the continuation of our diving operations had been compromised - consumed in flames, or gone up in smoke, as it were. On the other hand, we were fortunate that the fire had not started an hour or two earlier, when decompressing divers were adrift. Had the boat been disabled at sea the consequences could have been severe, perhaps fatal. At the very least it would have instigated a major Coast Guard rescue operation.

Gotto took the situation in stride and immediately began to rewire the engines. Richard Tulley remained on the boat to help. Additional assistance came in the form of Rich Field and Rob Royle. Polly had invited them to dive with us for a day or so, and they were waiting for us at the dock when we pulled in. Royle was an electrician. He wasted no time in rolling up his shirt sleeves and getting to work.

The rest of us went straight to the folk house to begin the daily grind of filling tanks and maintaining equipment.

## Trouble Compounded

We left late the next morning because Gotto couldn't get the incinerated starter motor rewound overnight. He made arrangements to borrow one, so we were forced to cool our heels until it was delivered. The replacement was installed within fifteen minutes of its arrival. Then we were on our way.

Lander and I were dive marshals. She wanted to set up the dec station, so I stayed on the *Sundancer* to help kit the divers. Because slack tide lasted only three-quarters of an hour, it was important that the first team hit the water at the exact onset of slack, and that the other teams entered with timetable precision: preferably at no longer than five-minute intervals. By strict adherence to this schedule, the fifth team would hit the water twenty minutes after the first, giving them twenty-five minutes before the tide began to run.

Lander deployed the five dec station modules from the RIB, then got in the water and joined the modules end to end with carabiners. Five-foot lengths of PVC pipe kept the modules from bunching up and the oxygen cylinders from banging together.

Meanwhile, Weston was having trouble with the inflatable's outboard motor: it sputtered intermittently and would not run at high rpm's. Finally, the motor conked out altogether. Lander was under water, Gotto and I were alone on the *Sundancer*. We chased after the chase boat and took it in tow. No sooner did we get back to the dec station (which was still connected to the downline) when the port starter motor burned up (it was the starboard starter motor that had burned up the day before). Now the *Sundancer* idled on one engine, and we had no RIB for rescue work. The situation was serious because we had divers in the water.

By this time all the divers were on the dec station. The last team released it from the downline. Lander now had to complete the assembly of the dec station by looping module five back to module one and joining the PVC pipe extension on module five to a carabiner on module one - like a snake biting its tail. The five modules would stream in a line as long as one end was secured and there was current or tide running to keep the connecting pipes taut. But once the station went adrift, the modules doubled back like a carpenter's folding rule. Even after connecting tip to toe, the circle would collapse so that modules opposite each other would come together. To solve this problem, Tulley added rope cross members which ran diagonally from one corner to the other - like an x in the middle of a square. The extra position required by the fifth team was converted into a triangle on the end of the square. The completed construction looked like the home plate of a baseball diamond.

I discussed options with Gotto while he maneuvered the

*Sundancer* on one engine toward the drifting dec station and the eleven divers who were totally unaware of our predicament. I suggested securing a line from the dec station to the side of the *Sundancer*. This was the method I invented for high-current drift decompressions on the *Monitor*. Gotto liked the idea, but thought that the boat would track better if we secured the line to the bow. Lander surfaced as we nosed up to the dec station. By this time I was in my drysuit. I rolled overboard with the line in my hand, swam past her, and clipped the end of the line to the dec station with a carabiner. Lander climbed into the boat.

During this "changing of the guard," Paul Owen switched to his oxygen bottle and discovered that it was empty. Normally he would have signaled his difficulty to the safety diver, who would have handed him the spare. But Lander wasn't in the water, and I didn't have the spare oxygen bottle. I wasn't even wearing a tank. Owen sent up his safety sausage as a sign that he needed help. He continued to breathe his nitrox.

Lander was already out of her suit. I peered down at Owen from the surface and saw him pointing to his oxygen bottle. I swam back to the boat, hastily threw a single on my back, grabbed the spare oxygen bottle, and dived overboard. I handed the bottle to Owen, unable to apologize for the delay and to explain the hazardous situation topside. I then checked each person in turn and exchanged okay signs. There wasn't enough oxygen in the spare bottle for Owen to complete an hour's decompression, so I swam back to the boat, got a full replacement bottle, and took that one to him.

Gotto wasn't sitting idle. He was effecting repairs. He started the starboard engine. With the engine still running, he disconnected its operational starter and installed it on the port engine in place of the burnt-out starter. This enabled him to start the port engine. As long as he kept the engines running there was no need for starters.

Weston wasn't sitting idle, either. He removed the cowling and worked diligently on the inflatable's outboard motor, checking hoses, fuel lines, electrical connections, and so on. Eventually he got the motor started. He fixed it so well that it even ran the way it was supposed to at high rpm's. He speculated that water had somehow gotten into the portable fuel tanks, perhaps through condensation.

The tide and a light breeze pushed us along nicely: dec station, *Sundancer*, and inflatable all tied together. If the wind had been blowing harder against the boats - acting as sails - the resultant speed would have lifted the dec station weights and caused the vertical lines to trail behind like streamers in a storm, taxing the divers' strength as they fought to hang on. Instead, the *Sundancer's* blunt stern presented a broad surface to the water in the direction of travel, and the dec station acted as a sea anchor. No one under water experienced undue strain.

I ascended with each diver when his decompression was finished. On the surface I explained the situation and gave instructions to follow the rope to the boat and then to swim along the side to the stern. Lander was waiting for them at the ladder. She removed sling bottles, helped people up the ladder, and guided them to a seat on the now-closed engine cover, beneath which the engines idled softly.

Ironically, the dives had gone smoothly and with no jeopardizing incidents. Surface support personnel underwent more stress than the divers. The ride back to the dock was uneventful.

## Troubled Waters

Despite the workload, the project would have been a delight for everyone involved had it not been for Bemis's continued harassment. He was making waves that were more tumultuous than the ones we encountered at sea. Since his empty threats failed to intimidate the members of The Starfish Enterprise, he tried to instigate trouble through official channels.

He began the placement of obstacles in our path by retaining a solicitor in Cork. His name was Richard Martin. Martin contacted Polly via the local police, who informed her that Bemis was threatening to file an injunction that would keep us off the wreck. Polly promptly sought the advice of local counsel: Grattan Roberts, also in Cork. In Roberts' opinion, Bemis could not get the High Court to issue an injunction because, as he wrote to Martin, "your clients Title to the wreck is seriously disputed." Once again, Bemis's claim was worthless without proof - proof which he consistently refused to provide.

Roberts wrote a letter to Martin which he also sent to the Court. "Our clients are quite entitled to dive in the vicinity of wrecks off the south coast of Ireland, Kinsale being a traditional base for divers who for many years have been coming here to dive in and around the area of wrecks. Diving is a very popular sport in the Kinsale area and makes a substantial contribution to the tourist industry in the take up of accommodation, boat rental and general spending in the town.

"We believe that your clients threats to obtain an injunction are ill-advised and if an injunction were granted would have serious repercussions for the tourist industry . . . Our clients have already won the hearts of the people of Kinsale and Cork in that they were instrumental in the saving of three lives when a fishing vessel sank off Sandycove Island at the entrance to Kinsale and the occupants became entangled in nets. In particular, one fisherman, Mr. Simon Quilligan of Cork, was recovered in an unconscious state and our clients ability to provide oxygen from their diving tanks and warm clothing saved Mr. Quilligan from a certain death. He is now recovering in the Regional Hospital in Cork. Rather than driving such humanitarian visitors from our coast with dubious applications for injunctions, we should be welcoming them with open arms."

The local police told Polly that even if an injunction were issued they could not enforce it because they had no jurisdiction at sea, and no way of knowing if, after departing the dock, we were actually headed for the *Lusitania*. In fact, our original plan included diving the *Minnehaha,* an ocean liner which lay nearby at about the same depth as the *Lucy*. It had never been dived, and would therefore provide a unique opportunity to dive a virgin, if less well known, shipwreck.

In the event, the application for injunction was never filed. Nor did we dive the *Minnehaha*. We became so swept up by the majesty of the *Lusitania* that the thought of sacrificing a single dive to visit another wreck became abhorrent.

## Receiver of Wreck

Having failed to set the local police against us, and declining the attempt at filing an injunction, Bemis tried to get the Receiver of Wreck involved in his machinations. This brings to mind another curious story. Once again, the focal point was Philippe Beaudry and his delusional claim to the *Empress of Ireland*.

Beaudry, in his rabid and irrational bids to prevent others from entering what he considered to be his exclusive diving domain, reported our activities to the Canadian Receiver of Wreck. (By "our" I mean Dave Bright, Bart Malone, John Moyer, and myself.) We were making annual excursions to Rimouski in order to dive the sunken liner.

In the nineteenth century, the Receiver of Wreck was a full-time government position. The Receiver was charged with the responsibility of protecting the property rights of owners whose ships and cargoes came ashore. In those days, "wreck" referred to flotsam, jetsam, and lagan (cargo buoyed for later retrieval), as well as to a vessel's hull, rigging, and appurtenances. Salvors could legally salvage goods and cargo that either remained within the hull or that washed up on the beach, with the proviso that all "wreck" had to be turned in to the Receiver and stored in the Receiver's warehouse until the owners were notified and an equitable distribution was made.

If an owner wished to reclaim his goods, he had to pay a reasonable salvage award. The award was based upon the actual expenses incurred by the salvor, the risk that was incurred by the salvage, and a profit for the salvor's services.

If an owner chose to relinquish ownership, the "wreck" was returned to the salvors, who could keep, sell, or auction the salvaged items in order to defray the costs of salvage and, hopefully, to turn a profit for their voluntary efforts.

If no owner came forward, the Receiver held on to the "wreck" for one year, during which time he attempted to locate the owner. In cases where no owner could be found, the "wreck" was given back to the salvors to do with as they pleased.

In addition to protecting ownership rights, the Receiver was authorized to prosecute illegal salvors - that is, those who salvaged "wreck" and did not turn it in to the Receiver. In other words, those who sought to relieve a legal owner of his goods in order to increase the profit margin. This was at a time when seaside residents looked upon shipwrecks as manna from heaven and did not respect the property rights of ship owners, and who chose to believe - because it was in their best interests to believe - that a ship owner relinquished all rights of ownership when his ship was accidentally wrecked.

Thus the purview of the Receiver of Wreck related to current events and to the disposition of property that held a quantifiable resale value: items that could be returned to the stream of commerce. The Receiver's authority was never intended to encompass wrecks that lay abandoned on the bottom of the ocean for some seventy-five years. The Receiver of Wreck is therefore equivalent to the human appendix - an organ that once had a useful function but which has since become vestigial.

The situation in Quebec was complicated by the fact that the Provincial authorities didn't know what had become of the position of Receiver of Wreck because such services had not been invoked in modern times. The law was on the books, but it no longer held any relevance. After some bureaucratic scrambling, a patient researcher in the provincial legal department determined that the Receiver's authority had been vested in the Customs and Excise division of the Revenue Department.

This came as a shock to the local agent when he was apprised by his superior officers of his new responsibility. Denis Provencher, wearing his uniform, badge, and name tag, met us at the dock upon our return one day and told us about the complaint that Beaudry had filed. In a sociable manner, Provencher explained how the job had landed on his desk. He readily admitted that he knew nothing about the Receiver's functions or authority, or how it pertained to artifacts recovered from a long-lost shipwreck. He promised to look into the matter and to pay us a visit the following afternoon, hopefully with some information that would clarify matters.

The next day Provencher showed up with a handful of documents that he had photocopied from some antique tomes in his office. Statute citations revealed the history of the delegation of authority, the scope of his responsibilities, and the enforcement options at his command. He quoted relevant passages aloud. Then he adopted a realistic approach that would address the conduct of our activity, satisfy the law, and protect his discretionary discharge of duty.

Since the "wreck" was no longer a recognized commodity, the only justification that Provencher could find for official intervention was, in his opinion, tenuous. The *Empress of Ireland* lay in international waters beyond provincial or federal jurisdiction. By an exaggerated interpretation of customs regulations, artifacts brought ashore could technically qualify as imported goods which might therefore be subject to import duty - if they had any value, and if they were intended to be sold in Canada. Provencher determined that the one-time commercial value of the goods had been diminished by their age, and that because the goods were merely passing through Canadian territory on their way to a foreign country, import duties did not apply.

Furthermore, no warehouse facilities were available for storing our "wreck." He asked that we make a list of the items we recovered. That night we laid out the artifacts in our motel room, and he took Polaroid photos of the display. We gave him proof of identity and contact information. He gave us his business card. If we were stopped by police or customs agents at the border, we were covered and he was covered. Everything was legit. And, on the off chance that the owner of the items in question laid a claim on them, he knew where to get in touch with us and could ask that the items then be placed in his care. If we didn't hear from him within a year, then title to the items was transferred to the salvors without further process. Thus the situation was resolved to the satisfaction of all. All, that is, except Beaudry, who it will be remembered had no claim of any kind to the wreck except that which existed in his warped imagination.

The situation was similar with respect to the *Lusitania*, except that in Ireland and the U.K. the position of Receiver of Wreck was extant and still officially recognized. However, the Receiver's jurisdiction extended only as far offshore as Ireland's territorial limit. Bemis's position that the Receiver had authority to act in his behalf was based upon two premises: that Bemis's claim to exclusive title had been established, and that the wreck lay within Irish territorial waters. The Receiver had already notified Polly that no notice of ownership existed on record. Now, Polly ascertained, the Receiver was unclear about the *Lusitania's* location relative to territorial boundaries. (I will return to this latter point in Part 10.) The best advice the Receiver could offer was that, if no artifacts were recovered, then Bemis had no grounds to prevent diving on the wreck.

## The Tide of Events

With the tides running later each day there was not enough time at night to complete preparations for the next day's dive. This meant that some of the tanks had to be filled in the morning for an afternoon slack. A new starter motor arrived and was installed, again by Rob Royle, who hoped to make another dive - but not this day.

Weston was having more trouble with the inflatable. Once again he found water in the fuel, causing the motor to run unevenly. He disposed of the bad fuel, filled the jugs from an untainted supply, and got the outboard motor running smoothly again. But then he discovered another problem: the steering linkage was severed.

One might almost believe that we had chartered the two most run-down boats in the country. Were these breakdowns the result of an abnormal string of bad luck, or did both captains lack diligence in conducting proper maintenance?

We lounged on the *Sundancer* while Weston effected repairs. Polly took the opportunity to make a formal announcement. With advices from counsel and the Receiver of Wreck, with the precarious balance of the threatened injunction, with potential repercussions from political circles, with the uncertainty of action from law enforcement agencies, and in the interests of completing our primary goal - the exploration of the wreck until our allotted time was over - she declared that there would be no more recovery of artifacts.

Wreck-divers love souvenirs, and already several items had been brought up from the wreck: a soap dish, a deck light, a porthole. Polly spoke in her most serious mien. Her words were greeted with silent assent if not with approbation. Nor did we doubt the possibility of being prohibited from further diving.

Polly's proclamation did not damper anyone's spirits, but it did arouse some anger toward a millionaire in New Mexico, whose long and threatening bark continued to nip at our heels like a mad Chihuahua.

The weather continued to be wonderful: warm and sunny with hardly a cloud in the sky. Tourists and locals strolled past the quay, ogling the pretty boats. We glimpsed them apathetically, waiting for Weston to complete his repairs, eager to get under way. Someone pointed to a man in a car, remarked that he looked familiar. Someone else agreed: "He was watching us yesterday as we unloaded the boat." Then came growing concurrence - the same man had been observing us the day before that as well.

A spy!

I affixed my telephoto lens to get a closer look at his face. I zoomed in, snapped a picture, but he was too far away to capture his features distinctly. Other cameras came into action. The complaints were the same as mine. Finally, after discussion and debate, I said I knew how to get a perfectly clear image. I replaced the telephoto with a wide angle lens. I jumped onto the dock, walked along the floating ramp, climbed the stone steps to the street, sauntered indifferently along the roadway, paused alongside the car, stared hard at the man, leaned over to the open window, held the camera in front of his face, and snapped the shutter.

The man looked up at me with total indifference. He didn't smile, he didn't grimace, he didn't even look curious. Without a word I walked behind the car and took a picture of the license plate. He continued to watch me impassively. I made eye contact as I strode past the car on my way back to the boat. Still he expressed no reaction. A few minutes later, he drove off. We never saw him again. Perhaps because his cover had been blown?

Weston was still trying to mend the broken steering linkage. By this time it was too late to catch the slack, so the dive for the day was canceled. As a precaution against tampering, Weston took the fuel jugs off the boat and locked them up for the night. Somehow that solved the problem of water in the fuel.

## Deep Discovery

I was particularly excited about the forthcoming dive because of what had been discovered two days before: the docking telegraph mentioned above. Tully and Powell promised to run a line to the area so that everyone could find this precious piece of history. They showed me on the drawings where the object was located so I could orient myself with respect to the shot line.

Lander rolled into the water at the same time I did, and we began our descent together. Again she lollygagged. She had become erratic, unreliable, and noncommunicative. I wasn't about to waste my valuable bottom time waiting for her. I hit the hull and never stopped swimming.

Visibility was an awful eight feet. The floating plankton that had previously hovered at mid-depth had descended to the wreck, blanketing the hull with fine living particles that simulated a desert sand storm. I was thankful for the guide line to the stern bridge wing - I never would have found it otherwise.

This time I was confident that the camera would work on full automatic as it was supposed to. The extra time afforded by the previous day's cancellation enabled me to read the manual yet again, and to experiment with various settings in my room. Imagine my bitter disappointment when I depressed the shutter release lever and nothing happened!

I felt like screaming. I couldn't understand why the camera worked in my room but not on the wreck. I was disgusted. Then, as I ruminated over what I had read in the manual, I had a flash of insight. According to the book, in autofocus mode the firing sequence progressed from focus lock to shutter release to film advance. Now suppose . . .

With my left hand I aimed my spare light at the center of the telegraph. With my right hand I depressed the shutter release lever. Flash! went the strobe, brighter than my insight. Now that I knew the trick, I detached the strobe and held it and the light together. This enabled me to vary the distance of the strobe and, consequently, the amount of illumination reaching the subject. I congratulated myself on figuring this out at 300 feet. I could never have done it on air.

Lander drifted past my peripheral vision while I was making photographic discoveries and images. She didn't hang around, and I didn't see her again until I got to our module on the dec station.

That night I poured over relative sections in the Nikon instruction manual. Nowhere did it state anything about the camera's inability to function in no-light conditions, or that the focusing motor must actuate before the shutter mechanism would trip, or that the focusing motor could not engage without adequate light on the subject. No wonder I had missed it!

At first I blamed myself for not practicing with the camera at home, or blamed the weather for preventing me from diving in North Carolina, where I intended to familiarize myself with the camera under water. But then I realized that the camera would have functioned in the bright Gulf Stream water. I still would not have learned about its unwritten weaknesses.

Forever after I have shot the camera on manual mode.

## The Changing Tide

So far we had followed the progression of the tides, diving fifty minutes later each day until we didn't get into the water until mid-afternoon. By the time we completed our decompression and rode back to the dock, it was quite late. So we jumped tides; that is, we moved ahead to the preceding slack which occurred some six hours earlier.

Instead of having fifty minutes longer between each dive and the next, the changeover shortened our preparation time by more than six hours - similar to the loss of time that occurs in the autumn during the overnight shift from daylight savings time to standard time. We worked harder, faster, and later into the night in order to make the transition without missing a dive.

## A Shot in the Dark

After spending a week on the stern we decided to explore the bow of the wreck. We didn't remove the stern shot. Gotto used it as a reference, then dropped another shot at the opposite end of the wreck. The shot weight fell through a hole in the hull and had to be pulled out. Nor was there any doubt where the weight had landed: on the high side just in front of the port bridge wing at 270 feet. This was a perfect location, one that enabled us to reach a host of fascinating areas over the next six days.

The shot line was secured in the middle of an indented crease from which the hull plates sloped upward in all directions. In the dark and limited visibility it was difficult to determine which way led to the gunwale and which way led to the turn of the bilge. From the tie-off point I felt as if I were climbing out of a hole dug into loose sand. Once out of the crease the hull was approximately level. The waterline side sloped down and eventually dropped vertically to the sea bed. The gunwale side dropped off fairly plumb at first, but went down in steps to the sea bed, where the debris field began.

The crease was longitudinal and about twenty feet long. The gash in the center of the crease was three feet across at the widest point, and tapered to a point at either end. Aft of the crease some twenty or thirty feet from the shot was a large hole inside of which lay a boiler. This boiler appeared to be intact. If cold sea water had rushed in through this crack and washed against the boiler's hot sides, it probably would have ruptured. Thus the crack most likely developed after the ship had sunk, probably as a result of natural collapse.

Above the bulwark farther aft stood two lifeboat davits about five feet apart, each from a different pair of davits. Below this area in a jumbled heap lay round portholes and rectangular windows, water tanks, bathroom fixtures, tiles, tubs, shower stalls, toilet bowls, sinks, and chinaware. The entire midship section was free of nets.

We continued to be plagued by poor visibility: eight feet on that first day on the bow. Succeeding days were better: ten feet, then fifteen feet, then twenty feet, and finally thirty feet. Guide lines were indispensable, as were marker strobes flashing on the downline.

## As I Pondered, Weak and Weary

The technical diving aspects of the *Lusitania* project were not the only obstacles that we had to overcome. We all felt the strain of the strict diving regimen, the self-imposed workload, the frustrating mechanical breakdowns, and the imminent threat of legal action. Compounding our anxiety and fatigue was, ironically, the blessing of uncommonly good weather.

No blow-outs due to wind meant no days off for rest and recuperation. By the beginning of the second week I was positively exhausted - yet I was unwilling to yield to my fatigue if it meant sacrificing a dive. On one occasion I bent down to pick up a set of doubles in Haskel Land to carry them to Compressor Country. I got the tanks only a couple of inches off the ground when a wave of weakness swept through every muscle in my body. The tanks banged on the macadam. I poised over them in a crouched position, feeling the strength drain out of me, but steeling myself for another attempt to get them in the air.

Simon Tapson placed a hand on my shoulder. "Don't feel bad, Gary. We're all tired."

His insight bolstered my sagging ego. At least I wasn't alone in my lassitude.

## INTERNAL DISHARMONY

On a couple of days we went out in seas that were marginal. After one particularly rough trip, with whitecaps breaking four to five feet over troughs and with occasionally larger waves, Chatterton complained in retrospect that the dive should have been aborted. His comment led to a general discussion that became somewhat heated.

Several people elected not to dive that day: that was their choice - diving was strictly voluntary. Chatterton argued that if he had been running a dive at home in similar circumstances, he would have called it off. All the Brits were against him, including those who chose not to dive. Their argument was that high wind conditions were common off the British coast, and that if they habitually aborted trips every time the weather was marginal, they would dive only half the time.

I was learning that these Brits were a tough breed. They certainly earned my respect by not even considering the option of turning back - the question never arose. They bore the brunt of the pounding seas without a murmur of protest. I have to admit that I didn't like the ride and I didn't want to be there - a warm cozy bed that didn't pitch and roll was uppermost in my mind - but I was driven by the irresistible urge to explore the *Lusitania*. My ambition to dive overruled my instinct to settle for physical comfort.

The position I took was that under normal circumstances I would have agreed with Chatterton. But these were not normal circumstances. The *Lusitania* was a rare experience that was not to be taken lightly - an opportunity that shouldn't be missed. I was willing to suffer some discomfort for the privilege of diving the wreck.

Tulley made the final riposte by noting that Chatterton's argument would have been more convincing had he chosen not to dive.

Despite overall good group dynamics, a few other incidents occurred that marred the prevalent congeniality. One time Hope wouldn't let me use his oxygen analyzer. My analyzer, which I had shipped to London when the Brits were in training and before the group had purchased one, had been used so heavily that the fuel cell died, making me dependent on the group analyzer. Simon explained that it was Hope's prerogative to keep his own analyzer for his personal use, and that he didn't have to let me use it. I suspected that Hope didn't like me.

Howard Woston accused the group of stealing a knife from the shop, to which we had open access. Polly was quite "cross" at Weston for making the accusation.

Lander exhibited growing disaffection. She often refused to take meals with the group. She became alternately aggressive and taciturn. She wouldn't stay by my side under water, and if I went after her she swam away. Her behavior finally forced me to ask bluntly whether she intended to dive with me as a buddy or to go off on her own. She indicated that she was diving alone. Once I knew where I stood I could conduct my dives accordingly. We were a team only in that we rolled into the water at the same time. She made no attempt to justify her behavior: not then, not during the flight home, not ever.

Chatterton incessantly needled Wilkins, whose unassuming manner belied his competence. Wilkins suffered Chatterton's taunts silently and refused to rise to the bait.

The most explosive disagreement erupted over supplementary divers. Rich Field and Rob Royle joined us at Polly's invitation, as did Tony Hillgrove, whose pleasant personality and unflagging assistance during the preliminary weekend trip endeared him to all of us. Both were accepted without qualification.

Polly also invited Chris Reynolds to join us for a dive. Reynolds was a navy diver stationed nearby. In his capacity as chamber operator, he was Polly's contact at the navy's hyperbaric facility, and he made himself personally responsible for ensuring that the chamber was available to The Starfish Enterprise should anyone require recompression treatment. He met us at the folk house a couple of times, then went out on the boat as an additional deck hand and surface supporter: roles in which he did yeoman's service, for that day had been particularly rough, with divers ricocheting about the deck like loose billiard balls. He was a likable person and a valuable ally.

The reason that he was not accepted unanimously was that he was a member of the Irish Technical Diving team that was slated to dive the *Lusitania* later that summer. This association was perceived by some with negative connotation, due largely to the attitude of the ITD's rash and outspoken ringleader, Desmond Quigley.

Quigley readily acquiesced to Bemis's demands - with deference if not with pusillanimity. He and his group signed Bemis's contract, thus ceding all their rights in the endeavor to Bemis and paying for it to boot. Bemis invested nothing in the ITD trip, accepted no liability, but expected to reap all the rewards - if any rewards accrued. Quigley had a nasty habit of boosting his ego by making snide remarks and derogatory statements about his perceived competitors, The Starfish Enterprise - a character flaw which earned him no respect in the wreck-diving community.

Reynolds' credibility suffered by association. Some believed that since he was a follower of the enemy camp, he might do us harm by reporting our activities. We had already ousted one spy - the unidentified man who kept our actions under surveillance from his car - we certainly did not want to invite another into our midst. On the other hand, Reynolds was instrumental in obtaining access to the chamber owned and operated by Ireland's Navy Diving Service. Should an accident occur, he would be responsible for scrambling a helicopter to transport the stricken diver to the chamber. He would then supervise the operation of the chamber.

Reynolds maintained that he had aligned himself with the ITD not out of allegiance to Bemis or to Quigley, but because he very much wanted to dive the *Lusitania*. He viewed the ITD trip as his sole opportunity to do so. And since he wasn't interested in artifacts or photographs, he had nothing to lose by agreeing to Bemis's conditions. He did admit, however, that although he had passed Robert Palmer's technical diving course, his total in-water mixed-gas experience was limited to his check-out dive. His skills might be weak but his enthusiasm wasn't lacking. Survival, of course, depended upon skill and experience, not enthusiasm.

Polly's motivation for inviting Reynolds to join us was pure: she wanted to repay him for his efforts in our behalf. The one most vehemently opposed to his participation was Christina Campbell. Her arguments were persuasive and not without foundation. Others took sides with less volubility. In the end it was resolved that Reynolds could dive as long as Polly held the reins and took responsibility for his safety. They had a rewarding and trouble-free dive together.

## NET SCAPE

In diving the bow we had some brief encounters with nets. Considering the paucity of fish we had seen, a means of catching them seemed out of place. Ballard's description of the nets as "treacherous" seems exaggerated to wreck-divers, although for submersible and ROV operations the possibility of entanglement loomed large. Partly this is because of the poor maneuverability characteristics inherent in underwater vehicles due to their bulk, mass, momentum, and remote control performance difficulties, in comparison to an untethered diver.

In open water a skilled operator can overcome these potential liabilities. But in close proximity to shipwrecks, where special conditions exist, a powered vehicle that is self-propelled or one that is tethered by a cable runs a greater risk of accidental contact than an experienced wreck-diver runs. In large measure this is because wreck-divers understand shipwrecks.

Moving water that hits a wreck is diverted around the structure. This diversion creates eddies as well as local concentrations of accelerated flow streams. Rust holes create siphons with a pronounced venturi effect. Wreck-divers understand these environmental conditions and can anticipate their occurrence. Wreck-divers, because they are physically present in the water, can *feel* the flow and can respond instantaneously - generally on an unconscious level.

The only large, buoyant net that we observed on the wreck was snagged on the starboard bulwark in front of the wheel house. It extended some twenty to thirty feet forward, billowing like a disembodied ghost, and covered a broad expanse of the lower forward deck just abaft the starboard capstan. Floats held it up to a height of twenty feet above the wreck.

I approached to within inches of this floating net without feeling threatened by its presence. I hovered and avoided contact, then turned and swam away. This is an example of what a diver can do that a submersible or ROV cannot.

Other nets lay draped across the upper hull and partway down the deck, but their filaments hugged the wreckage like a hair net clutching a woman's bun and were easily avoided. I was the only one who reported any entanglement. I was swimming close to the forecastle hull when a strand of aged netting caught my fin strap. I wasn't concerned because I sensed from the stretch that I could easily break the line. At that moment I noticed bright lights approaching from behind: Chatterton with his video camera. Instead of pulling free, I waited for him to get footage of the incident. I knew that it would make a dramatic scene.

As soon as Chatterton saw my predicament he reached for his knife. I bent my knee and snapped the rotting cord, then exchanged okay signs. Later I asked him how the footage came out. He said he didn't get the scene on tape because he lowered the camera when he pulled out his knife.

I berated him for his poor choice of priorities. "A photographer always gets the shot first."

If I had truly been trapped then I couldn't have gone anywhere. He should have finished taping, then cut me free. He defended himself by claiming that he was too concerned about my safety to think about videotaping.

## What's in a Name?

At the time the above incident occurred, I was photographing the letters that spelled out the *Lusitania's* name. Each brass letter stood about a foot and a half high. It appeared as if each letter was backbolted onto a steel plate which outlined the letter like an all-around shadow, and that the steel plate was then secured, possibly welded, to the hull.

A close-cropped net covered the last two letters: I and A. Yurga later cut the rotten strands away so as to better see and photograph the letters.

The upper hull where the letters were located was no longer flat but began to curve upward as if the hull were flared, which originally it was not. The letters themselves lay on about a 30° slant, the top of the letters being higher than the bottom.

## Forecastle Deck

Forward of the letters, the bulwark continued to bend upward until it appeared nearly vertical at the stem. This appearance of verticality might be due to a thickly encrusted net which was held upright by its floats. (This was not the same net referred to earlier, which was situated farther aft: just in front of the remains of the wheel house structure.)

The forecastle looked very much as it did in topside photos of the ship, complete with the railing, capstans, winches, mooring bitts, mast, and anchor chain. The forward hatch cover was gone. The Tapsons peered inside with their lights and spotted two navigational lanterns about two feet tall. The lanterns were buried in mud and only partially exposed.

The deck was angled perhaps as much as 75°. Some observers thought it was 60°.

## Below the Shot

A series of tall "steps" began about ten to fifteen feet below the bulwark adjacent to the shot. These were not true steps, but folded decks which gave the appearance of risers and landings. The route down passed through what used to be the lower level of the superstructure. Here lay portholes, a brass stand, a window with a ventilator top, and other odds and ends. Ten feet below this lay the sea bed: an admixture of dark sand and gravel.

At the juncture of the lower edge of the hull and debris field lay a section of plating pierced by three portholes - an arrow pointing the way back to the shot for anyone traveling along the juncture. Scattered about the area were stair treads and many loose portholes.

Imagine my shock when I arrived alone in the debris field to find a length of white PVC pipe lying across a loose porthole on the sea bed. I quickly determined that the pipe was part of the dec station, complete with carabiners at either end. Jouncing seas had forced open the carabiner gates on one of the cross pipes. I took pictures, then tucked the pipe under my left arm and charged toward the shot. I must have looked like Sir Lancelot in a jousting match. The six-foot length of pipe was too unwieldy to carry single-handed - in addition to my camera rig - so I abandoned it.

Forward of this point lay miscellaneous wreckage, including half of a china bowl (which I photographed). Safe progress was eventually halted by the vertically hanging net mentioned above.

## Wheel House Wreckage

About twenty-five feet abaft the plate with three portholes noted above, and which pointed the way to the shot, lay the bridge wreckage. This was perhaps the most extraordinary sight I could ever hope to see on a shipwreck as historic as the *Lusitania*: the carrot on the stick that kept me going farther and staying down longer than planned.

The upper level of the wheel house had sloughed off and collapsed like a steel spring slinky - the child's toy, no longer in vogue, that coiled and uncoiled its way down the stairs. If the individual coils represent the stories of the superstructure, picture the spring angled sideways like the leaning tower of Pisa, then crushed flat.

No decks or bulkheads stood upright here. Steel plates and sections of decking lay like dealt playing cards that overlapped each other on a table top. Dispersed helter-skelter lay the bronze remains of the navigational equipment - lying about like abandoned parts in a junk yard: telemotors, telegraphs, and annunciators, complete with pedestals and stands. In multiples!

Never in all my years of wreck-diving had I beheld such an awesome sight. The sea bed was littered with rare artifacts, like prime exhibit pieces in an ocean liner museum display. Controlling my excitement, I aimed the camera and framed my shots. I detached the strobe in order to offset the angle of light hitting the subjects, a trick which not only cut down the backscatter which resulted from light reflecting off floating particulate matter, but which added shadows to the subjects and gave the images a sense of dimension. I fired as quickly as the strobe recycled.

## Cast in Bronze

I didn't know it then and didn't find out until several weeks later when my film was developed, but my pictures solved one of the mysteries of the liner's last minutes afloat: had the engines been reversed in order to reduce the ship's speed so the lifeboats could be launched without capsizing?

The long-sought answer to this poignant question was literally cast in bronze. I captured the proof on film. The indicating arrow on the wheel house annunciator pointed to the side marked "ahead." The ship was going forward.

Captain Turner may have rung the telegraph to order the engines to be reversed, but if so the order was not acknowledged from below, perhaps because the engine room gang were chased from their stations by the rising flood. With the *Lusitania*'s engines maintaining full speed ahead, the huge torpedo hole scooped up brine like the bucket of a gigantic water wheel, causing the ship to settle more quickly.

Had the engines been reversed and the ship's speed reduced, the lifeboats possibly could have been launched without capsizing. Perhaps the extra time afloat would have given the passengers and crew a better chance at survival.

## Sabotage!

After we returned from the wreck on day twelve, Denis Kieran took out the RIB for an early evening jaunt. He was under way for only a short while when he detected an aberration in the motor's performance: a barely audible murmur that aroused his mechanics' sixth sense. He stopped the motor and pulled off the cowling to have a look. The full measure of the cause of the motor's malfunctioning was staggering.

The oil had been dumped out of the oil reservoir, the oil feed lines to the motor had been severed, and the electrical wires between the motor and the console warning light had been cut. Clearly this damage was no accident.

Whoever had sabotaged the motor knew how to do the job effectively. The broken circuit disabled the low-oil alarm that ordinarily alerted the operator to an oil-feed problem. Without advance warning, an operator might not know that anything was wrong until he heard the motor sputter just a moment before it seized. By that time the motor would have been ruined beyond repair.

There was no longer any doubt: some malicious person or persons unknown was trying to sabotage our trip. If a member of The Starfish Enterprise got killed in the process, so be it. Now there was an explanation for the other mysterious breakdowns - the ones that we had chalked off to coincidental happenings. The more suspicious of those among us (or perhaps the less ingenuous) suspected sabotage all along. My lot was cast with naivety.

The daily mobilization and the performance of the dives demanded exceptional concentration and physical exertion. Even so, the expedition was fulfilling because of the achievement of our goal. But what should have been a unique and exciting adventure whose only dangers were those imposed by the insensitivity of nature and by lack of personal experience, turned into a struggle for survival against the perilous intrigues of antagonistic schemers who were imbued with deadly intent.

Kieran notified the Garda (police) about the vandalism. A month later, I sent copies of my pictures of the spy and his automobile's license plate to Kieran, to give to the police. Nothing ever came of it.

Dogged by shysters, spies, and saboteurs, plagued by threats and intimidation, we plunged ever onward and downward.

## Winding Down

Kieran obtained a fiberglass skiff to replace the inflatable as a chase boat. It worked marginally well but could not carry the weight of the dec station, which then had to be stowed on and deployed from the *Sundancer* on the penultimate day of the trip. On the last day we dispensed with the chase boat altogether because of the group's sudden loss of enthusiasm and, perhaps, because of fatigue.

Some wanted to pack, some wanted to rest, some wanted to shop in Kinsale - or all of the above. Instead of ten divers we departed with only five. The support crew was shuffled so that I wound up sharing the job with Chatterton. Once we got on site, Nick Hope and Barb Lander opted not to dive. That left Jamie Powell, Simon Tapson, and Richard Tulley. They dived as a threesome.

The wind was howling so loud that we had to shout in order to communicate on deck, and even then many words were whipped away by the din. The seas were short and choppy and averaged five to six feet in height. After the trio entered the water, Chatterton jumped in and secured two modules of the dec station to the downline. He was still under water when a safety sausage broke the surface.

The *Sundancer* went after the safety sausage, which was drifting with the accelerating tide. Gotto maneuvered the boat upwind of the sausage. Hope jumped in with my snorkel. (I stayed on board in order to direct operations.) The divers were so deep that he couldn't see how many there were. From the bubbles we couldn't determine whether one, two, or all three divers were on the line. I gave Hope the spare dec station module and an oxygen bottle. He held the sausage line and the dec station line together while the *Sundancer* returned to the shot line buoy.

Chatterton had surfaced and found himself abandoned. He couldn't see the safety sausage because of the high waves, but he could see the boat in the distance and figured that we would come back for him. Shouting across the spume as we drew alongside, I explained the situation and gave him the possible scenarios. I needed information before I could make any decision on how best to proceed. He went down the shot line to count heads. A moment later he surfaced and reported that he saw no sign of anyone. I had to work on the assumption that all three divers were hanging below the safety sausage. I signaled for Chatterton to climb aboard, then we recovered the two buoys and the three oxygen bottles.

This took far more time to do than it does to tell.

Meanwhile, in mountainous seas that showed no signs of diminishing, Hope was still holding together the safety sausage line and the spare buoy. Gotto steered the boat back to the safety sausage. We dropped a dec station module in the water, and Chatterton jumped in to secure it to the safety sausage line. I breathed a deep sigh of relief when he surfaced and said that all three divers were down below.

Hope got out of the water, complaining volubly about how much he hated snorkels. (He didn't have one, which was why he borrowed mine.)

We retrieved Chatterton once again and took him back to the shot line. This time his job was to cut the marker buoy free from the line. (Powell, Tapson, and Tulley were supposed to have unshackled the chain from the wreck at the end of their dive.) Chatterton went as deep as he dared on a single tank that was half-filled with air - 200 feet. He couldn't surface immediately because he had to decompress. Then he climbed on board for the final time. He had gone up and down lines and ladders so many times that he must have felt like a yo-yo.

He and I switched roles. I jumped into the water and went down to watch the divers decompress - easy work compared to Chatterton's vertical peregrinations. After another hour the trio were ready to surface. Powell and Tulley ascended and started drifting away.

Simon signaled when he was ready to go up. I rose with him, maintaining eye contact. He grimaced and squeezed his left elbow - it throbbed with a dull ache. The closer we got to the surface the more painful the elbow became. He shook his head, signaled that he wanted to go back down. The pain went away at depth.

After a few minutes we tried once more for the surface. Again came the pain, so we went back down the line. Then his oxygen supply ran out. I signaled for him to hold on. I surfaced and swam to the boat, which was drifting nearby, and obtained an emergency oxygen bottle. I handed the regulator to Simon then clipped the bottle to his harness. We hung around for another twenty minutes or so before he decided to make another try for the surface.

We ascended gingerly, creeping upward by slow degrees, Simon breathing oxygen and bending his arm all the time and feeling for any sensation of discomfort. We hovered on the surface and waited for the symptoms to return. When they did not, we got on the boat. After Simon got out of his gear he went back to breathing oxygen. Now the problem was to keep him from talking and to concentrate on inhaling.

He didn't want Polly to find out about his problem because he thought that she might worry. We promised not to tell her, but word leaked out anyway.

From Powell and Tulley we learned about the dive. The trio swam away from the shot, reaching approximately midship. When they turned around they found the tide against them. They struggled against the rushing water at first, but soon concluded that they couldn't make it back to the shot in the time they had remaining. Rather than chance overstaying their time and depleting their mixed-gas supply, they made a controlled ascent.

They kept each other in sight at all times. They conducted their deep decompression stops by hovering with neutral buoyancy. At 100 feet Tulley sent up the safety sausage. Then they used the wreck-reel line for vertical orientation.

## Forlorn Farewell

Gotto checked the coordinates and ascertained the *Lusitania's* heading. The latitude was 51° 24.727' north, the longitude was 8° 32.866' west. The wreck lay with the bow pointing northeast, the stern southwest. According to the chart the closest point of land was Brow Head - 11.5 miles. Old Head of Kinsale lay 11.75 miles away.

Out of fourteen days we managed to dive on twelve. One day was lost to weather, another to deliberate evildoing. Of the collective experience of some one hundred twenty dives, I made nine - and yearned for more.

At one time the *Lusitania* was a mariner's elegant mistress. Now she was a woeful ruin whose collapsed and collapsing hull provided mute testimony of the destructive force of the sea. Yet to me - and to any wreck-diver - the condition of a shipwreck at the moment of experience is the image that will always be remembered. Comparisons with former glory are meaningless. We accept what we see as the current and temporary state of an ever-changing reality, of evolutionary decay.

In a sense, the presentation of a shipwreck is like a portrait of a child. The image of that moment in time is frozen. But shipwrecks and children change: wrecks break down, children grow up. These processes are inevitable and unstoppable. The brief span of exploration is a one-time event that can never be relived. It can only be remembered and captured on film.

My ecstasy over experiencing the wreck yielded to sadness from having to leave it. So much was left to explore . . .

At dinner I expressed my sentiments to the Brits for their outstanding achievement in the planning and conduct of the project, for their dedication in training, for their expertise in mixed-gas technology, for their knowledge of wrecks in general, and for their gracious hospitality.

I said, "There is no team of divers I would rather dive with - anywhere, any time."

The voyage of The Starfish Enterprise was over.

# Part 10
# After Shocks

### Life Goes On

Event plateaus such as the *Lusitania* expedition seldom represent the highest plane of attainment - they are merely steps along the ascending staircase of life.

Seven days after my final *Lusitania* dive found me tying in on the *Doria*. Considering how much travel time lay in between - the ferry from Cork to Swansea, the flight from London to Philadelphia, the drive from Philly to Montauk - I didn't have much time at home.

My dives on the *Doria* were commonplace, but events that occurred during the three-day trip altered one aspect of my life significantly. The *Wahoo* was chartered to a group from Italy of whom only one spoke recognizable English. Another could communicate after a fashion. The rest spoke only Italian. The Italians had a fair amount of deep diving experience and had trained for the *Doria* by diving in a fresh water lake of equivalent depth. They were all breathing air.

The experience they lacked was ocean-diving in general and wreck-diving in particular. If conditions on the wreck were good they shouldn't incur any problems. But if anything went wrong they were seriously out of their depth. None of them carried an emergency decompression reel - nor, if given a reel, would they have known how to deploy it. Their only option should they fail to locate the anchor line was to make a controlled ascent and to decompress adrift. I was concerned.

I was even more concerned when Steve Bielenda informed me on the dock that he was sending us out without an operational chase boat. The outboard motor for the inflatable wasn't working. At the last minute he found an outboard motor which he told me couldn't be mounted because it was too large for the boat - this was not a matter of weight but of speed: sudden acceleration might cause the inflatable to flip end over end. Why he bothered to provide a motor we couldn't use, I have no idea.

Bielenda stayed home and let Janet Bieser run the trip. Bieser made no provisions for the safety of the passengers in the event of an emergency. What were we to do if a diver went adrift? How could we retrieve him? The *Doria* lay in a location that was notorious for tides and current. Poor visibility, loss of ambient light, unfamiliarity with the wreck and with routine wreck-diving techniques, could easily combine to disorient a diver on the bottom.

I discussed these matters with the other two crew members, Billy

Deans and John Moyer. We worked out a method to manage the oversized motor on the inflatable. We could offset the weight by laying a set of doubles on the floor by the bow. During operation, one person would steer and run the motor from the stern while another would maintain a position forward. We would run the motor at partial speed so the bow wouldn't lift up during acceleration.

I thought we had the problem solved when Bieser burst into the cabin, screaming. (She had been lurking in the dark outside the door, listening to our conversation.) "I'm the captain of this boat and I said we're not gonna use the chase boat. I don't give a *fuck* about the Italians. They can fuckin' *die* for all I care. Once they're off my boat they're on their own."

This statement was remarkable. I don't understand how any humane person could have such callous disregard for human life. Bieser's statement was all the more remarkable for a licensed boat captain to make because she was charged by the Coast Guard with responsibility for the safety of her passengers and crew. She stormed out of the cabin, leaving the three of us speechless with mouths agape.

Bieser's inattention to detail was well-established. Once she left the *San Diego* without taking a head count. A diver surfaced in an empty sea. A harrowing couple of hours passed before he was picked up by a fishing boat headed for New Jersey. Bieser didn't know that a diver was missing until she intercepted a broadcast from the fishing boat captain. The castaway diver was left on his own to obtain transportation to the *Wahoo's* dock, more than one hundred miles away - and this without a wallet or suitable clothing on a journey that took him through New York City.

On a previous *Doria* trip Bieser forgot to deploy the trail line. Billy Campbell surfaced next to the boat in a ripping current and was instantly swept past the stern. There was no rope or buoy to grab, he couldn't make headway, and his shouts couldn't be heard above the noise of the compressor that was running on the after deck. Not until Dave Zubec climbed aboard and asked about his buddy did anyone know that Campbell was missing.

A check of the divers still decompressing revealed that Campbell was not among them. A search of the boat proved fruitless. A visual search of the surrounding sea was frustrated by fog. Campbell was either dead on the bottom or adrift. Bieser called the Coast Guard. The Coast Guard dispatched a helicopter to search downcurrent of our position.

At first the low-hanging fog militated against dispatching the inflatable - not because an observer couldn't see very far from sea level, but because those in the inflatable might not be able to relocate the *Wahoo*. Aluminum foil wrapped around a makeshift crosstree became a portable radar reflector. A pair of hand-held radios enabled constant communication. Gary Gilligan steered and ran the motor while Zubec concentrated on scouting. From the wheel house, Bieser gave course directions by watching the inflatable's blip on the radar screen.

Observation from the helicopter was hampered by fog. The inflatable was two miles downcurrent when Gilligan and Zubec spotted Campbell afloat. He was alive! A loud collective cheer practically overwhelmed the *Wahoo*. With true gallows humor - now that we knew that Campbell was unhurt - we started taking bets on how much gear he had remaining. (For some reason, scared divers adrift ditch their equipment.)

Gilligan followed a compass heading back to the *Wahoo*. Campbell had ditched everything except for his light and video camera. He was about to let go of the camera, but said that he decided to keep the light as a nighttime signaling device. He was never reimbursed for his lost equipment - *and* he still had to pay full fare for the charter.

Returning to the present situation, once Deans, Moyer, and I regained our composure after Bieser's unexpected outburst, we continued our brainstorming session. After all, we reasoned, our primary concern as crew members was for the safety and enjoyment of the passengers - even if the captain's philosophy differed. We kicked around some possible scenarios that didn't rely upon the chase boat that Bieser wouldn't let us use.

We could toss a life ring to a diver close to the boat, then haul him in with a line. To catch someone who surfaced out of tossing range, one of us could swim after him dragging a line unspooled from a decompression reel, held by someone on deck. But what if someone surfaced out of reach? Or was dragged away by the current? One of us would have to catch up with the diver and stay with him until the *Wahoo* slipped its moor and effected recovery - an uneasy prospect, but we each carried signaling devices to aid in the locating process: mirrors, strobes, and flares.

But Bieser didn't return to the wheel house. She lingered outside the cabin door and burst in upon our safety talks again. Her second tirade was worse than the first. After she stormed out of the cabin a second time, co-captain Hank Garvin picked up where she left off by delivering a speech on "She is the captain of this boat." I couldn't believe what I was hearing. Had they both gone insane? Or were they both so insecure that they couldn't stand to have their authority questioned?

In the event, the Italians adopted a cautious approach to the wreck. They did not stray far from the anchor line, they did not extend their bottom times, and hopefully they returned to Italy with a sense of fulfillment at having changed a long-held dream to a reality.

The reality for the crew was less fulfilling. I was banned forever from the *Wahoo*, Moyer's subsequent charter was canceled, and Deans's charter boat in Florida was sunk.

Bielenda called me the night before the following week's departure and, in a one-sided conversation that lasted less than a minute, ended a relationship that went back ten years. He did not want to hear my side of the story - he agreed unquestioningly with his subordinate's outlook. I could have shrugged it off philosophically by reasoning, "Oh, well. I've been thrown off better boats and by better people." But I was appalled by the injustice of the situation.

Moyer had chartered the *Wahoo* for another trip dedicated to looking for the bell. Plans that were a year in the making were dashed, and a dozen divers who had scheduled vacation time so they could participate in the venture had to be given the bad news. Bielenda didn't seem to care about that - or about losing thousands of dollars that he would have been paid for the charter.

The sinking of Deans's boat more than a thousand miles away was just a far-out coincidence that would have seemed contrived in a work of fiction.

## WINDFALL

My gear was packed and I was ready to go. Even if I couldn't dive the *Doria* I still wanted to dive. I called Gene Peterson to find out what trips his shop had scheduled. As luck would have it, he had a spot open on the *Down Deep*, Captain Bob Meimbresse.

The first day we went to an unidentified wreck that I had christened Bob's Cold Spot. The name stemmed from the fact that two years earlier we had intended to dive a recently discovered wreck called Homer's Hot Spot. Meimbresse had obtained the loran coordinates from another boat captain. When Peterson and I explored the wreck, we realized that it didn't fit the description that we had been given. We had discovered a different wreck!

This time I dived with Tom Packer. We were skirting along the edge of the wreck when I noticed a dark mass at the limit of visibility some fifty feet away across an empty patch of sand. I went to investigate. I found a piece of wreckage that was twenty or thirty feet across and that was partially covered by an ancient trawler net. Several loose portholes lay about. When I examined the wreckage more closely I spotted a helm stand. I had found the remains of the wheel house!

I attracted Packer's attention with my light, and when he reached my side I pointed out the helm stand. We determined that the pedestal wasn't bolted down and that the long shaft was disconnected, then ran out of time and had to leave. On the boat we discussed the find with Peterson. He was excited because the rudder indicator might be stamped with a manufacturer's name that could lead to the wreck's identity.

The wreck lay some fifty miles offshore in 160 feet of water. We hadn't counted on making two dives there, and were already underway to an inshore destination. We resolved to return the following day.

The inshore wreck was another one that had recently been discovered and was yet unidentified. Eight minutes into the dive I spotted a triangular corner of brass sticking an inch out of the sand behind the engine. Curious, I pulled and wriggled the object free from where it was buried. The flat length of brass was a couple of inches wide and a foot long. Lettering cut into the sheet identified it as a stencil. When I held it up to the light from the surface I read "S.S. CLEOPATRA."

Positive identification! Exuberantly I held up the stencil for Peterson, who was nearby, and watched his eyes grow to the size of silver dollars. Packer, also nearby, cheered through his regulator. I had written about the collision between the *Cleopatra* and the *Crystal Wave* in *Shipwrecks of Delaware and Maryland*.

The next day we returned to Bob's Cold Spot. Packer couldn't get off work so John Moyer took his place. We dived together and ran a line from the grapnel to the helm. Peterson and Lynn DelCorio followed the line and secured two 500-pound liftbags to the stand. The buoyancy broke the stand free from the encumbering wreckage, but did not send the stand to the surface. Instead, the stand hung about ten feet off the bottom, snagged by either the net or unseen wreckage. They secured another, smaller liftbag to the stand. It refused to budge. By this time they had to leave so they tied a decompression line to the stand, unreeled the sisal back to the anchor line and then up to the boat.

We hadn't planned on making a second dive on the wreck. We anticipated that two teams diving in succession could do the job. But we couldn't just leave the stand floating there, so Moyer and I went back down after a short surface interval. We followed the sisal line straight to the stand. I took photos and determined what was holding the stand to the bottom. Moyer tied a safety line to the pedestal.

When the helm stand initially floated off the bottom it dragged the surrounding net up with it. The trawler net now loomed high like a circus tent: taut strands encrusted with marine fouling organisms. The pedestal floated above the spot where the center pole of a big top would protrude. After a close examination I determined that neither the stand nor its shaft was caught on wreckage, but instead was simply enshrouded in netting.

I surveyed the circumference of the makeshift tent to determine how much slashing would be required to release the net. By the time Moyer had secured the safety line, I knew exactly where to apply the knife. The key connection was a single twisted rope.

I removed the lanyard from my wrist and laid my camera on the sand. I made sure that all dangling hoses and equipment were tucked beneath my body. I approached the base of the tent from a perpendicular angle. I stretched my arm to its fullest extent, and laid the edge of my knife on the strand I intended to cut. Before applying pressure I made eye contact with Moyer. He hovered in the clear and held the spool by its handles. He nodded that he was ready.

I sawed the knife on a forward stroke. Strands that were held under excessive tension parted and untwisted with a snap. I sawed back . . . and the pedestal exploded toward the surface like a Titan missile launched from a nuclear submarine. Neither of us saw the take-off. A huge section of netting was yanked off the wreck along with a few entangled beams. Left behind was a huge cloud of silt and mud that choked the area like the vapor of a rocket exhaust.

My heart rate quickened uncontrollably. The safety line spun from Moyer's spool like monofilament from a fishing reel when a thousand pound tuna has taken the hook. As soon as the line stopped unspooling, indicating that the liftbags had hit the surface, we backed across the sand, peering upward. We didn't want to be directly beneath the stand in case it dropped back down to the bottom.

The rest of the story is anticlimactic. The rudder indicator spelled out the name of the manufacturer, yet despite intensive follow-up research the wreck remains unidentified.

## Another Pair of Discoveries

A few days later I went on an exploratory trip that Chatterton had organized. The objective was to find the *Pan Pennsylvania*, a tanker torpedoed by a German U-boat in World War Two, and which I had covered in *Track of the Gray Wolf*. Hang numbers led us unerringly to the wreck, 250 feet deep. Yet the dive was a disappointment because the hull lay upside down, offered no means of entry, and rose only fifteen feet off the bottom. Enthusiasm for a second dive waned.

The group voted instead to check out a different site that we had passed on the way to the *Pan Pennsylvania*. The depth was the same. We anchored overnight, then hit the water early the following morning. As trip leader, Chatterton took the privilege of going down first. He tied in the grapnel, then released a Styrofoam cup to let us know the hook was secure. I went in as soon as I got the go-ahead.

The hull was amazingly intact. The grapnel was hooked in the port bow above the anchor. Bitts and winches reposed inside the bulwarks. The forecastle reached as high as 205 feet, with the main deck ten feet deeper. At the end of an exploratory sweep past two pairs of hatches, I approached the rise of the forecastle and intended to peek inside. And there lay the ship's bell!

At one time the davit stood on the after edge of the forecastle.

When the mounting bolts rusted through, the davit tipped over backward and landed upside down on the main deck, with the bell still affixed. In that position the bell looked like a large brass bowl, but to an expert eye there was no mistaking the distinctive flared lip. I snapped a series of photos.

A glance at my gauges disclosed that I had neither the time nor the air to secure a safety line and to fully inflate a liftbag. I attached a 100-pound liftbag to the neck of the davit as a marker, put in a puff of air to hold it upright, then dashed for the anchor line twenty feet away. During decompression, Chatterton hovered a couple of stops above me. With both hands I pantomimed the form of a bell - realizing as I did that my gestures could be misinterpreted as the curves of a shapely cheerleader. Chatterton got the right message.

Afterward, I asked Chatterton to assist me in raising the bell - not because I needed any help, but to include him in the act of recovery. The operation was straightforward. I replaced the 100-pound liftbag with a 250-pounder, tied my decompression line to the davit, handed the spool to Chatterton, then inflated the liftbag until the bell and davit slowly headed for the surface. We cut and tied the sisal to the wreck, and the job was done.

The bell and davit were hoisted onto the boat. After I climbed aboard, Brian Skerry eagerly chipped away the thin encrustation to reveal the name that was engraved in bronze: SEBASTIAN. I pride myself on my encyclopedic knowledge of local shipwreck lore, but in this instance I drew a complete blank. I wasn't familiar with any wreck named *Sebastian*. Could the ship have sunk under a different name?

I went through my records as soon as I got home. Within ten minutes I found the wreck in question. The *Sebastian* was a tanker that was consumed by fire in 1917. The recorded latitude and longitude matched perfectly with the loran coordinates.

### THE GOOFY GOLF BALL CAPER

The recoveries and discoveries related above were made during the span of a single week. They created quite a stir throughout the wreck-diving community - from cautious disbelief to caustic but pretended envy - and even appeared in the dive rags. The case could be made that getting thrown off the *Wahoo* was beneficial to me. It was certainly injurious to Bielenda because of the amount of business he lost as a result. My services for future *Doria* trips were quickly scooped up by Howard Klein, owner of the *Eagle's Nest,* and by Danny Crowell, who purchased the *Seeker* from Bill Nagle's widow, Ashley.

Three days after recovering the bell from the *Sebastian* found me collecting golf balls in water traps in seven feet of water. I didn't want the job, but Peter Hess was so insistent that I finally acceded to his importunities - not because of the potential monetary gain but strictly out of friendship.

Hess had a friend who was trying to do a favor for a friend of his who had the contract to reclaim golf balls at the Wilmington Country Club. He needed divers desperately. The golf balls were to be recycled by being cleaned and reconditioned.

In three and a half hours under water I recovered about five hundred golf balls. At the going rate of a nickel a ball I stood to make about $25 - barely enough to cover my expenses: air fills, lunch, and gas for the one-hundred-mile round trip. The highlight of the day occurred when Hess and I surfaced under a thick mat of vegetation that covered our heads so completely that we looked like monsters from a cheap horror flick. We laughed so hard that we very nearly drowned.

After one day I figured that I had fulfilled my obligation of friendship. To make matters worse, we got ripped off because the golf ball contractor gave Hess a bad check, then left the state. Only later did Hess discover that the man was a known fraud. We never collected our money.

## SNEAK PLAY

Contemporaneous with these diving exploits and shenanigans was an issue far more shocking. First knowledge of the underhanded plans then unfolding came from Paddy O'Sullivan. John Light's longtime friend had visited The Starfish Enterprise on more than one occasion at the folk house in Kinsale, and had invited the entire group to his home, where he showed us pictures and reminisced about his experiences with Light and the *Lusitania*.

To the Tapsons he sent a newspaper clipping from the London *Times*. Under Public Notices there appeared a warrant of Admiralty arrest filed by Gregg Bemis in the United States District Court for the Eastern District of Virginia, Norfolk Division, in which he claimed the right to be named the "sole rightful owner and possessor" of the *Lusitania*. The attorney of record was Richard Robol.

O'Sullivan highlighted one particular passage, pointing out in his letter, "Note in orange his bogus claim to the cargo."

The U.S. Federal Court ordered the publication of the notice on May 24, but its appearance was delayed until June 4 - the day *after* The Starfish Enterprise left London. As may already be apparent and will later become much more so, this legal foot-dragging ensured that we would be safely out of town when the notice appeared. Unwarned is unarmed and unprepared to act.

## 310 - The *Lusitania* Controversies

### PREPARING FOR BATTLE

As soon as O'Sullivan tipped off the Tapsons and put them on their guard, Polly spread the word to the other side of the Atlantic.

I was incensed. Cheryl has characterized me as having "a passion for justice," a sentiment which she saw manifested most prominently in my single-handed battle against the government over the right of public access to the *Monitor*. My idealistic temperament hasn't waned with the passage of time. On the contrary, my disposition toward perceived inequities has been bolstered by my growing disaffection for those who are dedicated to destroying what others have built or achieved - sometimes purely out of meanness, but usually in order to further their own base whims and desires. If this attitude brands me as a radical, so be it. I'm a helpless captive in a quest for the grail of romanticism.

I immediately offered Peter Hess $1,000 to initiate an action to intervene in Bemis's claim. He wouldn't take my money, but he agreed to handle the case pro bono publico (literally "for the public good," or for the good of the cause). I would have to pay expenses, of course, as well as court costs and associated expenditures. But he would not accept compensation for his services.

I wanted to enter the proceedings as an amicus curiae (a friend of the court) in order to disclose information which I believed Bemis and Robol neglected to mention. In other words, to submit the whole truth, not just that portion of the truth which was beneficial to Bemis's case.

There were complications. Hess was not licensed to practice law in Virginia, so he had to obtain the services of local counsel. Which brought up an interesting point: why would Bemis (who lived in New Mexico) bring suit against a wreck (which lay in Irish territorial waters) in the state of Virginia? Why not pursue the case in Ireland?

We had theories. Calvitt Clarke, Federal District judge for the Eastern District of Virginia, was partial to salvors. He had presided over other well-known and highly publicized shipwreck salvage cases, such as the *Titanic* and the *Central America*. Only an ingenue or an utter fool fails to accept the harsh reality that it is more important to have a favorable judge in your pocket than to have incontrovertible proof, especially when that proof is lacking. Furthermore, as I soon learned, Clarke was on a juggernaut to extend his authority and sphere of influence - reaching with "the long arm of the law" into far-flung international waters.

For years Robol had represented the Columbus-Ohio Discovery Group's claim to possession of the treasure from the *Central America*. Now he represented Bemis in the *Lusitania* case. Was this a coincidence brought about by the geography of his practice (in Norfolk, Virginia) or was there collusion between Columbus-Ohio and Bemis?

There was collusion, but although we only suspected it at the time, the reader will better appreciate the web of intrigue if the lowdown is revealed at this early point in the narrative.

My first suspicions of collusion were aroused by subsequent correspondence with Tommy Thompson, president of Columbus-America. I was tentatively thinking of including the *Central America* in my book on shipwrecks off South Carolina. I wanted to illustrate the chapter with underwater photos taken by the Group's robotic cameras. These pictures appeared multitudinous times in newspapers and magazines and on television. At first Thompson was enthusiastic at the prospect of having his pictures in my book. But after my involvement in the *Lusitania* case became known there was an abrupt change in attitude. He would not return my phone calls, would not respond to my letters, and he had underlings tell me in no uncertain terms that under no circumstances would I ever be permitted to use their images in my book.

After we obtained court documents relating to the case we learned that Bemis had represented to the court his quasi partnership with the Group by suggesting that artifacts from the *Lusitania* and the *Central America* were going to be displayed together in a traveling exhibit. (As of this writing no such exhibit has taken place.) We suspected that Bemis's affiliation with Columbus-America went much deeper. We suspected - although we were never able to prove - that Bemis wanted to sell *Lusitania* salvage rights to the Group at a profit. Columbus-America was looking hard at expanding their salvage operations worldwide, but the Group did not want to repeat the morass of legal complications in which the *Central America* was then embroiled - in which individuals, groups, and insurance companies surfaced after the gold was recovered and laid their own claims to the treasure - claims that cost millions of dollars in legal fees to prosecute and defend. We suspected that before consummating any deal, the Group demanded a free-and-clear title to the *Lusitania* and its contents. If these suspicions were real, then a great deal of money was riding on the outcome of the case.

We also speculated that there was little likelihood that Bemis could prevail in an Irish court without ironclad proof of ownership. By first winning his claim in an American court, which might be more lenient and forgiving, an Irish court might then be more conducive to rubber-stamping a previous opinion rather than to go through an exhaustive hearing process which it might perceive as repetitive. This sneak-in-the-back-door stratagem made sense for a case that had little or no merit.

It was my contention that if Bemis's allegations went uncontested, an indulgent judge might be hoodwinked by unsupported claims. I was scrapping for a fight and willing to pay to get into the ring.

## BALLARD AND BEMIS

The rumor mill was grinding away and there was plenty of grist. The rumor that suggested collusion between National Geographic and Bemis contained more than a grain of truth: it was *all* truth. Rather than report the rumor I will skip straight to the verification.

The National Geographic Society was a huge, multi-faceted company with numerous divisions. I wrote several times to Larry Nighswander, an editor, and asked for clarification about Bemis's relationship with NGS. He never replied.

I also contacted Bob Ballard, who worked at Woods Hole Oceanographic Institute, in Massachusetts. I first met Ballard in 1991 at the Boston Sea Rovers conference, where I gave a video presentation on the *Monitor* at the evening film festival. Ballard introduced me to an audience of thirteen hundred people.

For months and months my letters and phone calls went totally unacknowledged. Ballard was a VBP: a very busy person. He was busier than I was and carried a greater weight of responsibility. But that was no reason not to afford the courtesy of a reply - especially considering the cardinal importance of the subject matter to both of us.

When he finally did respond to my query it was because he needed something from me: information that only I could supply. He was planning a joint venture with the U.S. Navy that involved a new type of nuclear submarine that was 150 feet in length. The Navy wanted to test the research capability of this submarine by conducting a photographic reconnaissance on a sunken wreck. They chose the *Andrea Doria* for a trial run.

For all their virtues, submarines and submersibles have an inherent weakness with respect to operations conducted in close proximity to shipwrecks. That weakness is the likelihood of entanglement. Ballard was now aware of this stark vulnerability since one of the submersibles deployed to photograph the *Lucy* became hopelessly entangled in fishing net. The sub struggled up to a shallower depth where divers were called upon to cut it free. (This incident was left out of the television broadcast.)

Ballard would have demonstrated a stronger build of character had he called me on his own recognizance without having had an ulterior motive. Notwithstanding my opinion, before he posed his problem to me he addressed my prior query and launched immediately into a discussion of National Geographic's negotiations with Bemis. He confirmed every word that had been spread as hearsay.

When Bemis became aware of National Geographic's intention to produce a television show on the *Lusitania*, he put forth his claim of ownership and demanded compensation for his permission to grant access to the wreck. National Geo's lawyers investigated his claim. They believed that their position was legally defensible, but thought that ultimately it would prove less costly and take less time if they simply "paid him off" (Ballard's words). Bemis could tie up the case in court for years even though his claim was groundless.

National Geo had several million dollars invested in the project. Delay meant no return on revenues already spent. Legal fees could be prodigious. The company had its investment to recoup, money to make, a show to produce: important considerations for any profit-making organization.

National Geographic took the easy way out: they surrendered to a settlement - to the tune of some $50,000. Ballard said, "It was written off as an operational expense."

I told Ballard what he wanted to know about the *Doria*: the orientation of the wreck, its layout and state of collapse, the extent of the debris field, the hazards presented by nets and monofilament, the safest direction from which a submarine could approach the hull, why the superstructure should be avoided, and so on.

Afterward I was left with a disturbing thought: National Geographic had set a very bad precedent by yielding to the toll collector's demand. If The Starfish Enterprise had ever condescended to negotiating with Bemis, it was likely that this now established payola was the toll on which he based the value of his permission.

## Anecdotal Revelations

Meanwhile, Hess had yet to enter a plea in my behalf. His endless procrastination was frustrating. He plodded methodically through the legal protocol - getting all his ducks in a row, as he phrased it - but in my opinion, if we didn't shoot soon the ducks were bound to fly away. Notwithstanding my concerns about inertia, Hess was making strides in researching case law, obtaining evidence, and retaining local counsel.

Polly Tapson helped by sending directly to Hess the results of her own research, such as the Queen's Bench opinion and related legal documents. She also tracked down and spoke with two other people who had a vested interest in the case: John Pierce and Murial Light.

She wrote that John Pierce "was a part, if not the co-ordinating figure, of the Oceaneering Syndicate. . . . The Syndicate has rights to return to the wreck at any time - i.e. syndicate has rights to dive Bemis's hull. (If indeed it belongs to him). . . . He said the Syndicate could always make sure we could dive the wreck if we had further problems. I think he knew more about us than he revealed because one thing he did mention was that Bemis paid someone in Kinsale to keep an eye on us."

Paddy O'Sullivan acted as intermediary between Polly and John Light's widow. Then Murial Light agreed to let O'Sullivan give her phone number to Polly. Polly called Murial at home in Boston. About the subsequent conversation, she wrote, "Mrs. Light is currently motivated to be friendly to anyone wishing war on Bemis. She despises him. A month ago she received a letter from Bemis' lawyer stating if she didn't sign a waiver rights in the wreck she would run into legal fees because they would serve her. Last summer she received a telephone call asking if she was interested in providing documents that belonged to her late husband to contribute to a traveling *Lusitania* museum. She asked the caller if Bemis was involved. The man said no, said it was an organization with a track record for this sort of thing and they'd send her a video on a similar project to back up the story. Then a Mr Robol called to ask if she'd received the video!!! She received more correspondence (about the traveling museum) on unheaded paper and in cross-referencing this with the letter from Bemis's lawyer they were all the same company. . . . It's a very emotive subject for her. She firmly believes . . . that Bemis has always been driven by greed and is a very devious man. She knows we were divers on holiday and everything that Bemis did to try and ruin it for us. . . . It is not in our culture to be as proprietarial as it is in the US and Murial Light is originally from Kinsale and shares the same sentiment. Her husband hated Bemis too."

Murial Light's expressions about Bemis were her own. They have no validity in a court of law, but they go a long way toward understanding the charged emotional background that preceded Bemis's legal action, the timing of which begs another question: why didn't Bemis press his claim in court during the Oceaneering salvage operation, or after the Queen's Bench ruling which he blandished with such deference, or during his negotiations with National Geographic? He had ample opportunity, and certainly such an action - if found in his favor - would have bolstered his bargaining power. Why wait until 1994? Surely not because of The Starfish Enterprise.

Murial Light inferred a possible solution to this conundrum in her conversation with Polly Tapson. Until just a few months prior to that time, John Light was still alive. He would have vigorously protested any attempt that Bemis made to divest Light of the property which Light himself had purchased, and which Light could prove that he had purchased. Although it could be just a coincidence, the close chronology of events makes it appear that Bemis waited for John Light to die - for his testimony to be buried - before bringing forth his suit unopposed.

## Fifty Fathom Ventures

I was not alone in taking exception to Bemis's effrontery. John Chatterton, John Yurga, and Barb Lander were equally inflamed, and were willing to put up money in order to have their say in court. Rather than intervene separately as amici curiae, Hess thought our legal standing would be strengthened if we offered a united front. He suggested that the four of us form a corporation of which we would be the directors and the only shareholders. The corporation's sole function was to pool our financial resources in an effort to unmask the truth, and to expose the factual deficiencies in Bemis's declaration.

I wanted to get the Brits involved as well. My call to arms for allies from the other side of the Atlantic encountered a solid wall of resistance. They wanted no part of an American legal action. Their stance created no little disaccord between me and Polly Tapson, still the spokesperson for The Starfish Enterprise. Our friendship was strained to the breaking point as a result of our differences of opinion.

I have no one to blame but myself. Throughout life, complete contentment and peace of mind have eluded me because of my inability to accept the inequities that occur within my sphere of reference. I am tormented by injustice, misconduct, and unfair treatment, even when they do not directly affect me. Those who don't suffer such innate aversions and emotional inadequacies cannot fully appreciate how distressful they are. Conscience is an affliction that is painful to bear.

To me, the sting of injustice is like a touch on an open wound - a hypersensitivity which cannot be bandaged. I admire people who can shrug off miscarriages of justice, who can accept human baseness and evil, who are capable of ignoring all the wrongs in the world - and who can carry on blithely, their lives emotionally unaffected. They will never know how lucky they are.

Polly's points were well taken. It was humbling for me to learn how the people of other cultures viewed America's legal system as defective, perhaps obstructive, because grounds were not required for filing suits; the operative factor was money. By a curious quirk of the democratic form of government, tax-supported agencies and wealthy monopolists can afford to use the legal system to their advantage. They can win cases by investing more capital in an onerous, dilatory prosecution than an impoverished competitor can spend in his defense. This subsidized assault is the judicial equivalent of "right by might." The majority of cases are never tested on their merits because the lesser-funded litigant is forced to fold his hand in a game in which there is no limit on the bet. In the same vein, any bloated plutocrat can destroy an opponent financially or push him into bankruptcy if he is willing to sacrifice that portion of his wealth which exceeds the net worth of his opponent. In reality, the cause of justice or right-against-wrong frequently has little to do with the outcome of a case. Whoever has the most chips often wins by legitimate bluff.

She hit the nail on the head when she wrote, "Your 'knight in shining armour' approach to 'charging through the battlefront of civilization seeking change' and saving the *Lusitania* for all mankind is not going to work." My attitude - that of a blind crusader - must have seemed hopelessly childish and naive to a person of mean sophistication. And Polly was far above average in accepting the scandalous ways of the world.

It was stupid of me to assume that everyone felt the same way I did toward injustice. Furthermore, there were manifold other injustices ongoing in the world, yet I didn't campaign for all of them. No one person could, even if he dedicated his life in the attempt. My reasoning was flawed and presumptuous - since I chose which injustices to protest, it was wrong of me to expect others to protest the very same injustices, or to protest any injustices at all. In short, I tried to foist my character defects onto The Starfish Enterprise because we had shared an experience together. That the experience was an expedition to the *Lusitania* - the subject of the case - was irrelevant.

As Polly so aptly put it, "We just don't feel that participation is necessary in order to dive the wreck in future." To prove her point, several members of The Starfish Enterprise returned to Ireland in August and dived the *Lusitania* again.

Polly created the name The Starfish Enterprise. She didn't want the name exploited for a transitory corporation with which she was not aligned, in part because it suggested some connection with the diving expedition. During a brainstorming session, Hess and I originated the name Fifty Fathom Ventures, Inc.

Hess assigned each director to an office: I was President, John Yurga was Vice President, John Chatterton was Treasurer, and Barb Lander was Secretary. These assignments were purely arbitrary - no one had more or less authority or standing on the Board of Directors than any other. We were, in fact, a fictitious corporation but one that would be recognized by the court - an artifice, as it were, that was cumbersome but necessary in the American legal system.

Another legal expedient was the specification of a corporate goal. We couldn't appear in court as a collective amicus curiae. In order to intervene properly we had to advocate a competing interest consistent with legal propriety. Thus we entered a plea as competing salvors who challenged Bemis's appropriation of sole salvage rights.

To make matters quite clear, this judicial misdirection was just so much mumbo-jumbo in order to please the court. We had no intention of ever salvaging the *Lusitania*, and wouldn't accept title to the wreck if it were given to us. FFV's claim as a competing salvor was a legal position only, and nothing more.

## Davey Associates

Peter Hess found an attorney in Virginia who agreed to take our case on a pro bono basis. His name was Phil Davey. I drove to Norfolk to meet with him at his downtown office. Like Hess, Davey was a rarity among lawyers - strictly ethical, highly professional, an "officer of the court" in the purest sense of the term, and one who accepted his responsibility gravely. I would rather have a counselor like him at my side than all the shysters in the world.

He practiced law on the philosophy of absolute truth - the same philosophy on which American jurisprudence is founded (although not necessarily the philosophy on which it operates). Davey told me a parable about truth, the way he visualized it. Picture a cabinet full of cubbyholes. Present the truth to the court. Each truth is filed by the judge in a cubbyhole. Truths can be either helpful or damaging to a case. Davey's job was to guide the judge's hand, to place the truth in a context, and to have that truth filed in a cubbyhole that was beneficial to his client. The job of opposing counsel was to twist the truth, take it out of context, and have it filed where it did the most harm.

Davey instructed me to tell the truth no matter how damaging that truth may seem, and to let him control the guidance of that truth and to show it in proper perspective. He would be an abysmal failure in representing a client who couldn't abide by the truth, who had anything to hide, who had to rely on falsehood in order to prevail in a case. He was one of a vanishing breed. But so are honest clients.

He expected to spend only a couple of hours with me discussing the case, but so enraptured did we become in our philosophical digressions that we talked in his conference room until long after closing time, then went out and had dinner together. I came away with the feeling of mutual rapport.

## Foot-dragging

Peter Hess's greatest fault is procrastination. To him, a deadline is the time when he begins to think about filing for a continuance. By August he had not yet permitted Davey to file the intervention in FFV's behalf, claiming that he was still gathering ammunition against Bemis.

I argued - unsuccessfully - that sometimes you have to shoot the ammunition you have, then hope for resupply. The key to winning any battle is timing, especially when the enemy is on the move. If you wait too long for reinforcements to arrive, you might find yourself outmaneuvered. Many a battle has been lost without a single shot being fired.

Hess characterized my quick-assault strategy as "going off half-cocked."

## 318 - The *Lusitania* Controversies

We had a saying in the electrical business: you can't push rope. By this was meant that rope bunches up in the middle without the other end moving ahead. Hess could not be pushed.

### The Enemy Within

August brought another expedition to the *Monitor*. The first proposal to dive the *Monitor* on scuba had been submitted in my name only. My later expeditions were co-sponsored with Peter Hess. Because of our persistent suits against NOAA (four separate actions) over recreational diver access to the site, our names as signatories on subsequent permit applications raised the hackles of some of NOAA's bigwigs, particularly those whose dereliction of duty we called to political attention. They exceeded their authority in order to frustrate our efforts to explore and photograph the wreck.

Hess and I devised a plan to overcome this ongoing administrative resistance: submit the proposal in someone else's name - someone who was unknown and whose name would not wave a red flag in front of bullheaded bureaucrats. Barb Lander was hankering to take on a leadership role in wreck-diving. In 1993, when we offered to let her act as trip leader for the 1994 *Monitor* expedition, she jumped at the opportunity.

The operation of such trips was by now routine. During the years following my initial legal victory I systematized the logistics to the point where each expedition was basically stamped in the mold of the previous one. Already well established were the charter boat arrangement, oxygen rig deployment, floating decompression method, and so on. NOAA finally took my advice and installed a concrete mooring block next to the wreck. This facilitated securing a down line.

I gave Lander copies of our previous applications so she could retype them and bring them up to date. She made some minor changes, then submitted a proposal under her own by-line. Divers eager to see the Civil War ironclad still called me in order to sign up for the trip. I explained the new situation and the reason for the change in management, gave them Lander's phone number, and stepped out of the saddle to give Lander free rein. I didn't want her to be the leader in name only. I wanted her to lead. Only by doing the job herself could she learn the intricacies of conducting an expedition. My role was purely advisory.

The group assembled at our rented house in Hatteras. Cheryl spent the first week with me. She went diving on shallower wrecks with some of the other spouses, while I went out to the *Monitor*. Lander's husband and son were supposed to make it a family vacation, but those plans were changed.

Lander held a preliminary meeting at which she let everyone

know that this was *her* expedition, and that the presence of everyone there - including Hess and me - was tolerated on the assumption that we agreed to do her bidding. She wanted specific tasks performed, and it was our responsibility to perform those tasks to her entire satisfaction.

To be fair, the object of the expedition - as with our previous expeditions - was to conduct bona fide research that would increase our overall knowledge of the wreck, primarily by means of photography. Her goals fitted this established framework. What everyone took exception to was her heavy-handed manner and dictatorial attitude. She had suddenly become a control freak. From this first alienation it was all downhill.

On the boat she barked orders as if we were peons not worthy of respect. When it came to conducting the dive, she altered every part of a well-oiled machine that was based upon years of experience. She had her own way of doing things, and wouldn't listen when she was told that certain methods had already been tried and discarded. The result was a debacle with the decompression system in which we were unable to reach the oxygen hoses. Instead of deploying the system in the way that had always worked perfectly, she tried a new and unproven method. Those of us who were in the water were put considerably at risk. (Lander was on the boat at the time.) Danny Crowell could barely contain his anger.

That night there was general grumbling about the near disaster. Instead of learning by her mistakes, Lander took a defensive posture and refused to listen to criticism. Her arrogance got worse as the trip progressed, until hardly anyone would speak with her unless it was absolutely necessary.

Of the many examples of her wretched behavior I will cite only two. She wanted a drawing of a nest of mustard bottles in situ - this in addition to my own still and video photography. She didn't have a plastic slate on which to draw, so I loaned her my shipwreck survey slate - one of my own design which I featured in the *Primary Wreck Diving Guide*. After Lander photocopied the sketches, I asked for the slate's return. She refused to give it to me.

We argued off and on for a couple of days until I forced a major confrontation. She had the slate locked up in her car to make sure I couldn't get it. I explained that it held great sentimental value and that she had kept it from me long enough. I wanted it returned. She said, "I'm not giving it back."

I did not raise my voice but I was firm when I said, "You *have* to give it back to me. It *belongs* to me."

She scoffed at my appeal to reason and to the commonplace law of property rights. Among other things, she said that I didn't really want the slate, I wanted only to take it away from her, so I could destroy it.

The other members of the household listened from the sanctity of their bedrooms to Lander's shrill, irrational ravings.

She never did return my slate. I still wonder - did she want it for herself, or was her purpose to keep me from having it?

Others were the butt of similar mistreatment and evident hostility. By far the most serious breach of conduct occurred in the second week. Our charter boat, *Rapture of the Deep*, experienced mechanical difficulties that Captain Roger Huffman could not repair in short order. Lander made alternative arrangements with my long-time friend from the days of the Eastern Divers Association, Artie Kirchner, captain of the *Margie II*.

Before turning over the reins to Lander, I arranged to have Steve Lang and Uwe Lovas accompany the expedition in their own boat - a small center console job that could serve as a chase boat. It was good that I did. On one occasion Lander got lost and made a controlled ascent. I jumped in the chase boat with a back-up oxygen rig. Lang was at the helm. We caught up with the floating marker, dropped the rig into the water, then drifted during a long decompression.

Then came the day of iniquity. Lander went down with the first team. They made only a bounce dive because the visibility was bad - zero, in fact - due to a recent storm. She called off the second team's dive, then, without saying a word to anyone, she jumped off the boat fully kitted and swam to the nearby chase boat. I was aghast!

The two boats were in imminent danger of colliding as the wind and the current pushed them together. Lander was in the water between the boats and close to their sterns. Kirchner had to take the *Margie II* out of gear for fear of chopping up Lander with the propeller. Lang wasn't able to engage his motor because Lander grabbed the ladder. The boats pirouetted with their sterns almost touching. Those of us on the *Margie II* lined up along the port gunwale to try to shove the two boats apart. We barely averted collision. As soon as Lander had her fins on the ladder, Lang gunned the motor and veered the boat out of the way of the *Margie II's* drift.

The leader of the expedition had abandoned us!

Kirchner called Lander on the radio but she refused to answer. The chase boat raced out of sight we knew not where. Since Lander had left without issuing instructions or delegating authority, Kirchner headed toward the inlet. I discussed the situation with Harvey Storck and Greg Masi, both of whom could barely contain their anger - not that I was any less indignant over Lander's inexplicable and unconscionable behavior. Joanne Surowiec confirmed that the visibility had indeed been zero.

After we cooled down and mulled over the predicament that Lander had bequeathed to us, we concluded that if we couldn't dive the *Monitor* we could at least dive some other wreck on the way back to the dock.

Kirchner wouldn't hear of it!

He said that according to his interpretation of his understanding with Lander, he had been chartered to take us to the *Monitor* only, not to any other wreck. Since my friendship with Kirchner went back more than twenty years, since I had given him more than a week of my time to help him bring the *Margie II* from the building yard in Louisiana to Hatteras, since I had invited him to dive on previous *Monitor* trips, since I had helped acquire customers for his fledgling charter business, since I . . . and on and on. I elected myself the spokesperson for the group.

I argued that the boat was chartered for a full day's performance, and that if we couldn't dive the *Monitor* - for whatever reason - then he was obligated to take us elsewhere because the boat was still under charter. I don't want to belabor the point. I argued long and fruitlessly but - as Kirchner quickly pointed out - he made his agreement with Lander, not with me. That she had abandoned us in order to pursue her own personal agenda held no sway. And we had to pay for the charter to boot!

As the reader can well imagine, the household was in an uproar that night. Lander did not put in her appearance until after dinner, and then she wouldn't tell us where she had gone with Lang and Lovas. She said it was none of our business.

The next morning she told us that she was going out with Lang and Lovas to check the conditions on the *Monitor*. If conditions were favorable she would give us the go ahead. She didn't call until late in the morning. She told us that she was going diving with Lang and Lovas, and that we could go out on the *Margie II* if we wanted but that she would not authorize expedition funds to pay for the charter. (We had put up our money in advance.) We had to make our own charter arrangements with Kirchner and pay additional money to dive!

On such a sour note the expedition fizzled out. Yet despite Lander's conduct, the dives I made on the *Monitor* that year were among my most rewarding - for I recovered some of the bottles that I had discovered the year before. I didn't get to keep them - they were all surrendered to NOAA for preservation and conservation, and eventually for display. But the sheer act of recovery vindicated the point that I had been advocating for years: that wreck-reational divers could perform valid and useful functions in the activity that they loved best. *That* was sublimely satisfying.

## John Light's Heritage

Hess got in touch with Muriel Light and informed her of FFV's proceedings. Without representation - which the unprosperous widow could ill afford - and without delay, she sent a letter to the court along with supporting documentation. Her case against Bemis was far more valid than that of FFV.

She submitted copies of John Light's *Lusitania* purchase agreement, his last will and testament, and various letters. The purchase agreement was unimpeachable. Equally telling was correspondence that bore witness to irregularities - a deliberate understatement - in Light's erstwhile relationship with Bemis and Macomber.

Light to Macomber (July 12, 1972): "First off, as you are well aware, *all and any* 'supposed agreements' between myself, you, and F.G. Bemis Junior (as a third party) were illegal in their conception and were perpetrated upon me under false pretenses by you and Bemis, through your lawyer, Mr. F.W. Andres . . . who was at the same time representing himself falsely to me, and others, as my official 'Legal Counsel' and 'business adviser'! Furthermore, this illegal collusion between yourself and the other above noted parties, continued throughout our most regrettable relationship until it became apparent that I had recognized your completely unscrupulous intention to deprive me of all possible financial gain which I might realize through various projects connected with the 'story' of the 'Lusitania' . . . I recognized your chicanery, for what it was . . . F.W. Andres was working throughout our relationship in your and Bemis's interests, and not in mine as he had always claimed! . . .

"Secondly, you, F.G. Bemis, F.W. Andres and R.S. Hanson (as a 4th. party), entered into a premeditated conspiracy to illegally divest me, and my family, of all of my assets (real and unreal), as well as my potential future earning power to support my family in relationship to the 'Lusitania' and associated projects including 'Kinvarra Shipping Ltd.". That in carrying this out, you removed from my possession under false pretenses, all of the diving equipment which I purchased with funds supplied by the publishing firm of 'Holt, Rinehart & Winston", for the task of diving and photographing the 'Lusitania', therefore preventing me from completing that phase of the project! Furthermore, all of this diving equipment (as mentioned above) was subsequently sold by you . . . without ever obtaining from me, the rightful 'owner of record', any 'Bill of Sale' or 'transfer of ownership papers' . . .

"I am sure that I need not remind you that cases here in Ireland are still pending against you and your associates (in the High Court), in respect of work performed by 'Verolme Shipyard' and bills not paid by you in excess of $35,000.--: nor, that your 'Operational Plan" . . . was nothing more than a childish attempt to defraud 'Verolme" of monies due, for services rendered."

No doubt John Light would have testified effusively in the current proceeding, had he been alive to do so.

Light to Dan O'Connor, NBC producer (October 17, 1974): "Enclosed please find a complete set of the papers regarding the negotiations and purchase of the LUSITANIA: there *is* nothing else!! Following 'Bemis' taking over 'Kinvarra Shipping Ltd.', Andres wrote me and asked that I sign the rights of the LUSITANIA over to him. I point blank refused, despite a number of letters from him which went from pleading to demanding. As soon as I get these letters from Ireland, Ill get them copied and sent them on. Rest assured, however, no further 'transfer', what-so-ever, took place in regards to the LUSITANIA. My refusal was absolute, and my reasons well defined, as you shall see."

No doubt John Light would have elaborated on these points in court, had he been alive to do so.

## Again, Misrepresentation

Because of Muriel Light's interest and involvement in the case, FFV now received copies of the documents to which Polly Tapson alluded in her letter, with respect to Richard Robol of Huff, Poole & Mahoney.

Robol to Muriel Light (August 16, 1993): "I represent Columbus-America Discovery Group. The Group may be interested in obtaining files and materials, including historical data, related to the shipwreck LUSITANIA. I would like to call you to discuss this interest. I shall be calling you within the next several days."

Robol to Muriel Light (September 7, 1993): "Thank you for taking the time to speak with me the other day. As I mentioned, Columbus-America may be interested in discussing with you possible acquisition of any archives, papers, materials, artifacts and other items relating to the *Lusitania*. . . . I recognize that your materials may hold some value for you, and I am prepared to discuss possible compensation. Likewise, I think it would be appropriate for your family to receive appropriate recognition for their efforts (if you so desire) in any public exhibitions about the *Lusitania* (including joint exhibits with the S.S. *Central America*)."

Muriel Light to the court (July 30, 1994): "In the telephone conversation referred to in the September letter, Richard Robol in reply to a question as to who were the people involved in the Columbus America group, claimed no knowledge of F. Gregg Bemis. Yet just six months later, the same firm was indeed representing Bemis before your court. While it is certainly possible that Bemis hired Huff, Poole & Mahoney after September 1993, I must express my concern about what was otherwise a less than straightforward communication. This may not be a good start if one is concerned about the integrity of the process." Hats off to Muriel for understatement!

Muriel Light to Hess (July 31, 1994): "I am willing, should your attempt to stop this erroneous claim come to a court action, to help in any way possible. This includes a court appearance, deposition or using my name in conjunction with the other interested parties."

After discussing these letters with Hess, I offered to pay Muriel Light's expenses (travel, lodging, and meals) out of FFV funds if she would give live testimony in Norfolk.

## BEYOND THE DEADLINE

FFV's motion to intervene was finally submitted on September 7, nearly three months too late. I write "too late" because the public notice published in the London *Times* on June 4 stated that competing claims had to be filed "within 10 days after publication of notice."

In the real world, such a short response time is patently absurd. Hess argued that the ten-day stipulation was a legal convention and not a strictly exclusive limitation. He called the Clerk of the Court every other week or so to ask for an extension, and each time was told that since the case was not yet scheduled for hearing, a motion could still be submitted. However, the Clerk cannot speak for the judge, whose discretionary authority only the judge himself may exercise.

Phil Davey was not responsible for the overdue filing. His hands were tied by Hess's factual research. But once the motion was filed, Davey took immediate and constructive control of the case, devoting to it the time and attention it needed, notwithstanding that this reduced his time available for profit-making cases.

## A FOOT IN THE DOOR

Robol immediately filed a motion to dismiss FFV's intervention. His motion was flawed by partial truths and inconsistencies which drew attention to Bemis's "bad faith" or "unclean hands" (legal terms). Robol's tactic backfired, perhaps because he succeeded - where we had failed - in peeking the judge's curiosity about FFV's allegations of impropriety.

In citing FFV's tardiness, Robol wrote that the public notice appeared in various newspapers as the Court directed. But he failed to mention that the Court also directed that all potential claimants be notified personally.

FFV motion: "In his negotiations with members of the June, 1994, dive team, Plaintiff Bemis and his counsel repeatedly threatened litigation and injunctive proceedings in the courts of the Republic of Ireland, but never once mentioned the pendency of these proceedings, notwithstanding this Court's admonition in its Order of May 24th that potential claimants be given actual notice of the proceedings."

Robol also demanded of FFV the production of documents and

answers to interrogatories despite his position was that FFV was not a party to the action. FFV response: "The plaintiff is treating Fifty Fathom as a party which has the effect of waiving the objection to the filing of the claim." In other words, Robol wanted full disclosure of FFV's documents and information, but did not want FFV's appearance in court. Even in court you can't eat your cake and have it too.

Two other factors helped curry the court's favor. One was that "Fifty Fathom Ventures, Inc., divers and their colleagues are the only team to have safely and successfully dived the LUSITANIA since 1982." Bemis "does not have comparable capacity of the Claimant-Defendant's demonstrated ability . . . Upon information and belief, Claimant-Defendant denies that Bemis organized the 1993 expedition by National Geographic Society." The latter statement referred to Bemis's claim to have organized the NGS film shoot.

The other factor was jurisdiction. "Plaintiff's objection to Fifty Fathom's claim begs the question of why he filed his action in the Eastern District of Virginia, thirty four hundred miles away from the wreck when the wreck is within the territorial seas of Ireland and within twelve miles from Kinsale, Ireland."

FFV's Brief in Opposition to Plaintiff's Motion to Compel stated, "Bemis' motion to compel is premature. . . . Bemis has moved this court to preclude FFV from becoming a party to the action . . . Not until FFV becomes a party to the proceedings can it be held to answer discovery."

After all these claims and counterclaims, motions and motions to dismiss, and answers to all of the above, Judge Clarke decided to hold a hearing at which he let FFV's attorneys explain its position. FFV was represented by Phil Davey and his equally perceptive and ethical associate, Patrick Brogan. Bemis was represented by Rick Robol and one of the partners of the firm of Huff, Poole & Mahoney: Glen Huff.

At the hearing Clarke noted, "I don't know whether I have authority to enter discovery orders against a party that I have not yet permitted to enter the suit, but I can tell FFV that the burden is upon them, or it, and if they hope that I'm going to act favorably toward them, they are going to have to produce their principals in Norfolk for Mr. Robol to take their depositions."

Ironically, Clarke came down harder on Bemis than on FFV. "My actions in giving Mr. Bemis temporary custody of the vessel were based upon his representations to me that he had planned and organized a dive on this vessel for the summer of 1994. The summer has now passed and I haven't heard a peep out of anybody about that, and I expect an immediate report on whether it was done and, if so, when it was done, what was done, how it was done, and what was brought up, if anything, and where those things are. I also need a report from him in accordance with my previous order, I think, of where the artifacts previously brought up are now. I don't think that Mr. Bemis has helped himself a bit in this case in his activities since my order of May 24th. I think he has disregarded everything I've said."

Clarke also expressed his other major concern: "We're talking about a big ship and something that's 3,000 miles away from me that I don't want to have any jurisdiction over because I think the British courts ought to have in any event because it's a lot closer to them than it is to me. . . . It won't take much encouragement for me to dump it back out in the middle of the ocean. If your fellow wants me to hold onto it, he better tow the line."

## THE PAPER TIGER

Robol's next action was to wallpaper his office. The documents that he wanted FFV to produce took two pages to list, and included "All correspondence, telephone logs, memorandums, notes and other documents pertaining to any expedition by you or any other persons to the *Lusitania*. A copy of all dive logs, deck logs, engineer's logs, master's logs, medical logs, and other documents of any person or vessel participating or assisting in your salvage of the *Lusitania*. A copy of all photographs, videotapes, sketches, drawings, writings, records and other documents prepared during any expedition to the *Lusitania* by you, anyone acting with you, and/or anyone assisting in your behalf." And so on ad infinitum.

How we were supposed to procure documents from unnamed "other persons" Robol neglected to mention. But he did provide two pages of definitions. By "person" he meant "both natural person and corporate or other legal or business entities, including but not limited to partnerships, joint ventures, proprietorships, associations, trusts, agencies, boards, committees or commissions, as well as any of their officers, directors, agents, servants, or employees."

"Document" included "without limitation, the original and every non-identical copy of any written, printed, typed or other graphic material of any kind or nature and all mechanical, electronic or sound recordings, regardless of origin or location, whether sent or received or made or used internally in your possession, custody or control or prepared by or for you, including without limitation, letters, correspondence, memoranda, minutes of meetings, bulletins, reports, specifications, telegrams, notes, notebooks, scrapbooks, diaries, worksheets, summaries, transcripts, contracts, exhibits to contracts, records, interoffice communications, warranties, advertisements, fliers, promotional literature, computer sheets, cards, programs or printouts, drawings, graphs, tables, charts, photographs, sound recordings, disks, diskettes, tape or other electronical or electromagnetic records, writings and instructions, work papers, bills of lading, certificates, slips, policies, cover notes, bills, receipts, logs, books, diagrams, manuals, or other memorials in whatever form they may exist, and shall include drafts or non-final versions of any of the foregoing."

Davey informed me that Robol's paper punches were part of his style, an attempt to intimidate opponents into believing that compliance was prerequisite to either participating in a case or having any chance of prevailing. In one sense the demand to produce documents was a fishing expedition, but by and large the purpose was to make continued litigation unduly burdensome. For the cost of postage and the paper on which to print a computerized form letter, an attorney could hypothetically compel the opposition to spend eons of time and unimaginable fortunes in searching through files and photocopying possibly relevant documents. Worse than the abuse of the system of disclosure is the acceptance of this form of corruption by the courts.

Robol could ask for the sky, but until FFV was granted status as a litigant, we were obliged to produce only what was pertinent toward being recognized by the court. A more complete disclosure was out of place at this stage of the proceedings. Thus we produced evidence to validate our authenticity with respect to the *Lusitania*: dive log entries, photographs, videotapes, meetings of corporate meetings, and recovered artifacts. We did not produce our life histories, bulk mail, super market receipts, or other such trivia not germane to the case.

## SHOT FROM BEHIND

Barb Lander had personal motives which conflicted with the goals of FFV. She sent to Davey a copy of Polly Tapson's "R.M.S. Lusitania 1994 Expedition Notes." This 44-page report was intended for members of The Starfish Enterprise only, and was stamped "CONFIDENTIAL" and "FOR YOUR EYES ONLY" in red capitals on the cover. It detailed every aspect of the expedition's planning and procedure, including ways to deal with the Bemis threat (through the Receiver of Wreck, through injunctive relief, and so on).

FFV was not The Starfish Enterprise, and until FFV became a party to the litigation there was no reason to let Bemis have access to information that did not relate to the acceptance of our intervention. Submitting such a document was giving ammunition to the enemy.

Phil Davey agreed. Out of sight, out of mind. But once the report passed before his eyes, he could not in good conscience as an officer of the court deny release of the document to the plaintiff. In his opinion, his knowledge of the report's existence made it discoverable. He and I argued jurisprudence (I wanted the report withheld until FFV was recognized by the court as a litigant), but eventually I had to bow to his doctrine of firm legal ethics. The report was submitted as evidence despite my protest.

## Intellectual Property Wrongs

Shortly after the termination of The Starfish Enterprise's trip to the *Lusitania*, Robol sent letters to magazines and newspapers throughout the U.S. and the U.K. in which he intimated that he might take action against anyone who published *Lusitania* articles or photographs submitted by me or Polly Tapson, suggesting that we did not own the rights to our endeavors and achievements. This attempt to bully publishers, discredit Polly and me, and deny us potential earnings may not be morally justified, but apparently it was legal.

Every grade school student knows that the First Amendment guarantees freedom of the press. I had the Constitutional right to chronicle my adventures, and magazines and newspapers had the same right to publish them. Photographic images that I created were automatically copyrighted in my name, and remained copyrighted until fifty years after my death.

Bemis wanted to believe that his non-established ownership granted him intellectual property rights. Besides putting the cart before the horse, his claim of ownership was based upon a supposed chain of conveyances that went back to John Light's original purchase of the hull and fittings. The transfer of title from the Liverpool and London War Risks Insurance Association made no mention of intellectual property rights, nor were those rights ever assumed by John Light.

John Light executed a contract with Holt, Rinehart & Winston to provide film and photographic images in return for cash advances. This is called "work for hire." The results of Light's efforts - footage and stills - therefore belonged to the publishing company. Even if Bemis could prove that he had usurped Light's contract - despite Light's written statement to the contrary - the results of Light's efforts still belonged to Holt, Rinehart & Winston. Now it appeared that, through some process of creative reasoning or abstract logic, Bemis was claiming *future* photographic rights based upon Light's unfulfilled contract.

Neither Light's purchase of the wreck nor his publishing contract granted him universal rights to intellectual property, only rights to intellectual property that he *produced*.

Looked at another way, imagine taking a family photo in the street in front of your home. Your relatives are wearing designer clothes and in the background cars are parked in front of your neighbors' houses. Extending Bemis's theoretical position to the extreme, no one could ever publish or publicly display your snapshots because the objects in the background belonged to someone else.

The houses were deeded to various homeowners and came from the drawing boards of building architects. The cars that belonged to the neighbors or their visitors were designed by automobile manufacturers. Apparel was emblazoned with brand names.

It goes without saying that in a capitalistic society *everything* is owned by someone - whether the property is real or intellectual. Nor does it take an intellectual to understand that property rights necessarily have realistic limitations. (Is it considered trespassing when a letter carrier walks across your yard to deliver the mail, or when a jumbo jet flies over your house at 30,000 feet?)

A society which denied representations of itself would exist in chaos. Never again could a photograph be published. Books, newspapers, and magazines would contain only text. Students would not have the benefit of visual art. Cameras and film would be outlawed. The world would be usurped not by anarchy, but by absurdity.

Notwithstanding the rationality of the above, another truism asserts that only a dastard relies on intimidation and only a milquetoast can be intimidated. Most publishers jeered at Bemis's groundless threats. Cathie Cush published an illustrated piece in the newspaper for which she was the editor: *Underwater USA*. I was asked by three magazine publishers to write articles about the expedition and to illustrate them with my underwater photos. The first was Joel Silverstein, of *Sub Aqua Journal*. The second was Mike Menduno, of *AquaCorps*. Only the third changed his mind after receiving Bemis's letter, and decided not to publish: David Taylor, editor of *Rodale's Scuba Diving*; publisher Steve Blount agreed with Taylor's decision and paid me a kill fee instead.

U.K. publishers did not feel threatened. Christina Campbell wrote an article for a Scottish magazine. Paul Owen was interviewed about his participation in the expedition. Simon Tapson published photos of the wreck. Polly and Simon gave lectures. Chris Reynolds wrote a piece. And so on.

## Cargo

Davey unearthed information which contradicted Bemis's claim of ownership of the cargo. According to England's Treasury Solicitor, the *Lusitania* "carried a cargo owned by the Ministry of Munitions in which this office has an interest. . . . Being Crown cargo, the MOM cargo was not insured and remains the property of the Treasury Solicitor."

England's Department of Transportation wrote "our records show that the hull of the vessel was sold to a Mr J F Light in 1967. . . . As for the cargo, according to our records the title falls partly to the Department and partly to the Treasury Solicitor.

These letters were held in reserve, to be sprung at the proper time in order to have the most telling effect.

## The Interrogatory Battle

Robol was unhappy about FFV's partial response to interrogatories and the production of irrelevant documents. He wrote reams of papers in support of his "motion to compel." It was to his advantage to do so not just because he was being paid an exorbitant hourly rate for the gross generation of paper work, but because he needed to know how much information FFV possessed that he was trying to suppress.

Our strategy was to spring upon the judge Bemis's economy with the truth, without letting Bemis know in advance how much incriminating evidence we had uncovered.

FFV rejected Robol's fishing expedition "until the Court has formally accepted our Answer and Claim for filing." Robol not only objected to FFV's position, he went so far as to expand his interrogatories. For example, "State whether any of your officers or directors have ever been arrested for or convicted of a crime, infraction or offense, and if so, state the date, charge, circumstances and disposition of each."

This shameless tactic is akin to legalized blackmail, the purpose of which is to deceitfully dissuade testimony by implying that a witness's personal life will be exposed to view in the court room - or intentionally leaked to the media. A person's former doings, no matter how bad, have no relevance to his right to enter a plea in another proceeding, nor can it be used against him or brought into the open. But people with a sordid past will often retreat from scandal. The value of the stratagem is not always in the unveiling, but in the threat. The American legal system permits such unscrupulous enterprise.

None of FFV's directors had a criminal past, but we objected to answering this interrogatory on purely ethical grounds. We were not about to let Robol get away with such obvious exploitation. Furthermore, his request exceeded the threshold permitted by law. It isn't proper to ask if one has been *arrested*, only if one has been *convicted*. The lack of conviction substantiates innocence, but the knowledge of wrongful arrest can be used - dishonestly - to suggest involvement in criminal activity. And "infraction or offense" could technically include parking tickets and minor traffic violations.

After Davey ironed out all these issues of impropriety, he told Robol that he would let each director answer the question of conviction during deposition.

Among a dozen other red herring interrogatories were such as this: "Describe in detail and with particularity each and every occasion on which you have salvaged the *Lusitania*, the identities of each person who participated or assisted in such salvage and the identity of all persons having knowledge of any such salvage and all documents pertaining thereto."

This may have been attempt to ascertain the names of others who could be called in for abuse, but more likely it was an attempt to effectuate many hours of work at the cost of a minute or so spent in composing the question. Such are the oppressive wiles permitted by American law, and relished by attorneys who are unwilling to lay the facts before the court and try a case based solely on its merits.

## Sworn Statement

Depositions were scheduled for November. Mine was first. As usual when I travel to Norfolk or Virginia Beach, I planned to stay with Trueman and Nike Seamans and to do research at The Mariners Museum in nearby Newport News - this to make efficient use of the six-hour one-way drive.

Davey's pre-deposition instructions were simple: answer the questions truthfully. By this advice he meant that I should not be evasive and should not employ the dodge of memory loss. Certainly, if I didn't know the answer to a question I should say so. But he said that judges don't like to hear "I don't know" or "I don't recall" or "I don't remember." Too many such evasions demonstrated a lack of sincerity.

Robol did not appear at my deposition. Instead, I was deposed by Glen Huff, one of the partners in Huff, Poole & Mahoney. He seemed to have taken a personal interest in the conduct of the case. He didn't learn anything significant - not because I was cagey or withheld pertinent information, but because I had nothing significant to impart.

There were some questions which Davey directed me not to answer, either because they were irrelevant at this stage of the proceeding, or because they related to The Starfish Enterprise (which was not a party to the litigation and therefore not bound to give evidence in court), or because they required answers that called for speculation or hearsay (for instance, what did I think someone else thought or did).

Although I had only to answer questions truthfully - easy to do when you have nothing to hide - I had to watch for trick questions, innuendoes, and hidden allegations. Attorneys often ask questions in such a manner that the answer they want to hear is suggested or supplied in his language. This question-answer couplet could later be twisted to make inferences that the respondent didn't mean.

Many of Huff's questions dealt with the conduct of the dives and the technical aspects of decompression. These were easy for me to answer - in a conversational tone - because I did this all the time during wreck-diving workshops. I treated Huff as a novice, but not in a condescending manner - simply as one who knew nothing about the activity.

Huff spent an inordinate amount of time in establishing what I knew about the *Lusitania's* location. This was because Bemis and Robol had informed the court that the wreck lay in international waters, when in fact it lay within Irish territorial seas. As irrefutable proof I had a letter from Nic Gotto in which he gave the GPS coordinates, and Davey had a chart of Ireland on which the numbers were plotted. There was no doubt that the court had been misled. Such misdirection was necessary in order to convince the court in Virginia to accept Bemis's case.

It goes without saying that an American court cannot take jurisdiction over property that is located in a foreign country. That smacks of colonialism.

In order to establish that I had actually been on the *Lusitania*, and to demonstrate my competence as a photographer, I agreed to furnish prints of my photographs - but only with the understanding that by doing so I was not waiving my copyright and I was not placing my photos in the public domain. The photos were to be placed under seal, which means that only counsel and the court could see them.

I didn't have the prints with me that day. I sent them the following week. To ensure that the photographs could not be used commercially, I purposely took them to a store that did poor quality work. The prints were overexposed. Then I stamped my name across the face of each picture.

## Underhanded Scare Tactic

Davey cautioned us about collusion. He said that all too often the confederates in a suit collaborated to the point that their testimonies sounded like canned reiterations of a carefully rehearsed script. These word-for-word readings in different voices lacked the feel of authenticity. So instead of carefully preparing our statements, the members of FFV adopted a non-consensus policy: we did not try to get our stories straight. We even stopped discussing the events that occurred on the trip. That way, our testimonies appeared fresh and untainted by artifice and artificiality.

John Yurga and Barb Lander were slated to be deposed the next day. When the time came, Yurga arrived at Davey's office early so that he and Davey could have a pre-deposition conference. Yurga's deposition did not differ substantially from mine. He told the truth as he perceived it to be - through his own eyes - so his story came out essentially the same as mine except that it was expressed in different language and was conveyed from a contrasting viewpoint.

The differences in our testimony were minor. I reported the depth at which the breakaway line was clipped to the shot line as sixty feet; the day Yurga was dive marshal it was clipped at eighty feet, a function of the amount of loop in the shot line. We reported bottom depths that differed by two feet. Bottom times likewise varied. Yurga made no substantive revelations.

Lander showed up late. Her deposition was scheduled first that morning, but when she didn't make her appearance, Yurga took her place. Then she showed up in time to sit in on Yurga's deposition. After he completed his testimony, he went to The Mariners Museum to do shipwreck research. Because of Lander's lack of punctuality, Davey - who was meeting her for the first time - did not have an opportunity to offer advice and admonitions. As events unfolded, this didn't make any difference.

Attorneys for the plaintiff had another cowardly trick up their collective sleeve. Lander's deposition was barely under way when a process server pushed his way into Davey's office. He tried to serve Lander with a damage suit for interfering with Bemis's claim against the *Lusitania*. The complaint accused her of trespass, piracy, fraud, conspiracy, and other legal tripe that took nine pages to enumerate. Bemis claimed actual damages in excess of $60,000, and demanded treble that amount in punitive damages - nearly one quarter of a million dollars. Identical complaints were made out for me, Chatterton, and Yurga (whom the process server narrowly missed).

Polly Tapson was right: in America anyone can be sued at any time without any grounds whatsoever. It wasn't even necessary for the suer to show the court a shred of proof. He can sue on mere allegation. It has already been proven that anyone with a sleazy lawyer, financial backing, and an imagination for trumping up charges can get away with murder. No longer the land of the free and the home of the brave, America has become a country of disparity: one in which the rich get a slap on the wrist while the poor get a rope around the neck, in which the affluent serve time in Club Med luxury while the impoverished spend their sentences in penal poverty. What was there to protect the indigent from the onslaught of wealthy legal sanction?

Every American knows that it is easier to make big money by suing than by doing honest work. Did Bemis truly believe that he could increase his considerable fortune by a cool million dollars on such phony premises? Perhaps. But the suit contained a double-edged subterfuge that was far more sinister than the mere acquisition of ill-gotten money.

The purpose of a harassment suit - one whose pretext is baseless - is not necessarily to prevail in court, but to ruin an individual financially through protracted litigation and the exorbitant cost of defense. A defendant can win the case but go broke in the process. There are numerous cases on record in which people lost their homes, their life savings, and first-born child during the lengthy process of legal defense.

The more diabolical side of Bemis's damage suit was that it was filed in the State of Virginia. This slick legal maneuver would create a further financial burden on the defendants by forcing them to litigate far from home, and to retain local counsel. Bemis should have sued us in our home states - Lander and me in Pennsylvania, Chatterton and Yurga in New Jersey - but then he couldn't use the services of Huff, Poole & Mahoney, who were the driving force behind the legal but artful dodge.

While it goes without saying that Huff, Poole & Mahoney had nothing to gain from a suit that was filed outside the State of Virginia - that is, they couldn't charge Bemis their hourly rate if another attorney did the work - the insidious plot went much deeper.

According to judicial predicate, a person cannot be served by the forum State (the State in which a matter is being heard) unless that person has some contact with the state or can be caught within its borders; otherwise, the court has no jurisdiction over the person. In other words, Bemis could not serve us in Pennsylvania or New Jersey about a matter to be heard in Virginia *unless we entered the State*. We were not subject to service as long as we stayed outside of Virginia. The fiendish brilliance behind the damage suit was not merely Machiavellian, it was Mephistophelean!

If Bemis's attorneys had served me during my deposition, the damage suit would have lost its value as a threat. By letting me and Yurga go and serving Lander, they were sending the message that the next time we entered the State of Virginia - to testify at the hearing - a process server lurking in the court house would serve us with process, thereby entangling us (and Chatterton) in the crooked suit.

The scheme can be viewed not only as a means to extort money from those who opposed Bemis's doubtful declarations, or who aspired to compete against him as salvors, but to keep out the truth that we represented and the facts and the evidence that we were going to submit and that contradicted what Bemis had perpetrated upon the court when his word went uncontested. The damage suit was bogus both in context and in conduct - a pretext to thwart having to litigate Bemis's *Lusitania* claim on its merits.

Davey went ballistic. His usual calm exterior burned off to reveal a seething legal lion opposed to unethical chicanery. A shouting match erupted between him and Huff. Not only was the service of process unethical under the circumstances, it was illegal. Lander had appeared in the State of Virginia under a court directive - Judge Clarke himself had ordered the depositions. Every lawyer knows that a person cannot be lured into a court's jurisdiction by an order of the court only to be slapped with a civil suit. Davey reminded Huff of his hornbook law, and in no uncertain terms.

Davey grasped another implication, equally as insidious, which he stated to Huff and repeated for the record: "It is not proper to serve a witness when that witness is in the jurisdiction under a court order. I think this is an outrageous attempt to intimidate this witness. And I will not allow it. I simply will not allow it. I have asked Mr. Robinson to leave." (Robinson was the process server hired by Huff, Poole & Mahoney.)

Incredibly, Huff denied knowledge of what action his colleague, Rick Robol, intended. "I don't know exactly what the pleading or process was that was to be served."

After a lengthy argument, Huff backed down and noted his objection for the record. To make the situation quite clear, Davey later drafted an order which stipulated that the directors of FFV could not be served with process in Virginia if they were present in the State at the court's direction. Judge Clarke fully concurred, and let Bemis's attorneys know that he would not tolerate such behavior.

Of course, that didn't prevent Bemis from lodging complaints in our home states. But what would be the point?

During the subsequent course of Lander's deposition, Huff harped on the artifacts that had been recovered. He spent a great deal of time giving her the third degree about two ordinary bathroom tiles, one of which was broken in two. (Each rectangular tile was about three inches long.) They appeared one day on a shelf in Kieran's dive shop, somehow got stuffed in Lander's gear bag, and were brought back to the States with her luggage. No one admitted to recovering them or to knowing anything about them.

The value of *Lusitania* artifacts on the open market was established when Oceaneering auctioned off the booty that was recovered by the company in 1982. Common items such as portholes and dishes sold for only a few hundred dollars. Measured against that, the value of the two tiles - assuming that a gullible buyer could be found - must be gauged in pennies.

Yet it was this kind of minutiae to which Huff devoted much of his questioning. If he was trying to make a mountain out of a mole hill, undoubtedly it was because no other issues existed on which to spend his client's time. After Lander disavowed all knowledge of the source of the tiles, he suggested that they might not even have been recovered from the wreck, but could have been trash from the folk house. Lander conceded that this was a distinct possibility.

Chatterton got off the lightest. As the hearing date approached and he was unable to make himself available according to Huff's mercurial convenience, his deposition was conducted over the phone.

## PROFESSIONAL DISCOURTESY

Robol's demand that FFV disclose its evidence in full, while at the same time objecting to FFV's participation in the case, was the kind irrational and self-serving assumption that typified and characterized the Bemis defense posture. What made Robol's demands all the more absurd was his absolute refusal to cooperate in any way with FFV's attorneys, to the extent of denying the most fundamental professional courtesies.

Robol and Huff comported themselves as if depositions and disclosure rode a one-way street, with no thought at all toward reciprocation. Not only did they refuse to produce Bemis for deposition, they refused to release court documents that had been placed under seal. (The purpose of placing a document under seal is to keep it from the public, not from the attorneys of parties who had an interest in the case.) They wouldn't even release Bemis's affidavit! Not only wouldn't they release any documents, they would not respond to written requests or speak to Hess on the phone.

Wrote Hess: "This is at least the fifth time that the sealed affidavit has been requested. We have no intention of producing John Chatterton for a deposition until Bemis has been produced. Nor will any copies of Gary Gentile's photographs of the *Lusitania* be produced until we see some measure of reciprocity from you. Moreover, I have not received an executed copy of the Stipulated Protective Order governing photographic and videographic images of the shipwreck. . . . Because of Bemis' complete lack of effort at cooperation, FFV, Inc. intends to bring the issue of litigating in bad faith to the attention of the Court."

Huff eventually responded by writing that Chatterton's deposition was being held "hostage pending your taking of the deposition of Mr. Bemis." This is hardly a fair assessment considering that Huff had already obtained three depositions to our none.

The above is only a small sampling of the largely one-way dialogue that transpired with respect to Bemis's attorney's refusal to cooperate in the spirit of legal convention and in the extension of professional courtesy. Finally, Hess was moved to write most vitriolic: "While Fifty Fathom Ventures, Inc. has bent over backwards to comply with the Court's unusual order expediting Bemis' discovery from a non-party, apparently, Bemis has chosen instead to treat FFV, Inc. as if it has bent over forward."

And two days later, Hess added: "Just because Bemis has been able to use his toilet royalties to browbeat meeker parties into supplication before his grandiose delusions does not make him the owner of the *Lusitania*."

## Putting Out the Light

Robol was playing hardball with Muriel Light, too. He submitted a long list of interrogatories that he wanted her to answer, demanded an extensive production of documents, and issued a summons for her to be deposed in Virginia. Did a quarter-million dollar damage suit await her arrival in the State?

Some of Muriel's responses are telling. "I do not feel that it is proper that I am being sued by Mr. Bemis in Virginia. I do not live there... Why can't you at least have the decency to sue me in my home state of Massachusetts? This is very unfair to me and I am surprised that the court is allowing it to happen."

She also complained about "how you are treating me in this case. I received your 'Notice of Deposition' on the day before I was supposed to be in Virginia for it. How can someone be expected to drop everything to travel hundreds of miles overnight?" She lived more than five hundred miles from Norfolk.

In her sworn statement, she wrote, "My husband planned a series of dives to the *Lusitania* on mixed gases in the late 1960's. These were never attempted because the financial backer, Mr. Macomber, refused to pay for adequate safety equipment for the divers. It seemed as if Macomber's interest in the shipwreck was as a tax write-off, since he travelled from the United States to Ireland frequently, but never paid for the equipment necessary to ensure save diving. It is interesting to note, however, that Mr. Macomber could afford a $1,000,000 (U.S.) life insurance policy on my husband, for which he (but not John Light's and my 4 young children) was the beneficiary. . . .

"All of my husband's papers concerning the misconduct of Mr. Andres were sent to his mother, Ruth Light, who has died. When I learned of the sabotage of the dive boat in Kinsale, Ireland last summer, I had all of my husband's papers related to the *Lusitania* sent overseas."

Remember that Robol had already tried surreptitiously to obtain these documents from Muriel Light under the guise of the Columbus-America Discovery Group. She wanted to make sure that none of John Light's research fell into Bemis's hands.

Again FFV came to Muriel Light's rescue. Our attorneys filed a Motion to Quash the notice of deposition. The rules required eleven days advance notice. Furthermore, we charged that "the proposed taking of Muriel Light's deposition is nothing more than an effort to obtain discovery of her private information."

## COURT PREJUDICE

The reader must be wondering why the judge let Bemis's lawyers get away with such shenanigans and blatantly unfair treatment. My best guess is that he looked upon FFV and Muriel Light as interlopers, perhaps even as outlaws. After all, so far he had heard only one side of the story, and a highly inventive side at that. Bemis represented himself to the court as an innocent, maligned, and displaced owner who was doing great things on the *Lusitania* - things which were being wrecked by those who stood in his path and who proceeded against his will, and, by extension, the will of the court.

We didn't learn until late in the game that at the hearing held in May, Clarke appointed Bemis as the *Lusitania's* substitute custodian. This appointment bestowed upon Bemis certain rights and responsibilities. Much like a deputy, he was delegated authority to act in the court's behalf. He was expected to comport himself in accordance with the principles and ethical standards of the court. And he was supposed to carry out specified orders for the court.

In the judge's eyes, anyone who disobeyed Bemis's whims by extrapolation disobeyed the dictates of the court - to which the judge took personal offense. And since FFV had been portrayed to the judge in the worst light possible, we entered the case at a distinct disadvantage. By diving the *Lusitania* we had technically violated a standing court order.

What must have slipped Bemis's mind - and slipped the minds of all the attorneys at Huff, Poole & Mahoney - was his obligation to inform The Starfish Enterprise of his court appointed position. He threatened us with suits, injunctions, and arrest - right up the very night before we left London for Ireland - but, as Phil Davey wrote, "never once mentioned the pendency of these proceedings, notwithstanding this Court's admonition in its Order of May 24th that potential claimants be given actual notice of the proceedings."

The judge was explicit on this point. The May 24 Order included a "Notice to All Known Potential Claimants," which read, "the Court is concerned that other, known parties may have unresolved claims to the *Lusitania*. . . . Therefore, the Court ORDERS that Bemis, besides publishing notice as required, also provide each of these persons or entities (or their personal or corporate representatives) with a copy of the complaint and this Order as notice of his pending claim before the Court. Proof of such notice should be provided to the Court within ten (10) days of the date of this Order. . . . Bemis is CAUTIONED that he must comply promptly and fully with all directions and orders stated herein. Failure to do so will cause the Court to re-evaluate its exercise of jurisdiction in this matter." (Emphasis expressed in capital letters was printed that way in the original Order.)

This clause, thoughtfully provided by the judge, is what enabled FFV to get its collective foot in the courthouse door. If The Starfish Enterprise apparently violated a U.S. court order (to which, I might add, the Brits were not bound), Bemis was at fault for maliciously denying the group the essential information.

During a discussion of litigation strategy, Davey explained to me what he thought the judge wanted in this case, and what we had to do in order to meet his demands. Right away my hackles went up. A judge who "wants" something cannot dispense justice. A judge is supposed to be an impartial witness, an unbiased listener, a person who above all others should not have his judgment clouded - or blocked - by prejudice or preconceived notions. An honest judge would disqualify himself on the lack of judicial detachment.

## DOUBTS AND DISTRUST

Despite the judge's jaundiced opinion of FFV, he was shrewd enough to have suspicions about Bemis's credibility so that his mind was not unalterably closed to a contrary point of view. The seeds of the judge's skepticism were planted by Bemis and Robol themselves, when they attempted to steamroller the case through court on April 26, 1994.

Clarke took exception to Bemis's flippant attitude toward artifacts that Oceaneering recovered in 1982 and which Bemis later purchased at public auction.

Bemis: "I have some of the artifacts that I acquired from that expedition myself in Santa Fe, New Mexico. . . . some of the plates and silverware I use to eat with." Bemis had every right to treat those artifacts as he chose. They belonged to him.

But Clarke found Bemis's culinary proclivities disrespectful to the heritage which Bemis claimed the *Lusitania* represented, especially since Bemis testified that he wanted to "try to preserve this historical ship." His actions belied his words.

Clarke was also disturbed over the fact that the artifacts had been auctioned off to pay the expenses of the salvage operation, and that they had then been disbursed to various buyers around the world. Selling recovered items was routine practice in the salvage business - was, in fact, the raisson d'être for treasure salvage ventures, including Oceaneering's search for gold in the *Lusitania's* specie room.

Bemis stuck his foot in his mouth by making the false claim, "I sponsored the Oceaneering expedition." Undoubtedly he intended to convey to the judge that it was upon his initiative that the salvage operation had been mounted. But in attempting to take credit for the ideas and enterprise of others, Clarke assigned him the blame for letting the fruits of the operation get scattered and for not making a record of the purchasers.

Worse, Bemis claimed that he was not seeking title to the wreck and sole salvage rights for any profit motive, but that he was hoping to collect enough data and artifacts to construct a traveling exhibit on the *Lusitania*. This imaginary concept was portrayed in a long but vapid and heavily padded and overwritten document that can best be described as a diarrhea of words and a constipation of ideas.

If that were true, the judge wanted to know, why had Bemis sold off the artifacts that had already been recovered, if he needed them for the traveling exhibit? Backed into a corner by his own illogic, Bemis fumbled through a weak explanation of how he had no control over the artifacts in Oceaneering's hands - leaving hanging the question of how he lost control of an operation that he supposedly sponsored with his own money. (As of this writing, the traveling exhibit has failed to materialize.)

One can only surmise that Clarke detected a certain hypocrisy in Bemis's promises and flights of fancy, and saw through Robol's slick presentation and legal sleight of hand.

Nor were these the only devices in Robol's strategy that backfired. Clarke noted on numerous occasions - and kept coming back to - three other concerns that he had and that needed to be addressed to his satisfaction: jurisdiction, Bemis's "financial ability to carry on what he proposes to do," and the Queen's Bench ruling.

Judge Clarke: "What is this court's jurisdiction? I have two problems. One of them is if - my understanding is I don't have jurisdiction unless the artifacts are brought to this court and under this court's control, and if I accede to his wishes - I'm certainly not going to do it if he is going to spread these artifacts all over the world to private parties. If they are going to be kept in some sort of order for presentation to the public in museums and things of that nature, fine, but if it is just that he is out to make money and distribute it all around, I would be inclined to wait for a better offer. But I don't get much continuity from his testimony about what he is planning to do or whether he has got the wherewithal to do it."

If this indictment seems unduly harsh, bear in mind that Bemis asked for it by making certain promises. The judge was simply cautioning Bemis that he was bound to abide by the promises he made - that he was not to use them as false pretenses upon which to achieve some greedy ulterior motive that had not been placed before the court.

Bemis never said that he was willing to put up the money to achieve the lofty if vacant goals that he was proposing to the court. He pooh-poohed National Geographic's submersible efforts by stating "we have done about as much as you can do using that process." And, "I doubt that I will be involved with Dr. Ballard in the future, because he is not a diver, he has no interest in diving. . . . He plans to write a book on this past summer's expedition, and what comes out in that book, I don't know. But I think I'm pretty well finished with work with Dr. Ballard, and I have to find a new sponsor, and I have several ideas in that respect who would be interested in live human beings doing the work."

He kept alluding to getting "sponsorship which would make it possible in '95 to go down with divers and do a more thorough job of evaluating some of the questions that remain open." Ironically, he claimed to be negotiating with "two groups, one in Ireland and one in England, that I have been communicating with. Both of these are sports divers, but highly qualified, highly technical sports divers . . . The British group is just a well-established sport diving club, and these clubs are always looking for the next challenge."

If the British group that Bemis referred to was The Starfish Enterprise, then his definition of "negotiating" differs from mine and Webster's.

Bemis tried to establish his ownership to the wreck by means of a typed list of conveyances to which the name Andres was appended. Andres, however, had not *signed* the list. So who wrote it? When Davey later showed me the list, I said, "Anybody could have plunked this out on an old typewriter and added a fictitious name. That doesn't make it real." Andres couldn't validate the authenticity of the list unless he rose from the dead to do so.

Bemis explained, "John Light, George Macomber, and I each owned a third of the ship . . . and it remained that way until we went into bankruptcy or declared insolvency or whatever in - I think it was 1970 or '71."

Clarke: "Who went into bankruptcy?"

Bemis: "Well, the ownership, the company that we had. What we did, sir, was we set up a venture company to build the saturation diving complex, and we pledged the *Lusitania* to that company as the only security that we all had, if you will, in assets, hard assets, against the dollars that we were advancing for the building of this system. And so when the project failed, the ownership transferred to Macomber and myself as the people who put up all the money and had the security interest."

Before the judge could ask what happened to John Light's third of the ownership, or establish if any of the putative transfers had been agreed to by Light, Robol applied misdirection like a practiced stage magician and prestidigitator, by stating immediately, "All right, sir. Now, let me direct your attention to the last two pages of this exhibit," which were two memoranda from George Macomber in which, according to Bemis, he "took the position that he wanted no further liabilities, and as such, he turned over his assets to me, his ownership in the *Lusitania*."

Did Macomber do this out of the goodness of his heart? And when had a valuable property like the *Lusitania* become a liability? Questions without answers . . .

This was not the only fancy footwork that occurred that day. The most crucial issue of all was the Queen's Bench ruling. Consider the song and dance routine that followed Robol's introduction of a computer printout of that ruling into evidence.

## 342 - The *Lusitania* Controversies

Clarke: "Why don't you tell me what it says, Mr. Robol, so we can move along."

Robol: "All right, Your Honor. These are decisions in which Mr. Bemis was involved out of the Queen's Bench Decision in the court in London, which recognized his title." To Bemis, "And is that not correct, Mr. Bemis?"

Bemis: "That's correct."

Clarke (to Robol): "Well, I'm asking you as a lawyer, is that correct?"

Robol: "Yes, Your Honor, that's correct."

Bemis elaborated: "The way it worked actually was that at the courthouse steps the queen's attorney recognized my title to the ship, its appurtenances, all its equipment, what have you, and only continued to contest the cargo, those aspects of the cargo that we had brought up from the salvage, and this case in effect resolved that in our favor."

Later, Robol emphasized: "Mr. Bemis is here today . . . to get recognition and enforcement of the foreign judgment, the English judgment, here in the United States, in a U.S. court of law." Still later, Robol tried to muddy the waters, "The British court gave him title to the shipwreck and presumably therefore everything recovered from the shipwreck."

After more misdirection, Judge Clarke observed, "Well, under international law if the British courts have given him title, then I have to follow that, and everybody else has to."

Robol: "That's right, Your Honor. We believe that is correct, and because Mr. Bemis is concerned about possible violation of his right and title in the United States in the future, it would be incumbent upon him to get his title recognized and enforced, to get the foreign judgment recognized and enforced."

This was a case of the toll collector claiming squatter's rights based upon the number of tolls he had collected, and the fact that no one had yet had the audacity to question his right to collect those tolls (except The Starfish Enterprise, knowledge of which Bemis withheld from the court).

It has often been said that lawyers don't deal in facts, but in appearances. And appearances, as we all know, can be deceiving. I have already given my interpretation of the Queen's Bench ruling: that the British court did *not* rule on the issue of ownership because the litigants withdrew the issue and settled it among themselves. The out-of-court agreement held no legal standing - it bound only the participants to the agreement.

Yet Robol represented to Clarke that the Queen's Bench not only vested ownership of the wreck to Bemis, but granted him title to the cargo as well: a creative reconstruction that could not be obtained from the published opinion. The basis of Robol's argument had no foundation in fact.

Robol made it seem as if Clarke's job was an easy one, that he had only to rubber-stamp the Queen's Bench ruling without reading it, and to render his opinion in language that spelled out Bemis's rights the way Bemis wanted his rights to be written. And Bemis wanted this done without furnishing documentary evidence.

Clarke must have seen through the charade. He wasn't about to be stampeded into making a binding decision "for all the world" by having the Queen's Bench ruling waved before his eyes, then snatched away. "I have to read these British cases and see just what title they do give to Mr. Bemis, and I have got to determine what authority they grant Mr. Bemis to exercise salvage rights on this vessel."

Since the judge couldn't be blinded by smoke and mirrors, Robol tried posturing - he inserted as many as three "Your Honors" into a single sentence. The tactic was too transparent to work.

Clarke added that he was "going to have to do research . . . to see just how much I can go on faith that he will be able to put something together and on faith that he is going to preserve these things instead of spreading them all over the world in private hands for people to eat off of and that sort of thing. You know, if that is going to happen, I don't want to have any part of it. I'll just let everybody go down there and take what they can pick up off the floor of the sea."

That sentiment would appear to have given The Starfish Enterprise free reign to do as it pleased - until the judge changed his mind after the fact and reneged on his own words.

## Courtly Concessions

As the hearing date approached - mid to late November - and FFV's attorneys complained to the judge about Bemis's lack of cooperation, Clarke began to see through Bemis's tangled web of intrigue, a web that was being woven by Robol and by Huff, Poole & Mahoney.

Robol wanted the deposition process to be a one-way street. He refused to produce Bemis so that Davey could depose him. We were fighting the battle blind because we didn't know what evidence Bemis possessed.

The judge finally took exception to the inequity of the situation. Reciprocation was an established custom in the legal process. He ordered Robol to produce Bemis for deposition. In defiance of the judge's order, Bemis refused to appear.

Clarke also declared that the production of documents was as incumbent upon Bemis as it was upon FFV. Only grudgingly and after long delay did Robol yield documents that were pertinent to the case.

The judge further demanded that Bemis submit to the court at once the report on Bemis's 1994 commercial salvage activities, as he had been ordered to do in the judge's motion in May. In retrospect, I believe that Robol was withholding the submission of this report until FFV was thrown out or driven out of the case. The reason will soon become apparent.

## Gathering Ammunition

While Robol feinted and applied misdirection - focusing the judge's attention on FFV's alleged misdemeanors - we worked on countering Bemis's misinformation and half-baked truths with the facts, and on supplanting false documents with supported affidavits and testimony.

Hess persuaded Ballard to either testify in court or to submit an affidavit denying Bemis's participation in the National Geographic film shoot. Ballard grudgingly agreed to travel to Norfolk to testify as long as FFV paid his expenses. I agreed with alacrity, astonished by Ballard's belated cooperation.

Hess took a more cynical view. "Don't you see the hypocrisy of a millionaire asking for reimbursement from paupers who are fighting a case that he should have fought in the first place?"

National Geographic also agreed to jump on FFV's bandwagon. Bemis informed the court that he himself organized the 1993 film shoot. NGS took exception to Bemis's bold assumption of credit, especially since he held up the television production for ransom pending payment in advance. If Bemis's assertions went uncontested, his unsupported word would stand as the only evidence on record.

Hess found a willing advocate in NGS staff attorney Angelo Grima, who welcomed the opportunity to represent National Geo's interests. Grima pledged his support as well as that of Bruce Norfleet, a production assistant who helped with the on-site filming of the television special. Grima suggested that if Hess filed subpoenas on both him and Norfleet, they would gladly appear in court *at National Geo's expense.*

But our real ace in the hole was Chris Reynolds.

## Leaking Expenditures

Davey told me that Huff, Poole & Mahoney charged $150 per hour for the legal services of any attorney in the firm. If two attorneys worked together on a case, the charge was $300 per hour. This hourly rate was charged for all time spent on research, memos, phone calls, rough drafts, preparation of documents, filing claims and complaints and motions and notices and counterclaims and so on, depositions, conferences, interoffice discussions, and appearances in court or before the judge for pre-trial briefings. And that didn't count expenses.

The *Lusitania* case cost FFV the princely sum of $8,000. Split four ways, that was nearly as much as each of us paid for the dive trip. No record was kept of the number of hours that our attorneys invested in the case, but it was surely many hundreds. I calculated that if we had paid our attorneys the hourly rate charged by Huff, Poole & Mahoney, FFV would have been poorer to the tune of $40,000 to $50,000.

In order to let Robol know that FFV had the resources to prosecute its case, he let slip the fact that both he and Hess were working pro bono. Robol's delaying motions and harassment suits might cost Bemis a fortune, but FFV not a penny.

## Tacit Admission

Robol's dilatory tactics forced delays of the hearing date. We didn't have Bemis's deposition, and we hadn't yet seen any documents that supported his contentions.

Finally, the "Periodic Report of Substitute Custodian on the RMS *Lusitania*" was submitted to the court in Bemis's behalf. *Who wrote the report we never ascertained.*

Davey obtained a copy almost at once. When I saw it, the first observation that I called to his attention was that, like the Andres list of postulated conveyances, the report was unsigned. Nor was any name appended to it. An oversight? I thought not.

Davey noted further that the "Preliminary Draft Report" dated April 25, 1994 and submitted in Bemis's behalf, was also unsigned. He asked Robol to put his signature on these reports. Robol refused to do so.

When a lawyer signs a document he is authenticating its veracity - a grave responsibility that is not to be taken lightly by an officer of the court. Robol's refusal did more than cast suspicions on himself and on the validity of the document, it was tantamount to an admission of untruthfulness.

In consequence of Robol's refusal, Davey submitted to the court an "Objection to Papers Filed by Plaintiff" in which he noted that neither report had been "signed by either Mr. Bemis, as Plaintiff, or his counsel of record, as required by F.R.Civ.P. Rule 11. Wherefore, Claimant-in-Waiting, FFV, moves this Honorable Court to require Plaintiff to sign these reports in accordance with the rules."

Despite Davey's repeated requests, these reports were never signed by anyone.

## The Great Flimflam

The Periodic Report was a masterpiece of implied imaginative fiction - or was it fantasy? - whose creative author(s) must forever go uncredited. That the report was submitted on paper instead of whole cloth doesn't diminish its allusive quality.

The language in which the report was written was pedantic, patronizing, and intentionally obfuscatory. For example, the Custodian claimed to be continuing a "multidisciplinary approach to the investigation and development of the shipwreck site. . . . During the past four months as Substitute Custodian, a dive was made to the site to add to the data being compiled about the site . . . Eight highly-qualified deep sea divers were selected to participate in the Summer 1994 Expedition. . . . The Expedition successfully laid the cornerstone for the *in situ* park at the site, when it placed a memorial to the victims of the *Lusitania* disaster.

A point-counterpoint analysis is appropriate. For verification of the counterpoint we have Chris Reynolds to thank. In response to a request from FFV's attorneys, he submitted an affidavit which contradicted many of the claims made in the report.

Reynolds: "I was originally part of an 8-man Irish diving team preparing for an expedition to the RMS *Lusitania* under the direction of Des Quigley with the permission of Mr. Bemis. That expedition was canceled. . . . Des Quigley elected to go forward as a personal (private) venture with four divers. They were Des Quigley, Brendan O'Runaig, Eugene Cahill, and myself. . . .

"There were two teams of divers. . . . That first day, Des and Brendan were tasked with placing the shot (descent) line. We believed we were in the midships portion of the wreck. Des was unable to secure the shot line due to the nets, and the dive was aborted after five minutes of bottom time. Eugene Cahill and I made a limited, 15 minute dive in the area where Des was working.

"On August 3, a second, successful effort, was made to place the shot line in the midships section. The first team tied the shot line off and surfaced. Unfortunately, the [second] team's dive was aborted when my dive buddy made an uncontrolled ascent from the bottom after only 8 minutes."

Reynolds "refused to continue diving" after that. "To my knowledge, there was no diving on the RMS *Lusitania* on subsequent days."

What did the Substitute Custodian report about this glamorous exercise in futility? "During the course of the Expedition a number of significant observations were made: A substantial amount of damage has been done to the shipwreck during past military bombing. This has included explosions internal to the hull. A very high percentage of the portholes and glass have been blown out. A very high percentage of hull plating is loose, and there appears to have been substantial galvanic action connected with this. . . . There is particularly extensive damage to the bow area, with evidence of internal and external explosions. . . . An extensive collection of nets was observed in the bow area."

Let me pause to remind the reader that the Irish team believed they were amidships. So how did they make observations on the bow? Ironically, the only Irish team member to have seen the bow was Chris Reynolds - and that was when he dived with Polly Tapson during The Starfish Enterprise expedition!

The report makes a dozen more observations. These were quite remarkable considering that only eight dives were made in all, and that only once did two team members stray more than a few feet from the shot line.

Point from the Substitute Custodian's report: "There was one mishap that could have had far more serious consequences, had the Expedition not been so well-prepared. One of the dive instructors, Eugene Cahill, became entangled in one of the many fish nets in the vicinity of the bow. For a less well-prepared or inexperienced diver, this could have resulted in fatality. Fortunately, Mr. Cahill and his buddy were able to successfully disentangle the net and Cahill was able to reach the surface. Because of the extreme depth, however, and the limited dive time available at these depths, Cahill was forced to make a rapid ascent, with the symptoms of the bends that traditionally accompany such situations, including extreme discomfort, deep muscle tingling and short-term loss of muscle movement. Mr. Cahill's recovery has been excellent with the hope of no permanent injury, and he will be participating in future expeditions."

Counterpoint from Reynolds: "These were the circumstances of the accident. Eugene Cahill's side tank disconnected from his harness. The change in weight distribution caused him to turn upside down in the water. He spiraled in the water and became entangled in my guide line, like a cat in a ball of wool. After I cut him free from the line, he inflated his dry suit and began an uncontrolled ascent, feet first, to the surface. I tried to stop him, but could not. He accelerated to the surface as the gas in his suit expanded. I had to stay behind to decompress on our planned decompression schedule or suffer the same fate. He was taken to the hyperbaric facility at Haulbowline. . . . Mr. Cahill was hospitalized for an extended stay suffering from quadriplegia due to Central Nervous System involvement. He has been moved and is now [November] at the National Rehabilitation Institute of Ireland in Dun Laoshaire in Dublin. Against expert medical expectations, he is improving gradually, and it is hoped that he may even be able to walk again."

Point from the Substitute Custodian's report: "Each diver was required to have participated in a series of at least fifteen technical dives in excess of 90 meters."

Counterpoint from Reynolds: "We completed eight (8) dives in the 50-75m range and two (2) dives in the 75-80m range as our teams worked up for diving the RMS *Lusitania*."

The report stated that "the project received direct technical support from the British Oxygen Company." This is a highfalutin way of stating that the Irish team bought oxygen and perhaps other gases from the British gas supply company.

The report was not even consistent within itself. Consider this contradiction: page 3 noted an "abundance of marine life at the site," whereas page 6 noted "the general absence of marine flora."

Elsewhere the Substitute Custodian took credit for film work that was done by National Geographic, for a book that was being written about the film shoot, for the archiving of videotape and photographic stills at NGS expense, and so on ad nauseam. I don't want to belabor the idiocies and inconsistencies in the report, but I can't help but point out that no documentation was submitted to back up any of the assertions: no dive logs, diaries, correspondence, memoranda, telephone logs, sketches, drawings, photographs, or any of the items that Robol demanded FFV to disclose.

Bemis expected his report to be accepted on faith.

The final hypocrisy was the contradiction between Bemis's testimony on April 26 and the execution of "his" 1994 expedition: "Scuba divers . . . bottom time will be limited to ten or fifteen minutes, which is not enough to do any real work. So that's why you use divers under saturation." When you want work done, he said you "hire professional divers."

My biggest peeve: "News stories about the shipwreck and . . . over five major magazine articles have appeared." Ironically, the majority of the newsprint and magazine pieces to which the report referred were written by members of The Starfish Enterprise.

After reading the report of the Substitute Custodian, I understood why the plaintiff and his attorneys were so desperate to keep FFV out of the courtroom. The evidence we held and the truth we represented would pull the foundation out from under a carefully crafted house of cards.

The snow job concluded with: "During the past four months, a great deal has been achieved. In substantial measure, this has been due to this Court's efforts in providing a convenient, credible, neutral forum for determination of issues related to the shipwreck."

Such obvious fawning might have fooled an insecure grade school student, but I doubt that Judge Clarke was duped for a moment.

## Till Death Do Us Part

The case heated up dramatically just two days prior to the hearing. That was when a fire bomb was detonated outside my bedroom window.

It was still dark at 4 a.m. when the sudden blast rocked the normally calm neighborhood. Immediately following the explosion came a whoosh of flames that leaped skyward some twenty feet. The car parked next to mine was engulfed in a blaze that consumed the interior in a matter of minutes. By the time the fire engines arrived, there was little left of the car but a smoking, blackened frame.

I was lucky. My vehicle and house escaped unscathed. Not so my neighbor, who also parked in the communal driveway that made access possible to the garages of the row of two-story twins. Fire roared up the walls of his house, melting windows, burning wood, and singeing bricks, but doing no structural damage.

In the two decades that I had lived in the neighborhood, nothing even remotely like this had ever occurred.

Police and fire investigators examined the remains of the car. Because of the intense heat and nearly total destruction, they were unable to determine if the bomb had been tossed through a window or if it had been planted under the chassis and fitted with a delayed-action fuse. They were sure only that the fire was no accident; it was a deliberate and wanton act of destruction.

My neighbor admitted to having no known enemies, and certainly none who would commit a felonious criminal act as a way of doing him harm. Whereas I . . .

I did not let him know of my suspicions. He was already in tears over his loss, and to burden him with the notion that he might accidentally have gotten in the line of fire aimed at me wouldn't make him feel any better. Indeed, the car bombing could have been an incident unrelated to my activities. But I had learned to become wary of coincidences. I make no accusations, no allegations. No connection with the *Lusitania* case was ever established, much less proven.

Nevertheless, I had to consider the possibility that my life was in danger, that I was the intended target of an assassination, that the wrong vehicle had been bombed, that a timed device had gone off prematurely, that a trigger wired to the ignition had short-circuited, that the close call was intentional and was given as a warning of my vulnerability, that . . .

When in doubt, I automatically go into survival mode. I reasoned that it was best to act as if the worst case scenario were true.

I had intended to leave that morning for Virginia, do research at The Mariners Museum, sit in on Bemis's deposition the following afternoon, and attend the hearing the day after that. The car bombing put a whole new tenor on the case. Murdering me might have been a rabid act of vengeance, but it wouldn't halt FFV's intervention unless the other directors were scared off by my death, or by the threat of death. In any event, they had to be warned. I postponed my departure by a day.

I theorized that the person whose liquidation would most likely prevent FFV's appearance in court was Phil Davey. As FFV's local counsel he was the most at risk. I called his office as early as I thought he might have arrived. I should have expected his response - he could hardly lend credence to an event that sounded like a bad plot device from an old B movie.

A bomb in my back yard didn't sound real or serious to someone three hundred miles away. After warning Davey I called Peter Hess. He sounded equally incredulous, but was not as unmoved as Davey by the information. Then I called Chatterton and Yurga. I wasn't on speaking terms with Lander, so I asked Yurga to pass along the warning.

I asked Chatterton - a fellow Vietnam vet - if he thought I sounded paranoid. He said, "If you think someone is trying to kill you, that's paranoia. If someone actually *is* trying to kill you, that's reality."

I also called Nike Seamans, who was expecting me for dinner that night. I told her about the incident and the reason for my delay. We had a long talk about life, philosophy, and the implications of the event. I wound up by telling her that I was not depressed, I was in good spirits, I was excited about my writing projects, I had bright plans for the future, and if I should suffer a fatal "accident" on the highway she should relate our conversation to investigative authorities. I was *not* feeling suicidal.

Finally, I called the police department and told my story to a detective: that I was slated to testify at a federal hearing, that my testimony could be crucial to the case, and that I might have been targeted for termination. I expected more sympathy than I received. The police detective could not dispute the fact that a car had been blown up in the driveway behind my house - the report was already on file - yet he pooh-poohed the conspiracy theory as a contrivance of Hollywood screen writers. Apparently, such gimmicks were beyond the pale of routine police matters. Nevertheless, on the off chance that the next bomb might be tossed through a window into my living room, he promised to have my house watched during my absence.

The person I intentionally did *not* call was Cheryl. There was no need to worry her unnecessarily.

## The Lion's Den

I arose early the next morning. I prepared for the six-hour drive to Virginia by inspecting my vehicle's undercarriage and engine compartment. I didn't see anything out of the ordinary. Still, it was with great trepidation that I turned the key in the ignition. The usual grind of the starter diminished as if by a temporary power diversion. I barely had time to envision a trigger being armed when the engine roared into life, then settled down to a purr.

Later, Chatterton snickered when I told him about the alarm I felt at the moment of ignition. Shortly thereafter he bought a new vehicle which was equipped with a remote electronic starter. He laughed when he spoke about starting the car from around the corner of a brick building, but his laugh rang suspiciously hollow.

I was intensely alert during the drive. I kept a constant watch on my rearview mirrors. Whenever a vehicle overtook me, I gripped the wheel and prepared for action. More than once I wished for the comfort of a gun. But I didn't own a gun. I had *never* owned a gun. And I hadn't *fired* a gun since 1967.

The palatial offices of Huff, Poole & Mahoney occupied an entire multi-story building whose structure and parking lot filled a whole city block. The grounds were tastefully decorated with tall trees, pruned shrubbery, and a carefully cropped lawn. This was a far cry from the cramped handful of rooms that held Phil Davey's practice. I opined that honesty and integrity did not pay very well.

I walked into a lobby whose square footage totaled more than that of my entire house. The hallways were wider than my yard. The interior held more rooms than a hive contains cells. I announced to the receptionist that I was there to attend the Bemis deposition. She informed me that the deposition had been postponed yet again - but this time for only a couple of hours. Who should she say was calling. I made sure that she spelled my name correctly.

Judge Clarke's protection from Bemis's harassment suit applied only if I entered the state when ordered to do so by the court. My presence at Bemis's deposition was not only voluntary, it was a challenge.

The receptionist offered me a drink. I calculated the chances of receiving a poisoned beverage, and decided to go out for lunch. I told her I would return at two o'clock. That gave Robol enough time to call a process server.

After lunch I returned to the lobby where I met Phil Davey. We chatted for a few minutes until Robol showed up. Then the fireworks began. Robol had not expected me to be present at Bemis's deposition; he objected to my participation on the grounds that Bemis had not attended my deposition. Davey explained that as the client in these proceedings and the president of FFV, I had every right to be there; and that Bemis's choice to avoid my deposition did not bind me in any way.

We moved into a private room where the argument continued to rage. After many minutes no reconciliation evolved. Davey suggested that they call on the judge for a ruling. He was not in chambers at the moment, so we retired to the room in which the deposition was to be held. Bemis was waiting impatiently.

Bemis's behavior was curious. Although I stared directly into his eyes, he refused to look at me; most of the time he peered down at the desk in front of him. Never once during the course of the deposition, which lasted a couple of hours, did he make eye contact with me. I thought this was odd considering how much he had wanted to speak to me in May.

Eventually the judge got on the phone. A long dialogue transpired about the law and clients' rights. Clarke shot down every one of Robol's improvised objections. In desperation, Robol claimed that Davey might ask questions of a personal nature relating to Bemis's finances. Such answers should be available only to counsel (who were officers of the court and therefore bound by scrupulous allegiance to fidelity), and should not become public knowledge. Clarke compromised on this point: I must leave the room during any such questioning, and that portion of the deposition would be placed under seal. I agreed.

Robol's concerns were not based upon Davey's past conduct. Davey had no intention of delving into Bemis's personal matters, nor of pursuing issues extraneous to the case. Bemis had submitted a list of documents that purported to be the chain conveyances on which his claim of ownership was based, and he had made certain declarations in his substitute custodian report that we wanted verified by documentation. We asked for no new disclosures, only those that he had sworn to the court were already in his possession - disclosures which would be submitted under normal circumstances as a matter of legal routine. We just didn't want to be surprised on judgment day that he was withholding documents from FFV that he was planning to submit to the court.

Both Robol and Huff attended the deposition in order to defend a recalcitrant Bemis from Davey's simple quest for the truth.

I was not as calm as I appeared. The crux of the entire case depended upon the existence of the various title conveyances. Bemis had submitted only a *list* of these conveyances to the court, not the actual documents. These were the very same documents which Polly Tapson had asked him to produce the year before.

Davey framed his questions with precision. "Do you have with you the bill of sale from John Light to George Macomber for the 60% interest allegedly sold on January 10, 1968, as referred to in Plaintiff's Exhibit Number One which was offered to the Court on April 26, 1994?"

Bemis: "No, I do not."

Davey: "Do you have possession of it?"

Bemis: "I don't know."

This litany was repeated *twenty times*. Bemis did not have *any* of the purported bills of sale, agreements, transfers of interest, mortgages, financing statements, liens on "Light's salvage and other proprietary rights," copy letters from attorneys, "stock transfer books of

Kinvarra Shipping," the resolution "voting Kinvarra Shipping into liquidation," the bankruptcy petition, "a list of all the assets and liabilities of Kinvarra Shipping," a list of creditors, or assignment rights.

Davey: "Does such an assignment exist?"

Bemis reiterated his most frequent reply: "I don't know."

Bemis's arrogant attitude toward the legal proceeding that he himself initiated is exemplified by the following colloquy referring to the documents mentioned above. Davey: "Have you tried to find them prior to filing the lawsuit which is the subject of this entire action?" To which Bemis replied, "No."

I breathed a deep sigh of relief. Bemis possessed not a shred of proof that any of the alleged deeds or title transfers existed. No wonder he never bothered to look for them.

Speculating philosophically, when do you get rid of the deed to your house? When do you throw away the title to your car? If Bemis had invested as much money in the *Lusitania* as he claimed he had ($400,000 at that time, later amended to $800,000 - did that include legal fees?), and if the *Lusitania* were as valuable a property as he claimed it was, why hadn't he preserved any of the documentation to substantiate the legality of his ownership? Bemis didn't have an answer, and my suppositions held no sway in a court of law. The reader may draw his own conclusions.

The second major issue on which Davey wanted documentary validation was the substitute custodian's report. Davey first established that Bemis had no "correspondence, telephone logs, memoranda, notes, or other documents by and between yourself and the members of the Irish diving team incident to the planned August 1994 expedition to the wreck of the RMS *Lusitania*." Neither did he have "dive logs, deck logs, engineer's logs, master's logs, medical logs" or "photographs, videotapes, sketches, drawings, writings, records" or "any documents of evidentiary support for the report of the substitute custodian." Nor did he have any "documents related to your intention and effort to create an underwater park at the site."

It is tantalizing to recall that Robol demanded that FFV produce these very same types of documents: documents which Bemis refused to produce - or didn't possess.

Davey: "Have you authored or published any article, book, or technical studies regarding the RMS *Lusitania* or the wreck?"

Bemis didn't give a straightforward answer to so simple a question: "I would guess probably the answer is no."

About the substitute custodian's report, Davey asked, "Is it true and accurate to the best of your knowledge, information, and belief?"

Bemis: "Reasonably so, yes, absolutely. I won't attest to every adjective that's in there, but yes."

Davey: "Do you have trouble with some of the adjectives in there?"

Bemis: "No."

Davey: "Do you believe it to be the truth, the whole truth, and nothing but the truth?"

Robol: "I object to that as being ambiguous. . . ."

Bemis: "Well, I can't really answer. There are eight, nine pages."

Whereupon we took a break so that Bemis could read his report. Upon reconvening, Davey asked, "Are you willing to swear or affirm to the truth of everything in that report?"

Bemis was unwilling. "This represents the work not just of myself, but of other people as well as myself and their representations." He refused to verify the report in its entirety.

Davey took another tack. "I'm going to ask you to take this red pen and outline the areas that are representations made to you by other people to which you do not feel you can attest."

Bemis: "No, I don't choose to do that."

There ensued many discussions, objections, long speeches made by Robol, and the continuous refusal on Bemis's part to verify any aspect or part of the report. In the end, Bemis did not sign the report, nor did he encircle a single word or passage. The report had to be accepted on faith.

Davey tried to get to the bottom of the putative wreck observations, the conduct of the divers, the so-called success of the expedition, and so on. Bemis's most often stated response was "I don't recall," which he repeated so many times that he reminded me of President Reagan giving testimony at a Congressional hearing.

In short, Bemis could not or would not substantiate any representations that he had made to the court.

He did, however, retract some of the declarations made in the substitute custodian's report, particularly about the severity of Eugene Cahill's condition. We had intended to spring Chris Reynolds' affidavit on Bemis unannounced. But the original hearing date was postponed. Reynolds, who didn't know about the postponement, let it slip to Quigley that he made a sworn statement about Cahill's accident. Word immediately got back to Robol, who then had time to warn Bemis that the truth about the accident was known to the opposition. Thus we lost some of the impact that the Reynolds affidavit would have had by catching Bemis completely unawares. But Davey still got a lot of mileage out of it because Bemis now had to contradict statements made in the substitute custodian's report.

Bemis did not have - although he claimed to be compiling (after twelve years) - a list of all the artifacts that Oceaneering recovered in 1982. This despite his supposed involvement with the company and the salvage operation, and despite the fact that he was present at the auction in Aberdeen where he had bid on and purchased the *Lusitania* artifacts he currently owned. Yet *I* had a copy of the complete list (enumerated in Part 8).

An attachment to the substitute custodian's report itemized the artifacts that Bemis purported to have in his custody. I had information to the contrary. For example, Ken Marschall owned the steam whistle, all four deck lights, and the promenade deck window. Bemis knew this because he saw them at Marschall's home during a visit.

Bemis now contradicted his reported inventory by testifying that none of the eight portholes that he had previously claimed to own either belonged to him or were in his possession or control.

He also testified that he did not *invest* in the 1982 Oceaneering salvage operation. Instead, he was paid £10,000.

Asked if the name Brandon O'Runaig meant anything to him, Bemis replied "No." (O'Runaig was one of "his" divers - a member of the Irish team.)

I left the room so that Davey could ask questions which Robol believed would delve into Bemis's personal financial matters, the answers to which might embarrass him if they were made public knowledge. This portion of the deposition lasted only a few minutes. The questions were so innocuous that in a moment of generosity, Robol offered to unseal the record, which was done.

## FFV's Day in Court

Robol's refusal to produce Bemis for deposition until late in the afternoon before the hearing the following morning put FFV at a disadvantage. It left no time to develop a courtroom strategy based upon the weaknesses in Bemis's testimony and his economy with the truth. I took notes during the deposition so that Davey and I did not have to rely solely on memory about Bemis's carefully worded responses. That evening we discussed how best to handle the case.

I had little to offer beyond my innate sense of righteousness, which has no standing in an American court of law. The most important - perhaps the sole - criterion on which American jurisprudence is based is "procedure." Truth is forsworn as a subordinate adjunct.

It was left to Phil Davey and Patrick Brogan to develop the best course of action. Davey wrote a three-page trial outline that diagrammed the order in which FFV's directors would give testimony, the topics of discussion, the contradictions to be brought to the court's attention, and the evidence to be submitted: correspondence, affidavits, exhibits, and so on.

Davey put a rush order for the transcript of Bemis's deposition, which was transcribed overnight and delivered in the morning. FFV's directors and attorneys assembled in the courtroom along with boxes of files, folders, and paperwork. Ranged against us were Bemis, Robol, Huff, and their runners and assistants. Members of the press attended to cover the event.

Judge Calvitt Clarke brought the courtroom to order. For the record he announced the grounds for the hearing as well as the direction he wanted to session to take. It was obvious from the start that he was going to exercise complete control over the hearing, and not allow FFV to conduct its case to its own best advantage. Much of the evidence that our attorneys had accumulated had been submitted to the court along with pre-trial motions. However, the scope of our testimony would be limited to addressing the reason for our delay in filing the intervention.

Additionally, we were given a time limit based upon a social engagement the judge had that evening with his wife - he wanted to wrap up the case by late afternoon.

The battle was lost before the first engagement. Davey tossed aside his trial outline, then conferred with Peter Hess about how best to proceed. They decided to put me on the stand first, before the other directors.

## FRACTIONATING THE TRUTH

Although I swore to tell the truth, the whole truth, and nothing but the truth, on the stand I was permitted to tell only that portion of the truth which the judge wanted to hear. He didn't want anything on the record that painted his substitute custodian in a bad light. Whenever Davey or Hess asked a question that threatened to yield incriminating evidence about Bemis, if neither Robol nor Huff objected quickly enough, the judge objected and halted the entire line of questioning.

Davey and Hess had to word their questions creatively so as not to telegraph the answer before I had a chance to give it. And I had to be astute enough to know where they were going in order to answer the question promptly before Bemis's attorneys or the judge had time to object. I was certain that Clarke had already determined how he was going to decide the case. To that end, he didn't want any facts on the record on which his opinion could be overruled upon appeal.

In American law, the appeal process does not permit the admission of further evidence to bolster a claim. The record is frozen. Facts or evidence that were originally overlooked or intentionally excluded cannot be introduced at a higher court hearing; new evidence that comes to light cannot be presented. In the legal sense, an appeal has nothing to do with truth, justice, or the American way - it has to do with points of law and misinterpretations made by a lower court. A case that is properly rigged by a judge who controls the evidence that is introduced will not be overturned.

Under these constraints my testimony was not as revealing as it could have been. Upon cross examination, however, the judge let Huff and Robol ask any question they wanted, no matter how irrelevant, whether or not it had anything to do with FFV's delay in intervention. They asked questions about The Starfish Enterprise, about the expedition, about the history of the *Lusitania*, about my published articles and photographs, and so on. Furthermore, he let them ask leading questions in which the answer they wanted me to give was couched in the language of the question: "Isn't it true that . . ." Usually it was not true, but even so the insinuation became as much a part of the record as my denial.

These badgering, accusatory questions became so extraneous that despite my attorneys' silence, I was finally moved to state, "I don't know what that has to do with the delay in filing." The judge didn't even bother to look up from the paper he was reading.

Huff accused me of failing to return Bemis's phone call in May. I admitted that I was under no obligation to do so. The judge thought otherwise, and interrupted my testimony in order to fault me for ignoring the request of his substitute custodian. If I was in the wrong, then a lot of people who owed me money were equally delinquent.

The greatest injustice and indignity I suffered was the judge's reproof for not dealing with Bemis when I knew that the National Geographic Society had done so. I considered his imputation to be an insult. The company had set a bad precedent by acquiescing to the toll collector's demands, instead of having the guts to stand up and fight. I despised being cast in the shadow of National Geo's cowardice and stupidity, and I repudiated the judge's implication that we should have submitted to threats and blackmail because of National Geographic's weak will and spinelessness. Only with incredible restraint did I keep these condemnations to myself. Inwardly I seethed with indignation and contempt, but outwardly I expressed only calm deliberation. Yielding to emotion would have been self-indulgent and counterproductive.

After I got off the stand and resumed my seat at the table, Yurga - who was taking notes of the proceedings - leaned close and whispered that, after witnessing the grilling I had taken at the hands of Robol, Huff, and Clarke, he would do anything to avoid having to testify. He was only half joking.

## The Turncoat

Davey put Lander on the stand next. Her testimony didn't just shoot our case in the foot, it shot our case in the head. And her testimony didn't have to be threatened or tortured out of her - she volunteered it!

Under cross examination she claimed that Polly Tapson "instructed divers to sneak artifacts off the *Lusitania* and to lie in order to avoid trouble with Bemis." She also claimed that she "never would have taken anything from the wreck had she known that Clarke had awarded the rights to Bemis beforehand." Then she claimed that she "felt so guilty about how the expedition was being run that she withdrew from her fellow divers."

Lander thought that Polly did a poor job at leading the expedition. By contrast, I testified that "this expedition was the most well-planned expedition that I have ever been on. Better even than expeditions that I have planned myself."

Clarke was not so easily taken in by Lander's confession and her prayer for absolution. He asked her if she had felt guilty enough to return to the wreck the soap dish she had recovered. She said, "No."

These were precisely the kind of indictments Clarke needed in order to deny FFV's intervention.

## HUMBUG AND PERJURY

Lander's denunciations notwithstanding, FFV managed to win a few skirmishes by calling to the judge's attention certain improprieties and outright prevarications on the part of Bemis and his attorneys.

Clarke's May 24th order stipulated specifically that in addition to publishing the notice of arrest in the newspapers, *all* interested parties known or believed to have an interest in the wreck must be notified personally of Bemis's claim. Clarke was disenchanted to learn that Robol had failed to notify The Starfish Enterprise and Muriel Light: the very entities who, by dint of past and ongoing activity, had the greatest interest.

Worse yet was the fact that Robol had sent threatening letters and faxes to The Starfish Enterprise *without making any mention whatsoever of the proceedings*. This prompted the judge to reprimand Robol for his deception and deliberate inaction. Robol remarked weakly that he didn't think The Starfish Enterprise had an interest in Bemis's claim.

Worst of all, Robol finished his dim and anemic defense by bleating, "And I never even heard of this Muriel Light."

Robol's concocted assertion incited Davey to catapult from his chair, sputtering an objection as he fumbled through his files on the table in front of him. He produced copies of the two letters that Robol had sent to Muriel Light, ostensibly on behalf of the Columbus-America Discovery Group. Davey called attention to Robol's signature at the bottom of both letters.

Clarke glanced at the letters with quiet circumspection. Robol maintained a stance of audible silence. Since perjury is supposedly a serious crime, I would have thought that an officer of the court who swore a falsehood to a federal judge would receive judicial vilification that would lead eventually to disbarment. Instead, Clarke shrugged off the incident as if were an everyday occurrence in Virginia courtrooms. Clarke's innocuous comment: "I'm very disappointed in Mr. Robol for not disclosing that, because that should have been disclosed."

## Dismissal and a Chance at Redemption

Clarke decided that he did not want any more testimony or evidence placed on the record. Lander gave him all the ammunition he needed to exclude FFV's intervention. With a social engagement looming, he gave an off-the-cuff ruling from the bench: "I find that the claim was not timely filed, and the court dismisses the claim of FFV group for that reason."

*But -*

"The court also has before it the matter of jurisdiction . . . That issue can be raised at any time and . . . by anyone, whether that person is a party or not. . . . The court . . . can hear a nonparty argue the matter of jurisdiction. So if counsel for FFV tell me that they wish to pursue the matter, we will hear them at ten o'clock tomorrow morning."

All hell broke loose in the courtroom. At the center of the pandemonium raged four emotional attorneys. Two were ecstatic, two were bitterly distressed. I was totally confused.

Phil Davey threw his hands up in the air, and exclaimed, "We've won! We've won!"

Rick Robol slammed a sheaf of papers on the table. "We've been screwed."

I thought I must have heard wrong. It seemed that through transmogrification each attorney had mistakenly uttered the sentiments of the other. We just got our butts kicked. Davey should have been downtrodden, Robol should have been gloating.

For the record, Davey said, "We intend to call Mr. Bemis as our first witness."

Clarke said: "All right. That is fine. Mr. Bemis, you be sure to be here."

Comprehension dawned slowly - although admittedly, only after lengthy legal explanations from Davey and Hess. The judge's dismissal of FFV's salvage claim did not invalidate the rationale for initiating the claim. As I later told reporters, "We don't want to own the wreck. We never did. That was just the legal posture we needed to get into these hearings." We merely wanted to enlighten the court about the insidious nature of the case, to put *all* the facts on record instead of just carefully selected facts, to submit truths that were genuine instead of ingenuous, to contradict pretension with verity.

Just as Robol wanted FFV's documents but not FFV's intervention, so the judge wanted the result of FFV's research but not FFV as a disputant. We were being used by the judge to provide information that his substitute custodian had withheld. Since that was all we ever wanted in the first place, we had no cause to complain.

Robol's attempt at prestidigitation had already failed. Now the judge was offering FFV the opportunity - as amicus curiae - to prove that the emperor was not wearing new clothes but was naked in the eyes of the law. Robol was now going to have to fight the case on its merits instead of with unsubstantiated testimony and unsigned fictitious documents. His case was going to be screwed by the truth.

As I noted gleefully at the time, "We didn't have to win the case to serve the cause of justice, we only had to make Bemis lose."

## Bemis's Day in Court

FFV's day of infamy was December 7, 1994. As badly as FFV was destroyed in court, Bemis fared worse in the aftermath of destruction.

The judge took on a whole new attitude. Whereas previously he decried our attempt to introduce evidence that exposed Bemis in a negative light, now - although he didn't exactly welcome it - he reluctantly acknowledged that it was admissible. He didn't allow Davey to fully explore each and every exaggeration and falsification in the Substitute Custodian's Periodic Report - often cutting him short - but he listened to enough testimony to get the drift of where it was going. Davey was trying for the slam-dunk, but the judge permitted him only to make enough points to win the set. Our supporting documentation filled the gap where the testimony was at variance with the facts.

One issue on which Bemis changed his story was the matter of the spoon he claimed to have recovered in 1993 - on his single descent in the submersible during National Geo's film shoot - and which he had submitted to the court for arrest. *Now* he testified that the spoon he had showed the judge had been recovered in 1982, and that the spoon recovered in 1993 was at home. If he changed his testimony because he feared that FFV had information to contradict his claim, he was mistaken. Bemis's sudden attack of candor came as much of a surprise to us as it did to the judge.

From this admission it developed that Bemis had not brought all his *Lusitania* artifacts into the court's jurisdiction. When Davey asked him where they were, he said they were at home. The judge took over the questioning - he wasn't just angry, he was apoplectic. He quoted the portion of his May 24th Order that applied: "It is ORDERED that Bemis must immediately bring these artifacts, which are the basis for jurisdiction, within the Eastern District of Virginia and provide the Court with notice of their location." Clarke wanted to know how he could possibly have misconstrued what was written so plainly in black and white.

Bemis's attorneys tried but failed to placate the judge. Clarke said, "I don't deal in tokenism." Clarke wanted Bemis to turn over to the U.S. Marshal every artifact on the list he had appended to the Substitute Custodian's Periodic Report, the same as he had required of FFV. This was impossible for Bemis to do since, as I've already noted, some of the artifacts he claimed to possess he did not own or have access to - they merely padded the list to make it appear more impressive than it was.

The judge noted that his Order specified that "Bemis must also provide the Court with a complete and timely list of any and all artifacts salvaged and all artifacts brought into this district. Failure to comply with this will be grounds for dismissal of this action."

Huff looked like a circus performer as he backed away from the bench in a series of obsequious bowing motions while making promises of compliance. I couldn't help but laugh at his conciliatory histrionics and Bemis's obvious discomfiture. Bemis promised to pack and ship the artifacts as soon as he returned to New Mexico. Of course he didn't, because he couldn't. He eventually shipped about half the items on the list, still resorting to what the judge referred to as "tokenism."

Parenthetically, at a Ships of State exhibit then occurring in New York City, Bemis was credited as one of the contributors. What items he provided for the exhibit we never learned.

## False Report

It goes without saying the Davey ripped apart Bemis's Periodic Report for the sham that it was, especially highlighting the exaggerated nature of "his" expedition on which most of the report was based, and pointing out that its observations were either already known or could not possibly have been made in the few minutes that "his" divers had actually spent on the bottom.

Once again the judge took over from Davey, when he heard that at the time of the "expedition" Bemis was lounging at home in Santa Fe. At no time were "his" divers under his control. Clarke: "That's not being a custodian, as far as this court is concerned. . . . If Mr. Bemis expects me to treat his case with any sympathy, he's going to have to be on the scene."

Davey made sure to contradict Bemis's characterization of Eugene Cahill's "mishap" (one "that could have had far more serious consequences"). I doubt that Cahill, who faced spending the rest of his life in a hospital bed or a wheelchair, thought that his recovery had "been excellent."

Bemis writhed painfully on the stand under Davey's incessant assault against Bemis's thrifty display of frankness. Yet still the judge would not let Davey go in for the kill. Time and time again, just as Davey was thrusting forward with incriminating evidence, the judge interrupted and would permit no further development of the matter. It was frustrating not being able to twist the knife in the wound, no matter how deeply embedded the blade.

When Davey questioned Bemis about his "support" of the National Geographic film shoot, Bemis was on his best behavior. This was undoubtedly due to the fact that Angelo Grima and Bruce Norfleet were sitting in the courtroom. It would have been dramatic to let Bemis testify to organizing the film shoot (he had already taken credit for doing so in his report to the court), then spring Norfleet's disaffirming testimony unannounced. But Bemis spotted Grima and Norfleet in the lobby during a break in the proceedings, and in the event we did not have the opportunity to put Norfleet on the stand. Nonetheless, his presence alone kept Bemis on the straight and narrow, and as a consequence widened the credibility gap between Bemis's current testimony and his previous testimony, affidavits, court reports, and deposition.

My greatest vindication came when Judge Clarke explained for the record that the Queen's Bench court did not adjudicate the issue of title or ownership of the wreck, because that issue was settled out of court. Only the parties involved were bound by their agreement. I felt exonerated that the judge had seen through Robol's wretched razzmatazz.

## Where's the Wreck?

Finally came the subject of the *Lusitania's* location - the sharpened edge on which the issue of jurisdiction was balanced.

At the deposition Bemis said that he did not know the wreck's precise location. The original arrest papers defined a box approximately four miles on a side - sixteen square miles of ocean floor. The wreck was supposed to lie somewhere inside the box. The British Admiralty chart of the Irish coast showed three wreck symbols within the designated perimeter (one of which was probably the *Minnehaha*). Bemis said that he did not have the *Lusitania's* coordinates with him, and apparently neither did his attorneys. Bemis would not admit that the box was large enough to contain the wreck.

Bemis, coached by Robol, would not even accept the dimensions of the wreck, claiming that the main hull was surrounded by a debris field whose extent was unknown. Robol: "I object to the ambiguity;

what you mean by wreck, whether you are including the debris or main hull portion." When Davey asked him to define the wreck as the main hull whose dimensions were approximately eight hundred feet by two hundred feet, Bemis would not do so.

Bemis *thought* the main portion of the wreck - meaning the majority of the hull - lay not in the center of the box but in the northeast quadrant. It seemed strange to me that after twenty-five years of involvement with the *Lusitania*, Bemis could not talk about the wreck with more precision.

In court, Davey placed the chart before Bemis and asked him to indicate the wreck's position. Bemis squandered less attention on semantics before the judge than he had two days earlier. He indicated that the main hull probably lay in the box's northeast corner. Judge Clarke descended from the bench so he could see the chart. His astonishment was barely concealed by a grimace when he commented wryly that he couldn't believe that there was any doubt about the wreck's location, and that we - Bemis and FFV - couldn't agree on so simple a point.

Now came Bemis's jurisdictional justification. A small anomaly existed about one-third of a mile from the hull. John Light knew of this anomaly, and so did all the fisherman in the area because they often caught their nets on it. National Geo side-scanned the anomaly, but reached no conclusions about its nature. Bemis claimed that this anomaly *might* be associated with the *Lusitania*: either cargo, coal, debris, or a section of the hull that had fallen to the bottom when the ship had been torpedoed. He also theorized that a trail of debris *might* connect this anomaly with the main hull, thus establishing that *some portion of associated wreckage might lie farther offshore* than the collapsing hull.

The rationale for clutching at this tenuous straw was imaginative and ingenious, although contrived and ingenuous: in the event that it was established that the main hull did indeed lie within Irish territorial waters, a debris field that stretched beyond the boundary might insinuate that a part of the "wreck" (as defined by the Receiver of Wreck) lay in international waters. For this Hansel and Grethel fairy tale to fool any child, the proportions of the wreck had to remain undefined so that the bread crumbs - the "trail" of debris - could be arrested in American court. This attempt to wag the dog by the tail - to arrest property in Irish territory because a disarticulated segment might possibly have once been a part of it - stretched the bounds of plausibility to the breaking point.

Bemis's premise was based on pure speculation. He had conducted no studies of the anomaly and had no idea what it was - only what he wanted it to be. Nor did he know if the hypothetical trail of debris existed.

## Where's the Border?

Davey produced copies of several important documents that he had already submitted as supporting exhibits along with FFV's Motion to Dismiss. One of these was the Law of the Sea Convention, published by the United Nations. This internationally recognized source book included a map of Ireland showing the baseline from which its territorial limit was measured. The baseline was different from the coastline because of Ireland's rugged geological beach front: capes or headlands that protruded offshore like the tines of a giant fork, and small rocky islands. Instead of the territorial extension following these complicated convolutions, which overlapped and sometimes folded back upon themselves, the points of each headland were connected by lines to the adjacent headlands (or to islands considered to be extensions of the land), and the limit was measured from an artificial perimeter which stitched these far-flung points together.

The Law of the Sea Convention also included Ireland's Maritime Jurisdiction Amendment - passed in 1988 - which extended the country's territorial limit from three miles to twelve. Instead of embracing this publication as the unimpeachable treatise that it was, Robol objected to having it submitted to the court as evidence. Then he tried to have the document withdrawn. Next he alleged (without supporting documentation) that the United States might not recognize Ireland's territorial claim. Finally, he pretended that its accuracy couldn't be validated. His objections were patently absurd, and were overruled by the judge.

It is difficult for me to believe that Robol wasn't already aware of the Law of the Sea Convention and its ramifications on his case. An attorney does prudent research before submitting his case to the court. An officer of the court is honor bound not to hide facts from the judge nor to misrepresent the truth. As Davey explained to me, a lawyer's job was to try to make the judge view the facts in a light that was favorable toward his client. Willful misrepresentation was unethical.

Judges do not research the facts; they research and interpret the law. Factual research is the responsibility of attorneys and their clients. In that sense, a judge knew only what he was told: which facts were presented to him or, when a case went unopposed, what he was led to believe. If Clarke could be made to believe that Ireland's territorial limit was three miles - that is, if Robol and Huff could keep Ireland's territorial extension amendment out of the record - then the judge could exercise the court's discretion and extend his own jurisdiction.

That was the crux of Bemis's case. And that was what FFV had to oppose.

To pinpoint the wreck's location, Davey called Chatterton to the stand because he was a Coast Guard licensed "ocean operator" ("boat captain" in the vernacular) and an expert in the science of navigation and the reading of nautical charts. Davey produced the wreck's coordinates from Nic Gotto's affidavit and from my own records - I had taken the numbers directly from Gotto's GPS when the *Sundancer II* passed between the buoys that designated the bow and stern shot lines.

The *Lusitania* lay 11.5 nautical miles from the closest point of land, and 10.5 nautical miles from baseline between two headlands. By either criterion, the wreck clearly lay within Irish territorial limits. No amount of obfuscation could deny it.

It is curious to note that when Bemis filed his claim against the *Lusitania* in the State of Virginia, he informed the federal court that the wreck lay in international waters. When he sought an injunction against The Starfish Enterprise, he led the Irish Receiver of Wreck to believe that the *Lusitania* was within its jurisdiction. Now he wanted to *Lusitania* back in non-territorial seas. By this guile of fluid placement and creative chess board maneuvers, Bemis switched the *Lusitania's* location back and forth, to wherever it was to his advantage to have it at a particular moment in time.

## The Academics of the Jurisdictional Dispute

After FFV introduced all its documents, Judge Clarke was forced to acknowledge that the *Lusitania* clearly fell within Ireland's territorial sovereignty. He admitted as much in summation at the end of the day's evidentiary hearing. Despite his own admission, however, Clarke stated with a smile that he was taking jurisdiction of the wreck anyway. His interpretation of "hands across the water" was in direct opposition to international law. Such is the power of a federal court judge that he can do whatever he pleases.

The hullabaloo that followed raised waves and a tide of protest on both sides of the Atlantic. FFV offered resistance from the U.S. side, Polly Tapson dropped a bomb from the U.K. side, and the Irish bureaucracy reacted quickly and violently, as if the country had been invaded by a foreign aggressor.

Davey had kept the Irish consulate aware of the proceedings, hoping to inspire Ireland to file an intervention. The Irish consulate was blasé about the affair, as if it were of little consequence, or perhaps in disbelief that a judge would act with such effrontery. But once Clarke adopted formal jurisdiction, the Irish perception of the import of events changed dramatically.

Into this cauldron of Irish political wildfire Polly tossed another flaming brand - or was it a canister of napalm? She leaked to the press

## 366 - The *Lusitania* Controversies

that during The Starfish Enterprise's second expedition in August she spotted long metallic cylinders that could have been sealed lead tubes containing original works of art in the possession of Sir Hugh Lane, who died in the *Lusitania's* sinking. Lane was reported to have been conveying to Ireland several rare and valuable canvases painted by famous artists Peter Paul Rubens and Claude Monet, and possibly some works done by Rembrandt van Rijn. Lane bequeathed his art collection to the National Gallery in Dublin, of which he was the director.

The personal possessions of the passengers were not insured by the Liverpool and London War Risks Insurance Association, and therefore were not part of the assignment of the *Lusitania's* title to John Light. Lane's paintings - whether ensconced on land or at the bottom of the Irish Sea - were inherited by the National Gallery.

When these combined news items hit the fan, the Irish government kicked the U.S. court in the seat with a ponderous Irish brogue. Irish newspapers picked up the story and fanned the flames with foolscap. Within weeks, Ireland's Minister of Arts and Culture issued an Underwater Heritage Order to protect the *Lusitania* and other Irish property in the open sea from foreign invasion. Under this heritage order a license was required to dive the wreck, and under no circumstances could artifacts be recovered.

Against official Irish sanction, American legal rulings held little sway.

### THE GHOST OF CHRISTMAS PRESENT

During the Christmas holidays I noticed a uniformed officer approaching my house. I jumped to the conclusion that he was a process server. I opened the door eagerly before he reached the stoop, and invited him inside. If Bemis was willing to squander his money on a groundless law suit, I was willing to let him enrich the coffers of Huff, Poole & Mahoney. I had free legal service with which to defend myself - during the case and afterward.

I was disappointed to learn that the officer in uniform was the fire marshal. The car bombing behind my house was still being investigated because the culprit hadn't been found. He wanted to know everything I could tell him about the incident. I gave him same story I gave the police detective. Like the detective, the marshal thought my scenario was too far-fetched, but he listened and took note of my observations.

### THE *AQUACORPS* SYMPOSIUM

Attorneys for Bemis and FFV prepared written briefs for the judge. Here matters rested for several months while Clarke reviewed the case.

Meanwhile, in the real world, the *Lusitania* received a great amount of good press in the diving community. As the new year dawned, Mike Menduno milked the *Lusitania* for all it was worth. The January issue of the *AquaCorps Journal* was devoted largely to the wreck. In addition to staff-written fillers, feature articles were contributed by Polly Tapson and me. The issue was lavishly illustrated with topside and underwater photos. The magazine made its first appearance at the *AquaCorps* symposium held in San Francisco, in January 1995.

Half the members of The Starfish Enterprise attended the conference: Polly and Simon Tapson and all four Yanks. I gave an evening slide presentation to an audience packed with attentive technical divers, more than five hundred of whom filled the room. It's easy for an author and public speaker to derive more credit than he is due, because his name is constantly in the spotlight. I've always been sensitive to this. At the end of my presentation I projected a group photo taken upon completion of the trip, named the members of The Starfish Enterprise, and asked the attending members to come up on stage for their share of the recognition and applause.

I introduced them to the audience: Polly and Simon Tapson on my right, John Chatterton and John Yurga on my left. Barb Lander chose not to attend the evening festivities. (She snubbed my invitation by saying that she had "better things to do," then turned and walked away. That was our only interchange during the three-day conference. She avoided the others as well, especially Polly Tapson.)

## Opposition to Success

The success of the expedition was received with overwhelming enthusiasm by everyone at the conference except for two people: Gregg Bemis and Lad Handelman.

Bemis seemed out of place at a technical diving symposium: he wasn't a diver and he expressed no interest in learning how to dive. He attended for a different reason entirely. Initially he tried to bully Menduno into quashing all mention of The Starfish Enterprise's expedition to the *Lusitania*, both in the magazine and at the conference. Menduno wouldn't retreat a single step. The *Lusitania* was the technical dive of the year, and censorship went out with the American Revolution.

As a concession, Menduno offered to let Bemis tell his own story of the *Lusitania*, prior to the evening presentation. More than a dozen people crowded into a room large enough to hold half a gross. Nearly half of those who endured Bemis's vacuous rationalizations were members of The Starfish Enterprise. After a self-serving monologue about his chimerical claim to ownership - which he delivered with cool aplomb - he entertained questions from the patiently quiet gallery.

Polly Tapson fired the opening shot. Bemis responded with more restraint and self-assurance than he had demonstrated at his deposition or on the witness stand. Neither did he act defensive. He had never met Polly face to face, yet he identified her without hesitation. Polly was aggressive in making her points, and soon she was joined by Chatterton, who verged on the edge of open hostility. The discussion dissolved to a free-for-all when other entrants voiced their opinions.

I watched the diatribe but held my peace. I didn't think that anything could be gained by entering the fray. (Once at a wedding reception at which I refrained from engaging in a controversial exchange, someone said to me, "You don't say much." My reply: "I can't learn anything by talking. I already know everything I have to say.")

My friends were doing such an excellent job of lambasting Bemis that they didn't need me to back them up. Yet Bemis didn't retreat from their accusations. He fielded each penetrating indictment with a politician's cool reserve. No issues were resolved by the heated discussion, but the members of The Starfish Enterprise felt better for having voiced their objections in person. For Bemis the session was more of a contretemps than a fruitful airing of his viewpoint.

## Commercial Opposition

Bemis was not the only one who felt disdain at our success. Commercial divers were appalled at what technical divers were achieving. Perhaps their castigation resulted from what they perceived to be a threat to their livelihood: if technical divers could accomplish a task for a fraction of the cost that commercial divers charged, then commercial divers might lose jobs as a result of new competition.

Competition is the basis of a democratic society; monopoly is anathema. To meet the threat of competition, commercial divers sought to invoke regulatory sanctions that would disadvantage their competitors. Commercial divers were not the first group of elites to try to crush the competition through bureaucratic fiat; nor would they be the last. In America, where money can purchase political favors through mechanisms such as lobbying, it is easier to pass laws to suppress competition than to work harder and trim the fat that one has gotten used to supporting.

Lad Handelman was one of the more outspoken commercial diving dissidents. He owned a commercial diving outfit, and obviously felt threatened by technical diving expertise and encroachment. He approached me after the evening presentation in order to air his concerns about the conduct of the expedition. He told me that we should not have attempted the dive because he didn't believe it was safe. Nor did he condone the fact that we didn't have a recompression chamber on board.

The proof was in the pudding, as the old saying goes, but Handelman had his own platitudes to declare. I listened patiently without interrupting, then patronized him by refusing to be put on the defensive. It is impossible to convert ingrained beliefs with reason and rationality.

I always welcome considered advice on safety. But Handelman had his head in the clouds with respect to recompression chambers. If he were counseling mountaineers, he would have them stage helicopters along the route to the summit in order to evacuate climbers in trouble. A commercial diver advising a technical diver is like a trucker telling a racing car driver how to win the Indy 500.

## Fringe Opposition

A few others were envious of the achievements of The Starfish Enterprise. Chief among them was Des Quigley, Bemis's yes man and titular co-leader of the Irish diving team whose trip miscarried so disastrously. No one took seriously the words of a failure.

Some non-technical divers disparaged the expedition because of the example it might set for those who had less proficiency in the water. The rationale behind this short-sighted viewpoint was that inexperienced divers might be encouraged to attempt feats that were beyond their personal ability, that they might get hurt in emulation. Most of these espousers suffered from inferiority complex.

Their fatuous position implied that an individual should never raise his head above the masses, that he should not try to achieve any goals that were greater than his neighbor's, that he should mill in the middle of the herd, that he should glory in mediocrity - lest any demonstration of initiative embolden others to follow suit. These devolutionists would like to halt the flow of progress in order to prevent anyone from getting ahead of them. So they watch out for others and ridicule enterprise.

I'm so busy minding my own business that I don't have time to mind anyone else's. Each human being must make his own decisions, must seek his greatest potential. He must not be limited by the psychoses of the envious.

## Muriel Light's Day Out of Court

In July 1994, Muriel Light sent a brief letter of protest to the court in Virginia. She did not hire an attorney to represent her interests, she did not file any motions or interventions. She missed the court's deadline for filing a complaint just the same as FFV. Yet Judge Clarke let her enter the proceedings, using the same discretionary authority that he used to dismiss FFV. Davey thought the judge looked upon her favorably because of her status as a widow - a clear case of prejudice.

Muriel Light did not have the money to retain local counsel. Rather than see her lose her inheritance to a millionaire venture capitalist, FFV paid her legal expenses, and FFV's attorneys represented her pro bono. Davey's associate, Patrick Brogan, did most of the work in her behalf.

To make a long and sordid story short, Muriel Light chose to negotiate a settlement rather than fight for her rights in court. Bemis paid her an undisclosed sum for not appearing in court to testify, and for dropping out of the case. This left Bemis's case completely unopposed.

Afterward, I tried to get information from Muriel on John Light's background and his dives on the *Lusitania*. She refused to speak with me or to answer correspondence.

## JUDGE CLARKE'S RULING

Came the final hearing in March 1995. With all his opponents tossed out of the ring, Bemis looked like a shoo-in. Giving live testimony in Bemis's behalf was George Macomber, ex-partner in the Kinvarra Shipping Company. Macomber testified that "in the mid-1980's he withdrew from the operation and assigned his entire right, title and interest in the *Lusitania* to Bemis." (This assumes, of course, that he *had* an interest.)

The mid-1980's was a decade after Kinvarra Shipping had gone bankrupt. The time frame is suspiciously belated. Also unexplained to my satisfaction was why Macomber should give away for free his interest in a property which Bemis claimed to be so valuable, and from which Bemis had already extracted at least one substantial tribute ($52,500 from National Geographic, of which Macomber got none. How much Quigley paid Bemis is unknown to me.)

These and other puzzlements will forever go unresolved because the people who have the information aren't willing to let the full truth be known.

Without any live contestants to contradict Bemis's claim of ownership, the absence of full and authenticated documentation was moot. The court had no recourse but to find that Bemis had established his chain of title to the "*Lusitania's* hull, engines, tackle, apparel and appurtenances."

This meant that Bemis got to keep the artifacts that FFV had turned in to the court. Drat! Nor did he have to pay a salvage award - that was the price that we paid for intervention.

But -

"The Court finds that Cunard did not and could not transfer rights in the cargo or personal effects." Bemis had finagled title to the hull and appurtenances, but not to the wreck in its entirety. Cargo and personal effects could be salvaged by anyone who had the means to do so, and none of it could be taken away by Bemis.

Additionally, Clarke saw through the transparency of Robol's argument that the British ruling granted Bemis full ownership of the wreck. "The Queen's Bench found only that the Crown had no right to the property . . . Essentially, it awarded them possessory title. There was no adjudication as to rights in future artifacts." ("Possessory title" means title only to the artifacts that were actually recovered by the partners in the 1982 salvage operation, and which were therefore in their possession.) Judge Clarke found that the Queen's Bench made no ruling with respect to artifacts that might be recovered in the future, nor did it make any determination about Bemis's claim of ownership of the wreck that lay on the bottom.

## Artifact or Artifice?

Now came the issue of the juggled spoons. Initially, Bemis testified that he recovered a spoon in 1993 by means of the submersible's manipulators, and that this spoon was a tablespoon, ostensibly part of Cunard's silver service. Somewhere along the line, however, this spoon was misplaced and supposedly a different spoon was submitted to the court (purportedly by accident). The spoon that was submitted was either purchased by Bemis at auction or given to him as his share of the booty from the 1982 salvage operation, and could not be used to arrest the wreck for the purpose of claiming exclusive salvage rights.

Later, Bemis submitted his token artifacts to the court: a dozen china plates and sixteen silver-plated knives, forks, and spoons (three soup spoons and a serving spoon). No mention was made of the other items on his list, items which Bemis had previously claimed to own. *Now* Bemis contended that the spoon he recovered in 1993 (and there was no way to prove which spoon it was) was not a Cunard spoon at all but instead was part of the cargo. This sleight-of-hand gave the appearance of legal lip service, lending credence to the proverb "there's many a slip 'twixt the spoon and the lip."

Changing spoons in mid stream, so to speak, enabled Robol to take a different legal stance: one in which Bemis could extend his ownership to include the entire cargo and all the personal effects because he had reduced a portion of the property to his possession. His two-pronged thrust sought "injunctive relief so that he may have the sole right to continue salvage operations." One prong relied upon the Rule of Finds, the other upon the Law of Salvage.

Clarke ruled against Bemis on both prongs. He agreed that the property had been voluntarily abandoned. But, with respect to the

Rule of Finds, the recovery of a single spoon "does not demonstrate the possession, dominion or control sufficient to vest title to all the cargo and personal effects."

With respect to the Law of Salvage, "the Court finds that Bemis does not meet the requisite 'possession' requirements, has not conducted ongoing salvage operations and has not shown a fair chance of success in the future. To support this finding, the Court notes that in the past 18 years Bemis has participated in only three expeditions to the shipwreck."

In all that time, Bemis never initiated salvage operations on his own, nor did he financially support any of the operations that were conducted. Instead, he lurked in the background until someone else formulated plans, then he inveigled his way into the operations in order to collect his toll.

Robol tried to bolster his case by going to Massachusetts to depose Bob Ballard. (Expense must have been no object.) Ballard no longer had any use for Bemis because he had already gotten what he wanted - the film was in the can. Now that Ballard had nothing to lose, he was at last willing to take a stand against Bemis. Robol submitted Ballard's deposition to the court as proof that Bemis was instrumental in the creation and the conduct of the film shoot.

Clarke's response to this was compelling. He found that Bemis "did not substantially contribute to the operation. In fact, the Court notes that he was paid for the effort. In the words of Dr. Ballard, when asked if Bemis played a cooperative role in the 1993 expedition, 'Yes, in that he did not obstruct or in any way try to affect or alter our operational plan.'"

Clarke: "The most recent expedition, during the summer of 1994, fell far short of its expected success." After noting the expedition's shortcomings, "the Court concludes that Bemis is not in 'possession' of the contents for purposes of pursuing exclusive salvage rights. Furthermore, the Court finds that Bemis is not participating in ongoing salvage operations nor has he shown a fair chance of future successful salvage."

Ultimately, "The Court denies his request for exclusive salvage rights to the cargo and personal effects."

The conclusions reached above were all matters of law. But Clarke also noted that Bemis acted "in less than good-faith."

Even with the appropriate spoon, Bemis couldn't eat his cake and have it too.

## Case Closed

After the court's opinion was rendered, Peter Hess - with his seemingly infinite wealth of epigrams and catch phrases - characterized the hardships that FFV endured by noting wryly, "It's like taking cod liver oil. It tastes awful going down, but it solves the problem."

Bemis appealed Clarke's ruling. Robol's fifty-page brief was a pointless reiteration that was surfeited with exaggerations as well as with perversions and misrepresentations of the truth.

Robol wrote that Bemis "engaged in numerous salvage activities from the mid-1960s to the present." That Ballard "assisted Bemis' efforts." That the "depth makes scuba diving impractical" (despite the fact that Bemis's latest "expedition" was conducted on scuba). That Bemis developed a program whose "primary mission" was "the protection and promotion of the intangible values of the *Lusitania* shipwreck as a resource for information in the fields of history, archaeology and science. . . . His goal included deliberately leaving artifacts on the shipwreck *in situ*, within their historical provenance. . . . Bemis' efforts included substantial work to achieve these goals. . . . This included analysis of the data collected during at-sea operations, ongoing historical research, progress on magazine articles, films and books and preparations for further at-sea operations." (To my knowledge, Bemis has written no books or articles or produced any films.)

Robol portrayed Ballard and National Geographic as "subcontractors" who assisted Bemis in "his on-going efforts to recover, salvage, conserve and publicize the *Lusitania*." This statement contains two contradictions. In the previous paragraph he stated that Bemis's goal was to leave artifacts *in situ*; now his goal was to recover artifacts. Furthermore, instead of publicizing the *Lusitania*, Bemis labored to impede publication of magazine articles and to prevent public presentations from occurring.

Robol claimed that in 1994, "eight deep-sea divers" made "a number of significant observations . . . with respect to the extent of the structural change in the shipwreck, unusual condition of the portholes, evidence of internal and external explosions, degree of oxidization (*sic*) in the hull, surface conditions of brass, changes in the extent of silting compared to previous expeditions, and ambient light field. . . . Work also included further analysis of bottom conditions and mapping of the seafloor at the shipwreck site. . . ." All this by four scuba divers on a handful of bounce dives!

"Bemis has promoted major educational publications and events to disseminate the information gained during his programs for exploration of the shipwreck. . . . Bemis has encouraged public lectures and museum exhibits about the shipwreck." He also took credit for the television broadcasts in the U.S. and the U.K. "In addition, five major magazine articles have been published." (Two were mine, one was Polly Tapson's, another was Christina Campbell's.)

With reference to Judge Clarke's orders that *all* parties who might have a competing interest in the *Lusitania* be notified, Robol wrote with willful misconception, "It is undisputed that Bemis complied with these orders and gave the required notice." Perhaps his memory was remiss, and once again he forgot about Muriel Light and The Starfish Enterprise.

Robol bemoaned the fact that "the district court gave Bemis little or no credit for salvage activities of John Light during the 1960s, or those of Light and Macomber (which later included Bemis) during the 1970s." He failed to explain why Bemis should receive credit for the accomplishments of another, and forgot to mention that no diving or salvage was conducted in the 1970s.

Robol noted that by acquiring the rights ascribed by the "War Reclamation Board," Bemis assumed "all liabilities and expenses that may attach to the wreck." He didn't mention that Bemis's 1994 contract with the Irish divers stated specifically that "each participant . . . releases Bemis from any and all liability." Nor did Bemis pay any of Eugene Cahill's medical expenses.

Bemis's apparent ignorance (and that of his attorney) with respect to the 1982 Oceaneering salvage operation was reflected throughout the entire history of the proceedings. Bemis seems to have had no information contrary to the Queen's Bench observation that ninety-four artifacts were recovered at that time. The truth is that *thousands* of artifacts were recovered, but for the sake of convenience the artifacts were grouped in ninety-four lots, each of which contained a count and a description of the items. Clarke was led to believe by Robol's initial filings that ninety-four individual artifacts represented the entire haul. Robol never tried to correct this misnumeration, and even in his appeal brief he wrote that "Bemis . . . recovered many dozens of items." I am astonished that Bemis was not better acquainted with the facts. That he knew so little about the Oceaneering operation speaks for the depth of his involvement.

Perhaps Robol hoped that the appeals court was too busy to read the refuting documentation supplied by FFV, or too tired to review the applicable case law. If so, he was in error. The appeals court recognized Robol's pious fiction for what it was, and upheld Judge Clarke's ruling on every point.

Not content with that, Robol took the case all the way to the Supreme Court. That august body refused to hear the case, and let stand the rulings of the lower courts.

And there it stands. A Pyrrhic victory - or eternal limbo?

## Concurrence

Time heals all wounds, and wounds all heels.

I'm no longer afraid of my car blowing up in my driveway. I've gotten over the anguish of confronting a grave injustice. I've moved on to other projects. I'm working toward other goals.

With the soothing passage of time I have come to an understanding within myself - an acceptance, if you will - that Bemis was at least partially right in his assertions. I am willing to meet him halfway.

During the course of litigation, Bemis alleged that he purchased the *Lusitania* in the 1970's. I am forced to admit at last that he might possibly have purchased the wreck. Where we disagree completely is with the time frame. In my opinion he bought the *Lusitania* in 1994.

# Whither

## ANDREA DORIA FATALITIES

In 1995 I received a call from reporter Theresa Foley. Among other things, she wanted to know how many diver's had died on the *Andrea Doria*. I didn't know for sure, but told her that I thought the number was around half a dozen, perhaps as many as eight.

I was appalled when I read her technical diving piece in the *New York Times*. She inflated the figure to over sixty, claiming that more people died while diving the wreck than were killed in the collision and sinking. She and her kind of vulgar sensationalism are what equate all reporters with yellow journalism. I sent a staunch letter of protest to the *New York Times* and to Theresa Foley personally. Neither had the courage to respond.

Foley's crass, intentional misrepresentation was the kind of news picked up by the wire services, then spread like an epidemic disease. If such bloated distortions are repeated often enough, they become established as authenticated myths in the public consciousness.

My worst fears were realized. Within months an article appeared in *Scientific American* in which Foley's figures were repeated in similar language. Already Foley's falsehood had become source material.

One of the authors of the *Scientific American* article was Peter Bennett: president, director, and CEO of the Divers Alert Network, as well as chief editor of the organization's bi-monthly magazine, *Alert Diver*. According to the magazine's declared philosophy, "*Alert Diver* is a forum for ideas and information relative to diving safety, education and practice. . . . Ideas, comments and support are encouraged and appreciated."

I wrote to Bennett and outlined the situation. He responded quickly, claiming that *Scientific American's* staff writer Glenn Zorpette had added the damaging information to the article, and that Bennett hadn't seen the final copy before it went to press. He thought the correct number of fatalities was between eight and twelve.

I wrote to Zorpette. He, too, responded quickly. He didn't have faith in my argument, but neither was he defensive. He admitted that he had used Foley as his source, but said that Billy Deans, Mike Menduno, and others had concurred that the figure was at least as high as forty. The spread of misinformation was widening - was perhaps already self-perpetuating and out of control.

I couldn't understand how Deans and Menduno had the temerity to pass themselves off as authorities on the *Andrea Doria*; Menduno especially, because he had never even dived the wreck. I called them both. They were astonished by my pronouncement. Menduno was not persuaded to believe me, but Deans was willing to accept the fact that my count was closer to reality than his.

Zorpette didn't know who I was or what basis I had for my claim. He extrapolated from four fatality reports published in *AquaCorps*. "So in the preceding 30-some years - when people were diving the *Doria* on relatively primitive gear - there were fewer than four other deaths? It doesn't add up." This kind of logic assumes that if a person got two parking tickets within the span of a single week, by extrapolation he must have been ticketed more than one hundred times in the previous year. Nevertheless, Zorpette was open-minded because "it bothers me deeply to have published something incorrect."

He also noted that "Peter Bennett was absolutely mistaken" about not having had the opportunity to review the article prior to publication. "Bennett received a copy of the entire draft."

I sent a resume of my qualifications to Zorpette. Once he realized that I had written the book on the *Doria*, and became aware of my long-time experience with the wreck, he accepted my expertise on the subject. But I still didn't know the precise number of fatalities. So I decided to ascertain the facts.

By means of letters, e-mails, phone calls, and face-to-face conversations, I communicated with those who were involved in the recoveries or were on the boat when a fatality occurred. I spoke with dive buddies and boat captains. I compiled a list which included the name of the deceased, the date of death, the name of the boat, the name of the buddy, and the names of those who actually performed the recoveries. I cross-checked all the information.

After a couple of months I arrived at the precise figure: seven. And that was for the entire history of the wreck.

I wasn't surprised, but many of those I interviewed were shocked. Even experienced *Doria* divers believed the figure was higher. The dive community in general was flabbergasted. During the course of my investigation I discovered how the mystique of death originated.

On each occasion on which a diver died there was a boatload of witnesses. Each witness related his rendition of the event to his friends and various acquaintances. Those who heard these anecdotes over the course of several years attributed each separate version of a single event to a multitude of events. Every time a story was retold - and got farther away from its source - facts were dropped, mutated, or embellished to fill the gaps. Similarities blurred. The chronology was lost. The end result was a false perception that exaggerated the truth.

I sent the results of my research to Bennett, Zorpette, and Menduno, along with a "letter to the editor" that could be used in place of a retraction. Menduno elected not to print my letter, perhaps because Foley was then free-lancing for *AquaCorps*.

Zorpette passed my letter on to the letters editor, along with his recommendation. *Scientific American* declined to print my letter or to make a retraction. I suppose the magazine had a reputation to maintain - a reputation that I now perceived was a facade, like that of the *New York Times*. A magazine or newspaper could not maintain circulation by acknowledging poor research or by calling attention to its mistakes. Better to let subscribers believe untruths than to tarnish the imagery of esteem.

Bennett had someone at DAN corroborate my supporting documentation with the files at the National Underwater Accident Collection Center. The result was a perfect tally. The NUACC could find no *Doria* diver fatalities in its files other than those that I had established. I felt vindicated. I thought this confirmation would clinch the matter and that DAN would correct the error, but I was wrong. Or perhaps I was naive enough to expect Bennett to do the right thing.

I thought that publishing the truth in *Alert Diver* would be a good way to start quelling a rumor begun in malice. I shouldn't have to deal with red tape, personnel hierarchies, management issues, or adverse policy. I was in touch with the head honcho in the organization, on the same wavelength with the organization's stated philosophy, and Bennett was in full agreement with me on the evidence and the facts. Since he wrote an editorial in every issue, it would be a simple matter for him to dispel any misconceptions that had been formed in the diving community, where the perversion of the truth did the most harm.

I was both disconcerted and disgusted when Bennett refused to print my letter to the editor or to correct his own mistake from *Scientific American*. While privately he acknowledged the wrong that he had perpetrated, he wouldn't publish anything to help correct the situation. By his omission the public was led to believe a lie. So much for *Alert Diver's* bogus editorial philosophy and for the credibility of its leadership.

## THE UNCHARTED ADVENTURE

In my early teens I read about Roy Chapman Andrews' expeditions to the Gobi Desert in the search for fossil remains of ancient man and dinosaurs. Largely on the basis of that I decided to become an archaeologist or paleontologist. What I didn't realize at that age - and what took me most of a lifetime to discover - was that what attracted me the most about Andrews' exciting narrative was not the science, but the adventure: the physical challenge and the confrontation with nature as opposed to intellectual curiosity, or the sheer thrill of exploring the unknown instead of a yearning for scientific achievement. Subconsciously, I suppose, a quest in the guise of science was my way of justifying adventure.

Since then I've learned that all of life is an adventure that needs no justification. How each person perceives and pursues that adventure is what makes everyone different. I chose a path through life that was motivated by my nature. It is not the way for everyone - even for those who profess to envy my lifestyle. What many people fail to realize is what my own adventure has lacked. By that I don't mean just economic security and a comfortable family life.

I've missed out on experiences that others might consider to be essential ingredients for contentment. I've never been to an opera or a ballet; I've never attended a concert; I've never gone to any kind of sporting event (such as baseball, football, and basketball); I've never seen a race car rally; I don't hang out in bars and I've never had a beer; I've never had a smoke; I've never watched soap operas; I've seen none of the sitcoms that people always rave about; I don't recognize tunes made popular by television commercials; I've never shopped - I've gone to stores only to purchase specific essential items; I don't have a radio in my house and the one in my car I've never switched on; I don't delve into politics; I don't play video games; I don't go to amusement parks; I've never bought a chance or a lottery ticket; I don't gamble; I don't go to movies; I don't take vacations. I don't own any jewelry and don't have fancy clothes. My house is furnished with hand-me-downs or pieces I've found in the trash. And so on.

Based on all these don't's and haven't's my life might seem pretty drab. I suppose it depends on viewpoint. Would *you* find contentment in browsing through musty book stores; pawing through dusty archives; plunking away at a computer keyboard for ten or twelve hours a day, seven days a week; sleeping in a car, or on the ground, or in the cold and rain; being away from home for six months out of twelve; eating out cans and fast-food restaurants; coming back to an empty house; seeking personal satisfaction instead of acceptance?

What is sacrifice to one is nonessential to another.

Each person must choose his own adventure.

## THE PURSUIT OF FULFILLMENT

Due to the changing nature of progress, I fear that the book is going the way of the slide rule. This is of vast importance to me - as an author and avid reader - but of little consequence to others. If readership continues to decline I might have to find another occupation. Writing manuscripts to store in a filing cabinet has no value to me. I might have to seek some other ways to infuse my life with fulfillment.

It is no great feat to be wise after the event. I suppose I could bemoan the wrecks not found, the relationships not established. But it is more important to emphasize successes than to dwell upon failures and missed opportunities. As I look back along the path of adventure behind me, I see more than books and dives that have given me satisfaction. I see more than the misfortunes that I've suffered. I see a long line of people whose friendship has enriched my life beyond measure. Friendship is a precious if intangible commodity.

Given the controversial nature of the stories I've recounted - the sinking of the *Lusitania*, the chaos of Vietnam, the evolution of wreck-diving, the birth of technical diving, and bitter litigation - one might believe that contention reigns supreme. Not true.

In life, as in fiction, conflict generates interest. Newspapers tell of murder and mayhem, rape and pillage, political intrigue, crime run rampant - not because these afflictions predominate but because that is what people want to read.

My own story is peopled with characters to whom morality held no meaning, whose values were blatantly questionable, whose principles were flexible in order to suit their own convenience, who possessed no sense of social conscience, who committed crimes against decency, who were emotionally immature, who lacked competent self-awareness, self-analysis, and self-actualization. Some were more concerned with appearances than with substance. Some needed to wear a Kevlar vest in order to keep from breaking their ribs while they were pounding themselves on the chest. Others sought respect by taking credit for the accomplishments of others; they failed to realize that respect cannot be assumed or stolen, it must be earned in order to have any meaning. Some were so entrenched in the pursuit of self-advancement that they worked harder toward destroying their perceived competition than they did on honest achievement. And instead of doing what was right, some people did what was politically correct or personally advantageous. In the final analysis, some people are not symbiotic with society, but are parasitic - they give nothing in return for what they take; they just take.

Although the people who exhibited these disparate character flaws seem to have been pandemic in my story, in the comprehensive picture they existed in the minority. The vast number of divers I have met throughout the years have been good, honest, conscientious people - a genuine slice of all that is virtuous in humanity. They have not received their fair share of the credit for being who they are, for making mankind what it is. If I have not given sufficient credit to those who went unnamed, it is because my story has been recounted through my own eyes - a singular and exclusive point of view.

The friendships I have made throughout the years are what have made my life rewarding. All other experiences pale by comparison.

## Ever Onward, Ever Downward

Date: June 17, 1995. Dive boat: *Miss Lindsey*. Objective: German cruiser *Frankfurt*.

It was like old times: back where technical wreck-diving began, five years earlier.

Eight of us prepared to make the descent. I felt no tremors, no abject fear; I didn't shake with dreadful anticipation. But I still maintained a healthy respect for the potential peril of depth.

Ken Clayton was the indisputable leader of the trip: partly due to his energy and drive, partly due to his inner need to exercise control. But it was also partly due to my failure to take command. I've been a reluctant leader all my life. I've never felt the need to take the helm when someone else was willing to do the job. I've been more of a maverick, shrugging off the reins of responsibility whenever possible, taking up the slack when there were no other volunteers. If no one was headed where I wanted to go, I either went there alone or convinced others to go there with me.

While I was away diving the *Lusitania* and other deep-water wrecks, Clayton doggedly maintained the pursuit of his single-minded goal: to find and dive all eight German warships that were bombed and shelled in the 1921 Billy Mitchell tests. The discovery of the *Frankfurt* put to rest the uncertainty of what he and I had dived in 1992, at 350 feet. What we believed at the time to be the *Frankfurt* was more likely the *G-102*, a destroyer. Now we were diving the real McCoy.

Clayton descended with the first group of four: Brad Sheard, Doug Sommerhill, and Harvey Storck. Half an hour later I went down with my long-time friends: Steve Gatto, Tom Packer, and Jon Hulburt. We met the others on the line as they were making their ascent.

Gone were the days of travel mix and heliox. Now I breathed my nitrox deco gas for the first hundred feet, then switched to trimix-10/60 and breathed it all the way to the bottom. Hulburt had trouble with his drysuit inflator so he paused at 300 feet. The three of us left him to bring up the rear. We rocketed down through the dull green water like shoppers on a high-speed escalator racing to a bargain basement sale.

The wreck took form when we were fifty feet above it. The giant warship was clearly recognizable in bright ambient light which held a Caribbean quality, with lateral visibility reaching seventy-five feet or more. The sight was awesome. We touched down on the bottom six minutes into the dive. All was well.

I swam off the edge of the broken-down wreck into the debris field. I placed my depth gauge on the white reflective sand and took a reading. Then I returned to where Gatto and Packer were examining the wreckage. After a couple of minutes together, I separated from them and went on a brief tour by myself.

A projection of the wreck vaulted high overhead. I swam beneath an overhang, then rolled over and looked up - very much as I did in

1971 when I made my very first wreck-dive, on the *Persephone* in 60 feet of water. I felt the same churning queasiness in the pit of my stomach that I felt on that long-ago dive. A lot of water had flowed under the bridge during the intervening years. Yet I still thrilled with excitement about being under water, still throbbed with anticipation over what I might see and discover.

I swam into a beckoning compartment that was bounded by three existing bulkheads. Then I swam out on the far side of the wreck. I was fifty feet from the anchor line, and out of sight of my buddies. Instead of returning the way I had come, I circumnavigated the standing structure. A few minutes later I rejoined Gatto and Packer.

We explored around the grapnel for the remaining minutes of the dive. Poking through the twisted steel debris, I spotted what appeared to be a cow bell. I picked it up, examined it, then realized that I wanted to keep it. But in streamlining my equipment I had opted not to carry a goodie bag. Only with some difficulty did I convince Packer to put the bell into his bag - he couldn't believe I wanted it. Eventually he consented - rather as he had when he had put the lobster in his bag when we had searched along the bottom for the *Andrea Doria's* bell. This experience was similar, but deeper.

At fourteen minutes our bottom time was over. The eight minutes that we spent on the wreck was an intense emotional and sensory experience - far more pleasurable and much less visceral than my initial dive on the *Ostfriesland*. The gut-wrenching days of facing unknown elements lay in the past.

On the way up the line I glanced at my pressure gauge: 500 psi. Barely enough to reach the nitrox switch, I calculated, if I was conservative with my gas. I had yet to reach 300 feet when I felt a restriction in the gas flow. Instinctively I looked up and down: Hulburt was out of sight above, Gatto and Packer were a few feet below, looking at their gauges. From restricted flow to no flow took but one breath. I gulped. How stupid to die on the anchor when the danger of depth was past. Could I reach Gatto and Packer before my last breath gave out?

No time. I grabbed the regulator on my second sling bottle, pulled it free from the rubber thong, shoved the mouthpiece between my lips - and inhaled.

In planning for the dive and making allowances for gas consumption, we calculated that one set of doubles would hold enough gas for the projected bottom time. Nevertheless, with a pony bottle mind set and a back-up mentality, we all felt safer carrying an additional supply. My "pony bottle" in this case was a 120-cubic-foot single. I had hoped not to use it - because I didn't want to pay to have it refilled; I wanted to save it for another dive. As I breathed deeply, I thanked myself for planning for contingencies and for not being overly cheap.

The decompression went smoothly. I switched to nitrox and the appropriate depth, then breathed surface-supplied oxygen at the 20- and 10-foot stops. Two hours of decompression passed without a hitch. On the boat there was cheering and celebration.

Most gratified of all was Ken Clayton, for one of his wishes had come true. He hadn't yet found all the German warships - one of the destroyers was still missing. But when he and I compared maximum depths, my gauge registered 417 feet; his read 421. He finally went deeper than I did!

## Perspectives

Early wreck-divers cut their teeth on wrecks at a depth of 130 feet. At the time that was considered deep. Later they extended their range to wrecks that lay farther offshore and in the dark dismal waters of the Mud Hole, reaching 190 feet. Today these wrecks are considered routine dive sites.

The *Andrea Doria* was once a venerated shrine among shipwrecks, a Siren call to aspiring deep wreck-divers. Now the wreck is a training ground for technical divers.

The *Ostfriesland* was a watershed: a turning point in the advent of technical diving. In only a handful of years the Billy Mitchell wrecks have become a backwater: quaintly remembered for their historical significance - both to naval warfare and to technical diving - and visited by more and more divers every year: divers who cannot comprehend either the attitudinal obstruction or the blinding mental barrier that these deep wrecks once presented.

It took a quarter of a century of progressive depth attainment for me to make an exploratory dive to 420 feet. Today, with specialized instruction and modern high-tech equipment, an ambitious diver can make such a dive in less than a handful of years.

Each deeper wreck was a step on the ladder of progress toward greater knowledge and proficiency, toward a better understanding of human potential. Further advance is inevitable. Only those who accept the limitations imposed by others will lose the race before entering the running.

There are no culminations in the adventure of life. There are only plateaus whose level surfaces offer respites from advancement - way stations along the endless track of progress. Ever higher plateaus rise beyond.

The human race is a relay race. Each generation takes the torch from the previous generation, fuels the flame for a while, then passes the fiery brand to the next generation.

Now the turn is yours.

Take the torch.

# Index

Names in italics refer to ships, boats, books, songs, poems, newspapers, magazines, films, or television shows. Page numbers in italics refer to photos or captions.

Abandoned Shipwreck Act: 70-75
Aberdeen (Scotland): 222
Adams, Brock: 74
Addis, Jeff: 80
Admiralty: see "British Admiralty"
Admiralty Court (British): see "Queen's Bench"
*Advanced Wreck Diving Guide*: 59-61, 117, 148, 224, 228
*Alert Diver*: 376-378
*Alien*: 175
American Broadcasting Company (ABC): 216
American Revolution: 367
American Water Sports: 130
Amundsen, Roald: 167
*Andrea Doria*: 5-21, 43, 51-58, 77, 79, 86-92, 118-121, 130, 143, 146-147, 149, 161, 171, 173, *178-179, 180, 181*, 225, 229, 233, 239-240, 276, 302-305, 308, 312-313, 382-383
  arrest: 168-170
  artistic rendering: 251-252
  number of diver fatalities: 376-378
*Andrea Doria: Dive to an Era*: 16, 51, 84, 85, 153-154, 225
*Andrea Doria: Floating Art Gallery*: 168
Andres, F.W.: 322-323, 337, 341, 345
Andrews, Roy Chapman: 378-379
*AquaCorps*: 116-117, 156, 157, 245, 329, 366-368, 377
Aquazepp: 113-114
*Araby Maid*: 112-116, 118, 124, *181*
*Archimedes*: 217-221
Arena, Sal: 5, 7, 8, 14-15, 17-20, 28-31, 52, 93
argon: 106, 134, 135, 136, 138, 275
*Arizona*: 237
*Atlantic*: 40
Atlantic Alliance for Maritime Heritage Conservation: 71-73
Atlantic Divers: 147
*Atlantic Lady*: 112
Atlantic Wreck Divers: 85, 88-92
*Atlantic Twin*: 7

*Atocha*: 70-71
attention deficit disorder: 59
*Ayuruoca*: 48, 52
Bachand, Bob: 61
Bailey, Thomas: 214-215
Ballard, Robert: 231-232, 312-313, 340, 344, 372-373
Barnett, John: 5, 13-14
Bartlett, Harry: 39-41
*Bass*: 168
bathysphere: 34
Bauer, Ted: 144
Beasley, Risdon: 233
Beaudry, Philippe: 237-238, 285-287
Beebe, William: 34
Beesly, Patrick: 215
Bell, Ronny: 152
Bemis, F. Gregg: 234-236, 238-239, 249, 253-255, 262, 284-285, 287, 293-294, 309-375
Bennett, Peter: 376-378
Benoit, Frank: 99
Berman, Steve: 121, 149, 153, 155
Bielenda, Steve: 16-18, 52-53, 85, 88, 91, 119-121, 154, 302, 305
Bieser, Janet: 53, 120, 302-304
Billy Mitchell wrecks: 107, 152, 173, 227-229, 381-383
Blount, Steve: 329
Blue Hole, Virginia: 155
blue holes: 248
Blue Spring State Park (FL): 45-46
Bluett, Dave: 37-38, 80
Bob's Cold Spot: 305
Bohrer, Randy: 121
booster pump: see "Haskel booster"
Boring, Mike: 51, 93-94, 173
Boston Sea Rovers: 312
Bradley, Bill: 72-74
breakaway system (for decompression): 122-123
Brennan, Kevin: 146
Brielle (NJ): 46, 62
Bright, Dave: 144, 239-240, 285
British Admiralty: 209, 214, 215, 233,

## Index - 385

362
British Aircraft Carrier: 92
British Airways: 252
British Broadcasting Company (BBC): 214, 216
British Oxygen Company: 348
Broco torch: 47-49, 88
Brogan, Patrick: 325, 355, 370
Brow Head: 301
Brown, Gaye: 144
Buckley, Doug: 173
Buhlmann, A.A.: 51, 157
Burdewick, Ron: 5
Bureau of Mines (U.S.): 211
Butler, Jane: 60
Byrd, Robert: 74-75
Cahill, Eugene: 346-347, 354, 361-362, 374
Caloyianis, Nick: 15
Campbell, Billy: 303-304
Campbell, Christina: *182, 190, 206*, 256, 264, 266-269, 272, 294, 329, 373
Canadian Coast Guard: 39, 76
Canadian fisheries bureau: 39
Canadian Pacific Railway: 237-238
Canon: 253
Cape Charles (VA): 106
Cape Hatteras (NC): 69
Capitol Divers: 80
cave diving: 149, 153-155, *181*
*Central America*: 310-311, 323
certification: 163-166
Chatterton, John: 87-89, 92, 146-147, 174, *184, 189, 206*, 240-241, 242, 245, 255, 260, 273, 279-280, 292-293, 295, 299-300, 307-308, 315-316, 333-336, 350-351, 367-368
Chesapeake Bay: 92
Chidsey, Donald Barr: 214
*Choapa*: 157
Chowdhury, Bernie: 119-121
City Bank: 5
Clarke, Judge Calvitt: 310-374
Clayton, Ken: 80-81, 93-99, 106-112, 127-142, 147, 149, 151-153, 157-166, 173-174, *180*, 227-229, 381-383
*Cleopatra*: 306
Coast Guard: see "United States Coast Guard" or "Canadian Coast Guard"
Cobh (ex-Queenstown): 271
cocaine: 47
Columbus, Christopher: 167, 172
Columbus-Ohio Discovery Group: 310-311, 323, 337, 358
Compressor Country: *187*, 280, 291

Coney Island: 34
confrontational avoidance: 255
Congress: 72, 73
Constitution: 70
cordless telephone: 229-230
Cork (Ireland): 263, 284, 302
Cornell Maritime Press: 60
Cowardly Lion: 65
Craig, John D.: 210, 211
creativity: 58-59
Crowell, Danny: 146, 242, 308, 319
Crystal skull: 262
*Crystal Wave*: 306
Cunard Steamship Company: 209, 236, 370, 371
Cush, Cathie: 329
Dady, Ed: 153
*Danny & Rickie*: 39-41
Davey, Phil: 317, 324-374
*Day They Sank the Lusitania, The*: 214
DCAP: see Decompression Computation and Analysis Program
Deans, Billy: 41-45, 52-55, 58, 77, 78, 99, 102-104, 112-118, 124-125, 147-148, 156, 157, 166, 240, 302-305, 376-377
de Camp, Michael A.: 40, 41, 97, 113
Deco-Brain: 51-52
Decompression Computation and Analysis Program (DCAP): 103, 107, 124-126, 174
decompression computer: 51-52
decompression injury: 258-260
decompression station: see "dec station"
dec station: *180, 187, 194, 198-199*, 256, 263, 265-268, 275, 279, 281-283, 290, 296, 299
"defrayed prestidigitation": 35
*Delaware*: 27
DelCorio, Lynn: *back cover*, 306-307
DeLotto, Lou: 52-53
DeMaria, Don: 99, 102, 112-116, 118, 147
Demeter: 27
Democrates: 27
Department of Marine (Dublin): 234
Department of Transportation (British): 235-236, 329
Diamond Shoals (NC): 161
Disney, Walt: 265
Dive Equipment Manufacturers Association (DEMA): 156
Divers Alert Network (DAN): 376-378
Dive Shop of New Jersey, The: 15, 24, 26-27
Diving Unlimited International (DUI):

## 386 – The *Lusitania* Controversies

261
"Don't shoot the messenger": 27
*Doonie Brae*: 212
Doria, Admiral Andrea: 13
Douglas, Margaret: 158, 160
*Down Deep*: 305-307
Dreyer, George: 147-149
drift decompression: 122-123, *180*, 275, 283, 318-319
*Dr. Strangelove*: 101
Dry Tortugas: 78, 99-100, 102, 112-116, 118, 143, 147
Dudas, John: 22-26, 28, 47, 125, 244
Dun Laoshaire (Dublin): 347
Dupont: 112
*Eagle's Nest*: 308
Eastern Divers Association: 5, 320
*Eastwind*: 87
Edwards, Don: 76
Ehorn, Paul: 75
Emperor's new clothes: 247, 360
*Empress of Ireland*: 144, 149, 237-238, 285-287
Empress of Ireland Historical Society, The: 237
*Engineering: An Illustrated Weekly Journal*: 215
England, Kevin: 146, 242
*Ethel C.*: 98, *179*
Explorer's Club: 165, *180*
Extended Range Diving Organization (ERDO): 171-172
44 Fathom Wreck: 92
Farnquist, Tom: 75
Federal Register: 72-74
Field, Rich: *185*, 257, 281, 293
Fifty Fathom Ventures: 315-375
Filer, Ric: 130-131
firebombing: *207*,
First Amendment to the Constitution: 328
Fisher, Mel: 70-71
*Florida*: 171
Florida Keys: 52
Foley, Theresa: 376-378
Fort Jefferson: 99, 102
Frankenstein, Dr.: 29
*Frankfurt*: 151-152, 157-160, 381-383
Fraunhofer line: 210
Freeport (NY) : 7
*G-102*: 160, 381
Gamble, Steve: 114-116
Gambone, Guido: 170, *181*
Garay, Eric: 37
Garcia, Mary Grace: 146
Garda (Irish police): 298

Garvin, Hank: 120, 171, 304
Gary Gentile Fan Club: 224, 226
Gary Gentile Productions (GGP): 81-84
Gascon, Kenny: 51
Gas Station, The: 157, 164
Gatto, Steve: 15-16, 34, 52, 97-98, 146-147, 157-158, *178-179*, 241, 244, 381-383
*Gekos*: 38, 94, 107-108
Gentile, Michael: 67-68
GGP: see "Gary Gentile Productions"
Gilligan, Gary: 85-86, 89-92, 119-121, 146, *179*, 304
Ginnie Springs: 149
Gimbel, Peter: 15, 51
Gleason, Bill: 166
Glenn, John: 74
Gmitter, Tom: 34
Gobi Desert: 378-379
gorilla diver: 149-150, 167
Gotto, Nic: *184*, *197*, 233, 249, 265, 267-270, 275, 279, 281-283, 290, 299-301, 332, 365
Govier, Ray: 234
Grand Dame of the Sea: 5, 12, 21, 119
Graseby: 261
Great Lakes: 75-77
Grima, Angelo: 344, 362
Griswold, Lawrence: 131
Gulf Stream: 43, 94, 122, 290
Halifax (NS): 38-41
Hamilton, Bill: 103, 107, 117, 124-126, 174
*Hampshire*: 221
Handelman, Lad: 367-369
Hansel and Grethel: 363
Hanson, R.S.: 322
Haskel booster: *187*, 261, 263, 279-280
Haskel Land: *187*, 280, 291
Hatteras (NC): 38, 124, 161, 318-321
Haulbowline (Ireland): 347
Hayward, Joyce: 72-73, 75-76
Health Designs International: 87
helium: 99, 103, 107-108, 111, 127, 136, 149-150, 156, 157, 162, 170-171, 212-213
 attributes: 210-211
 discovery: 210
 Mark 8 backpack: 213
Hendricks, Sr., Walt: 117
Herrara, Lisa: 171
Herschenrider, Milt: 78
Hess, Peter: 69, 170, 176, 253-254, 309-310, 313, 315-319, 324, 350, 359, 372-373
Hickey, Des: 215

Highlander, Jim: 219
high-tech diving: 117, 150
Hillary, Edmund: 137-138
Hillgrove, Tony: *185*, 257, 259, 293
Hillier, Mike Jr.: 151-152
Hillier, Mike Sr.: 108, 152-153, 173
historic preservation: 70
Hoffman, George: 26-28, 244
Holmes, Sherlock: 262
Holt, Rinehart & Winston: 210, 322, 328
Homer's Hot Spot: 305
Hope, Nick: *182, 187, 190, 195, 206*, 256, 261, 264, 267-269, 272, 279, 280, 292, 299-300
Horan, Dave: 71
Hotforge Ltd.: 235-236
hot mix concept: 174
House of Representatives: 71
*How Much is That Doggie in the Window*: 160
Huffman, Roger: 92, 123, 320
Huff, Glen: 325, 331, 334-336, 352, 361
Huff, Poole & Mahoney: 323-374
Hughes, George: 78
Hulburt, Jon: 4, 19, 28-32, 38-39, 41-42, 44-46, 48, 60, 62-63, 64, 88-89, 97, 112, 126, 146, 157, 381-383
Hulburt, Judy: 44
"if you can hook it, we can dive it": 40
*Illustrious*: 93
inflatable boat: 6, 39, 76, 121, 123, 302-304
intellectual property rights: 328-329
Internal Revenue Service: 35
*Ioannis P. Goulandris*: 47-49, 52, 88, *178-179*
Irish Technical Diving (ITD): 248-249, 293, 346-347, 353, 355, 369
Isle Royale: 76, 144
*Ironclad Legacy: Battles of the USS Monitor*: 69
Italia: 16
*Jackpot*: 241
Janssen, Pierre: 210
Jarvis, Ray: 151
Jaszyn, Rick: 55-58, 146
JFK airport: 252
Johns, the: see "John Chatterton" and "John Yurga"
jonline: 60-61, 122
*Joss*: 144
*Kamloops*: 144-145
Keatts, Hank: 99
Keen, Larry: 38, 94, 98, 107-108
Keller, Hannes: 211, 212
Kessler, Dennis

Keystone Kops: 8
Key West (FL): 41-45, 58, 99, 102, 112, 114, 116, 147, 157, 240
Key West Divers: 77
Kieran, Denis: 233, 263, 264, 298-299, 335
Kieran's Folk House Inn: *186*, 232-233, 263, 309
King, Jim: 78-79, 156
*King Kong*: 129
Kinsale, Ireland: *188*, 212-213, 232-233, 263, 271, 284, 299, 309, 313, 314, 325, 337
*Kinvarra*: 212-213
Kinvarra Shipping Company: 322-323, 353, 370
Kirchner, Artie: 6, 11, 47-48, 51, 161, 320-321
Kitchener, Lord: 221
*Kiwi*: 241
Klag, Glen: 130
Klein, Howard: 240, 308
Knowlton, Dan: 217-220
Koenig, Alan: 222-223
Koonce, Don: 173
Lachenmeyer, John: 5-12, 17
Lake Maggiore (Switzerland): 211
Lambertson, Chris: 21
Lander, Barb: 147-149, 159 (family), 160-163, 171, 174, *184, 189, 206*, 229, 239, 245, 246, 252, 255, 257, 266, 272, 280, 282-284, 289-290, 293, 299, 315-316, 318-321, 327, 332-335, 350, 357-358, 359, 367
Lane, Hugh: 366
Lang, Steve: 320-321
Law of Salvage: 371-372
Law of the Sea Convention: 364
LeBlanc, Ryan: 75
Library of Congress: 37, 73
Lichtman, Norman: 24, 27
*Life* magazine: 211, 214
Light, John: 209-216, 225, 232-234, 236, 278, 309, 314, 321-323, 328-329, 337, 341, 352, 363, 366, 370, 374
Light, Muriel: 234-235, 254, 313-314, 321-324, 336-337, 358, 369-370, 374
Light, Ruth: 337
*Lillian Luckenbach*: 93
Lister, Barry: 235-236
Liverpool and London War Risks Insurance Association: 209, 236, 328, 374
Llewellyn, Betsy: 144
Lloyd's Law Reports: 223

Lloyd's of London: 209
Lockyer, Norman: 210
London *Times*: see "*Times* (London)"
Long, Dick: 261
Louisiana Purchase: 72
Lovas, Uwe: 320-321
*Lusitania*: 176, 225
    artifacts: 220-223, 371-372
    cemetary: 271
    court ruling (British): 235-236
    depth charge: 218
    gold: 209, 220
    injunction: 284-285
    nets on wreck: 294-296
    ownership question: 233-236
    ownership suit: 309-375
    photos: *177, 189-193, 200-205, 208*
    purchase of wreck: 209-210
    wreck description: 217-220, 273-274, 277-279, 290-291, 294-298
"*Lusitania*": 215
*Lusitania, The*: 214-215
*Lusitania Disaster, The*: 214-215
*Lusitania File, The*: 216
Lyndon, Alf: 219
Lynnhaven Dive Center: 108
MacLeish, Kenneth: 211
Macomber, George: 234-236, 254, 322, 337, 341, 352, 370, 374
Madison Blue: 153-155
Magellan, Ferdinand: 167
Maine Coaster: 46, 233
Malone, Bart: 111, 144, 146, 157, 164-165, 285
Manchee, Pete: 86-87, 112, 127-128, 130-142, 152-153, 157, *180*, 229
Mann, Graham: 217-219
*Margie II*: 161, 320-321
Mariners Museum, The: 331, 333, 349
Maritime Jurisdiction Amendment (of Ireland): 364
Marschall, Ken: *177, 208*, 215-216, 251-252, 263-264, 355
Mars Rock: 40
Martin, Richard: 284
*Mary Catherine*: 38
Maser, Drew: 4, 49, 66 (family), 84, 253
Masi, Greg: 97-98, 173, 320
Mathews, Roy: 152
Mathewson, Duncan: 71
McGarvey, Mike: 28-29
McKenney, Jack: 87
McKinney, Charles: 71, 73
Mead, Margaret: 172
*Medina*: 260
Meimbresse, Bob: 305

Menduno, Mike: 116-117, 156, 157, 245, 329, 367, 376-378
Merseyside County Museum: 222
*Military History of the Lusitania, The*: 213
Miller harness: 54
Milligan, Joe: 78
Minister of Arts and Culture: 366
*Minnehaha*: 285, 362
Minto, Bruce: 71
*Miss Lindsay*: 108-109, 129, 131-134, 151-153, 173, 381
*Misteriosa*: 99, 114-116, 118
Mitchell, Billy: 143, 158, 161 (see also "Billy Mitchell wrecks")
Monet, Claude: 366
Moneyhun, Emmet: 75
Money Wreck: 27
*Monitor*: 69, 71, 84, 104-105, 110, 113, 118, 121-125, 143, 149, 158, 165, 176, *180, 207*, 224, 239, 275, 283, 310, 312, 318-321
Montauk (NY): 7, 12, 19, 302
Moore, Mike: 253
Morehead City (NC): 38
Mount Everest: 137-138, 143, 150, 232
Moyer, John: 40, 47-52, 144, 146, 168-170, 285, 303-307
Mud Hole: 15, 27, 43, 85, 157, 161, 383
Mummer's parade: 67
Munchkin voice: 160
Murtha, Jim: 5, 10-11, 39
Musee de Mer (Rimouski): 237
Museum of Man (London): 262
*Myra Vag*: 216
Nagle, Ashley: 63-64, 308
Nagle, Bill: 5, 9-13, 22-24, 37, 41-42, 45-52, 62-66, 85, 87-89, 146, 168-170, *178-179*, 233, 242-243, 308
*Nantucket* lightship: 11
Nantucket Shoals: 11
narcosis: 150, 210
    description of: 96
National Archives: 37, 106, 161
National Association of Underwater Instructors (NAUI): 166
National Broadcasting Company: 323
National Diving Center: 151
National Gallery (Dublin): 366
National Geographic Society: *177*, 230-233, 249-252, 254, 312-314, 325, 340, 344, 348, 357, 360, 362, 363
National Oceanic and Atmospheric Administration (NOAA): 71, 84, 104-105, 123, 149, 176, 318, 321
National Rehabilitation Institute of

Ireland: 347
National Underwater Accident Collection Center: 378
Naval Arms Limitation Treaty: 81, 92
Naval Historical Center: 37
Naval Intelligence Division (British Admiralty): 215
Navy Diving Service (Ireland): 294
Navy Table extrapolation: see "repetitive dive extrapolation"
*New Jersey*: 161-163, 173
New Orleans (LA): 245
Newport News (VA): 93, 331
*New York Times*: 214, 376-378
Nighswander, Larry: 312
*Night to Remember, A*: 215
Nikon: 253
*Nina*: 167
NOAA: see National Oceanic and Atmospheric Administration
Nonchalant Lamont: 19
Norfleet, Bruce: 344, 362
Norfolk (VA): 310, 331
Norgay, Tenzing: 137-138
*Norness*: 174-176, 229
Novak, Cheryl: 171, 224, 247, 253, 310, 318, 350
Nova Scotia: 38-41
Ocean City (MD): 38
Oceaneering International: 216-223, 234-236, 254, 277, 313-314, 335, 339-340, 354-355, 374
*Ocean Venture*: 62-64, 132-133, 144
O'Connor, Dan: 323
Ocracoke Island (NC): 37
Old Head of Kinsale: 267, 269, 301
*Oregon*: 16, 93
Ormsby, John: 42-43, 52-58, 77
O'Runaig, Brendan: 346, 355
OSHA (Occupational Safety and Health Act): 172
*Ostfriesland*: 80-81, 98-99, 102, 104, 106-112, 115, 124-142, 143, 145, 147, 149, 150-153, 156-159, 170, 173, *180*, 225, 228, 264, 382-383
O'Sullivan, Paddy: 212-213, 309, 310, 314
Owen, Paul: *183, 189, 197, 206*, 256, 261, 264-265, 267-268, 275, 283, 329
Packer, Tom: 34, 48-49, 51-52, 97-98, 119, 146, 157-158, *179*, 228, 241, 305-306, 381-383
Pagano, Jeff: 146
Palmer, Robert: 248-249, 294
*Pan Pennsylvania*: 307

*Panther*: 256
Parks Diving Services: 128
Parks, Mike: 127-128, 134, 144-145, 151
Pearl Harbor: 143
Penguin and the Albatross, The: 166-167
Peninsular & Oriental Line: 260
*Penylan*: 260
Periodic Report of the Substitute Custodian: 345-348, 352-355, 360-361
*Persephone*: 382
Persico, Frank: 17
Peterson, Gene: 4, 23, 24, 40-41, 146, 147, 305-307
Peterson, Joanie: 4
Peugeot: *186*, 261, 263, 264
Pickens, Slim: 101
Pierce, John: 233-235, 254, 313
*Pinta* (Columbus's ship): 167
*Pinta* (NJ wreck): 13
Pittsburgh (PA): 27
Pletnik, John: 13
Plokhoy, Glen: 87-89, 146
Pollack, Ross: 214
Popular Dive Guide Series: 61, 149
Potomac Air Gas: 108, 127-128
Powell, Jamie: *183, 189, 206*, 252, 255, 258, 267-268, 276, 279, 289, 299-301
*Primary Wreck Diving Guide*: 319
Provencher, Denis: 286-287
Queen's Bench (British Admiralty Court): 235-236, 313-314, 340-343, 362, 371, 374
Queenstown: see "Cobh"
Quigley, Desmond: 293-294, 346, 354, 369, 370
Quilligan, Simon: *189*, 284
Quinn (seaman): 217
Ramsay, William: 210
*Rape 'N Pillage*: 46
*Rapture of the Deep*: 320
Reagan, Ronald: 354
Receiver of Wreck: 234, 285-288, 327, 363, 365
Reef Raiders Dive Shop: 41, 77
Reekie, John: 121, 153-155
Rembrandt van Rijn: 366
remotely operated vehicle: (see "ROV")
repetitive dive extrapolation: 20-21, 86, 107, 157
*Republic*: 171, 229
Reyes, Alex: 153-155
Reynolds, Chris: *185*, 293-294, 329, 344, 346-348, 354
*Rhein*: 118

RIB (rigid inflatable boat): *188, 195-197,* 233, 265-269, 282-283, 288-289, 298-299
rigid inflatable boat: see "RIB"
Rimouski (Quebec): 144, 237, 285
*R.M.S. Lusitania: Triumph of the Edwardian Age*: 215-216, 251
Roberts, Grattan: 284-285
Roberts, Smokey: 138, 152
Robinson, Mr.: 335
Robol, Richard: 253-254, 262, 309-375
*Rodale's Scuba Diving*: 329
*Room 40: British Naval Intelligence 1914-1918*: 215
Rooney, Pat: 146
Rosenberg, Jerry: 28-33
ROV: 216-219, 231, 294-295
Royle, Rob: *185, 196-197,* 257, 281, 288, 293
Rube Goldberg: 51
Rubens, Peter Paul: 366
Rui, Romano: 18
Rule of Finds: 371-372
Rutkowski, Dick: 166
Ryan, Paul: 214-215
*S-5*: 77-78
Saccarro, Bernie: 87
*San Diego*: 16, 93, 303
Sandycove Island: 284
San Francisco (CA): 367
Santa Fe (NM): 234, 339
*Santa Maria*: 167
Sarlo, Lou: 146, 157
saturation diving: 211-213, 216-220, 277, 341, 348
Sauder, Eric: 204, 215-216, 251-252
*Scientific American*: 376-378
Scorpio: 216-219
Sea Fisheries Office (Ireland): 234
*Sea Hawk*: 17
*Sea Hunter*: 5-20, 30, 52, 93
*Sea Lion*: 26, 46, 78
Seamans, Nike: 47, 62, 63, 158, 331, 350
Seamans, Trueman: 47, 62, 63, 331
Sea Sports Publications: 61
*Sebastian*: 184, *206,* 307-308
*Seeker*: 46-52, 85-86, 88-89, 91-92, 146, *179,* 308
Senate: 72-75
*Seven Days to Disaster*: 215
Sheard, Brad: 28-29, 32, 146, 157, 381
Sheen, Justice: 235-236
Ships of State: 361
shipwreck research: 37-38, 77, 80
*Shipwrecks of Delaware and Maryland*: 306

*Shipwrecks of New Jersey*: 61
*Shipwrecks of Virginia*: 81, 227
Shumbarger, Gary: 75-76
side-scan sonar: 78
Silverstein, Joel: 171-172, 329
Simpson, Alan: 74
Simpson, Colin: 214-215
Sir Lancelot: 296
Skerry, Brian: 99-100, 308
*Skin Diver Magazine*: 166
Slate, Spencer: 52
Small, Peter: 211
Smith, Gus: 215
Smith, Stan: 13-14
Snap-On Tools: 46, 85
Snyder, Louis: 213
Sokoloff, Steve: 78
Sommerhill, Doug: 173, 381
*Sommerstad*: 22, 28
  diver fatalities: 22-26, 28-33
SOS meter: 51
Sotheby's: 222
South Pole: 167
Southport (NC): 253
Sparta: 27
*Sports Illustrated*: 222-223
*Squalus*: 211
Starace, John: 15
Starfish Enterprise, The: *177, 189,* 224-227-233, 240, 245, 248-249, 253-256, 259, 264, 270, 280, 285, 293, 298, 301, 309, 313-316, 327-328, 331, 338, 341-343, 347, 348, 357-358, 365-369, 374
*St. Cathan*: 86
St. Lawrence Seaway: 144
St. Nazaire: 255
Stock Island (Key West, FL): 77
Stoker, Kenneth: 19
Storck, Harvey: 152, 173, 320, 381
Stroh, Dave: 18, 20
Suarez, Ed: 87, 130, 152-153
*Sub Aqua Journal*: 171, 329
substitute custodian: see "Periodic Report of the Substitute Custodian"
*Sundancer II*: *184, 188, 195, 197,* 233, 249, 265-270, 275, 279, 281-283, 288, 299-301, 365
Superman: 5
Supreme Court: 274
Surowiec, Joanne: 320
Swansea (Wales): 263, 302
*20/20*: 216
Tapson, Polly: *182, 184, 186, 195, 206,* 224-226, 229, 232-235, 238-241, 245, 254-255, 258-262, 267-270,

# Index - 391

275, 279, 284, 287, 288, 293, 296, 300, 309-310, 313-316, 323, 327-329, 333, 347, 352, 357-358, 365-368, 373
Tapson, Simon: *182, 186, 187, 206,* 224-225, 232-233, 255, 256, 258-260, 262, 267-271, 275, 279, 280, 292, 296, 299-301, 309-310, 329, 367
*Tarpon*: 37
Taylor, David: 329
technical diving: 150, 157, 165-168, 248, 294, 341, 368-369
   standards: 170-172
telepresence: 231
Temple University Hospital: 84
Ten Mile Wreck: 257
Terry, John: 80, 128, 130
Thompson, Tommy: 311
Thomson, Elihu: 211
Three Stooges: 140
Thunder Bay, Ontario: 76
tide: 269, 290
*Time for Dragons, A*: 58
*Times* (London): 216, 309, 324
*Titanic*: 225, 237, 251, 310
Tolbert, Ron: 173
toll collector: 236-237, 254, 313, 342, 372
*Topcat*: 13
Toronto Colsolidated Ltd.: 222-223
*Track of the Gray Wolf*: 307
*Treasure Wars, The*: 70
Treasury Solicitor: 329
Trotter, Dave: 75
Tulley, Richard: *183, 189, 206,* 256, 258, 260, 261, 263, 265-268, 279, 281, 282, 289, 292, 299-301
Turner, Captain William: *193, 205,* 297
Tussey underwater housing: 252-253, 281
type XXI: 100
*U-117*: 161
*U-140*: 161
*U-2513*: 78, 99-102, *179*
*UB-148*: 161
U-boat: 62, 78, 100, 161, 174, 221, 307
*Ultimate Wreck Diving Guide*: 155-156, 164, 224, 227
Underwater Heritage Order: 366
*Underwater USA*: 329
United Nations: 364
United States Coast Guard: 11, 14, 27, 28-33, 46, 50, 303
United States District Court for the Eastern District of Virginia, Norfolk Division: 309-375

United States Navy: 104, 143, 211, 213, 312
United States Navy Standard Air Decompression Tables: 20, 30, 51, 86-87, 96-97
University of Pennsylvania: 21
University of South Carolina: 87
Valley Forge General Hospital: 76
van Name, Floyd: 241
Verolme Shipyard: 212, 322
Veterans Administration: 65
Vietnam: 64, 76, 350, 380
*Virginia*: 161
Virginia Beach (VA): 47, 62, 93, 108, 152, 331
Virginia Capes: 152
Wadsworth, Al: 38
*Wahoo*: 16-17, 52, 87, 89, 91-92, 119-121, 146, 302-305, 308
Warehouse, Kathy: 22-23
Warner, Pam: 73
Warren, Mark D.: 215
War Reclamation Board: see "Liverpool and London War Risks Insurance Association"
*Washington*: 81, 92-97, 102-103, 107, 130, 137, 153, 225
Washington, DC: 27, 37, 66, 71, 73-75, 77, 80, 149, 151
WAVY-TV: 158
Weston, Howard: *195, 197,* 233, 265, 266-267, 269, 275, 282-283, 288-289, 292-293
*Who Sank the Lusitania*: 214
Wilkens, Dave: *183, 206,* 256, 267-268, 275-277, 279, 293
*Wilkes-Barre*: 41-45, 52, 56, 58, 77-81, 102-103, 116-117, 147-149, 156
Woods Hole Oceanographic Institute: 312
World War One: 161, 215, 231
World War Two: 62, 93, 174, 215, 218, 233, 239, 307
*Wreck Diving Adventures*: 37, 253
Wright Brothers: 167
Yurga, John: 146, 174, *177, 184, 189, 197, 206,* 240-241, 245, 252, 255, 260, 273, 295-296, 315-316, 332-334, 350, 357, 367
Zorpette, Glenn: 376-378
Zubec, Dave: 303-304

## Books by the Author
### Fiction

Vietnam
*Lonely Conflict*

Action/Adventure
*Memory Lane*
*Mind Set*

Supernatural
*The Lurking*

Science Fiction
*Entropy*
*Return to Mars*
*Silent Autumn*

The Time Dragons Trilogy
*A Time for Dragons*
*Dragons Past*
*No Future for Dragons*

### Nonfiction

*Advanced Wreck Diving Guide*
*Ultimate Wreck Diving Guide*

*Track of the Gray Wolf*
*Shipwrecks of New Jersey*

---

Available (postage paid) from:   GARY GENTILE PRODUCTIONS
P.O. Box 57137
**Nonfiction**   Philadelphia, PA  19111

$25 *Andrea Doria: Dive to an Era* (hard cover)
$20 *USS San Diego: the Last Armored Cruiser*
$20 *Wreck Diving Adventures*
$20 *The Nautical Cyclopedia*
$20 *Primary Wreck Diving Guide*
$30 *The Technical Diving Handbook*
   Civil War ironclad **Monitor**
$25 (hard cover) *Ironclad Legacy: Battles of the USS Monitor*
$25 (video tape - VHS or PAL) *The Battle for the USS Monitor*
   The ***Lusitania*** Controversies (hard covers)
$25 Book One: *Atrocity of War and a Wreck-Diving History*
$25 Book Two: *Dangerous Descents into Shipwrecks and Law*
   The Popular Dive Guide Series
$20 *Shipwrecks of New York*
$20 *Shipwrecks of Delaware and Maryland*
$20 *Shipwrecks of Virginia*
$20 *Shipwrecks of North Carolina: from the Diamond Shoals North*
$20 *Shipwrecks of North Carolina: from Hatteras Inlet South*
   **Wreck Diving Adventure Novel**
$20 *The Peking Papers* (hard cover)

Website
http://www.pilot.infi.net/~boring/gentile.html